Bioremediation and Biotechnology

Khalid Rehman Hakeem • Rouf Ahmad Bhat
Humaira Qadri
Editors

Bioremediation and Biotechnology

Sustainable Approaches to Pollution Degradation

Springer

Editors
Khalid Rehman Hakeem
Department of Biological Sciences
King Abdulaziz University
Princess Dr. Najla Bint Saud Al-Saud
Center for Excellence Research in
Biotechnology, King Abdulaziz University
Jeddah, Saudi Arabia

Rouf Ahmad Bhat
Department of Environmental Science
Cluster University, Sri Pratap College
Srinagar, Jammu and Kashmir, India

Humaira Qadri
Department of Environmental Science
Cluster University, Sri Pratap College
Srinagar, Jammu and Kashmir, India

ISBN 978-3-030-35690-3 ISBN 978-3-030-35691-0 (eBook)
https://doi.org/10.1007/978-3-030-35691-0

This Springer imprint is published by the registered company Springer Nature Switzerland AG
The registered company address is: Gewerbestrasse 11, 6330 Cham, Switzerland

To our Parents and siblings

Foreword

Environmental pollution is currently a problem of major global concern. The pollutants in general are accepted as unwanted additions for our living world. The large-scale industrial activities have ended up with serious contaminations of our soils and waters by adding several hazardous and toxic compounds together with undesirable xenobiotics, recalcitrants as well as other chemical compounds which differ much in their chemical structure from natural organic compounds. The highly toxic pollutants in our surroundings are mutagenic as well as carcinogenic for all living beings. They are diverse, versatile, and adapt to adverse conditions. Intensive work is being done to remediate the environmental contaminants in the biogeochemical cycles; however, we need to find out new pathways that might lead towards complete removal of pollutants.

One of the latest trends has been use of biotechnology in the bioremediation. It is currently becoming the most wanted field for restoration of our environment. Several bioremediation strategies are followed for treating polluted locations and wastes. For this purpose, investigators are undertaking studies on the biochemistry, bioavailability, and bioactivity. Several workers have started studies on the use of molecular genetics for biodegradation together with enzyme-tailoring and DNA-shuffling. Bioremediation in itself is meant to remove the pollutants and prevent pollution at global, regional, and local levels. Bioremediation as a biological process is regarded as a sustainable approach to clean the environmental pollutants. Huge numbers of researchers are exploiting application of bioremediation. Several taxa from algae, fungi, and bacteria can solubilize, transport, and deposit the pollutants and detoxify dyes and complex chemicals. However, technology varies according to the fact whether wastes under question are in their natural setting or removed and transported into bioreactor. In situ bioremediation is regarded as a complex phenomenon involving several contaminants affected by different microorganisms with different metabolic pathways.

The book includes 15 chapters from different parts of the world. First chapter by a group of authors from India is titled as "Concerns and Threats of Contamination on Aquatic Ecosystems." It presents information on the aquatic ecosystems as ultimate sinks for contaminants originating from urbanization, industrialization, agricultural activities, overuse of pesticides/fertilizers, and sewage from residential/industrial areas. All these are degrading our water quality, spreading many infectious diseases. Authors stress the need for regular monitoring and controlling of pollutant discharge into aquatic environs. The researchers from Saudi Arabia and India discuss the "Effect of Pesticides on Fish Fauna: Threats, Challenges, and Possible Remedies" in the second chapter. This chapter has tried to analyze the facts related to the use of pesticides, their historical background, classification, toxicity in fish and routes of exposure, together with the effects and threats posed to fish life as well as challenges in their monitoring. The remedial techniques that may replace the pesticides have been discussed. Third chapter by Indian group of scientists presents the information on the "Impact of invasive plants in aquatic ecosystems." Authors mainly deal with lakes which provide tremendous benefits to the living beings on global basis through their cultural, esthetic, socioeconomic, and ecological values. Their dispersal strategies, time period of perpetuation, and mode of invasion have been discussed as prerequisite for management of invasive alien species. In Chap. 4, a group of researchers from Mexico discuss the "Role of Modern Innovative Techniques for Assessing and Monitoring Environmental Pollution." They have included information on the detection of microsystems, control as well as automation of chemical processes, and an analysis of the data presented. Again another group of workers from India present data on the "Global Scenario of Remediation Techniques to Combat Environmental Pollution" in Chap. 5. Various remediation techniques devised across the world to tackle environmental pollution are discussed in this chapter, under ex situ and in situ remediation techniques besides those for remediation of air pollution together with emerging technologies like nano, microbial fuel cell, and ultrasonic. Chapter 6 prepared by the scientists from Romania is titled as "Biopesticides: Clean and Viable Technology for Healthy Environment." It refers to the use of living biological organisms or their metabolites for control in the agricultural production, mainly microbial pesticides. Another group of workers from Pakistan have pooled up information on the "Inoculum Addition in the Presence of Plant Rhizosphere for Petroleum Polluted Soil Remediation" in Chap. 7. It explores the ambiguity by providing a complete description on bioremediation as well as facts regarding the selection of plant rhizosphere microbial community, more suitable for degradation of pollutants than addition of microbial culture in soil. The rhizodegradation has been discussed at length. Researchers from India have pooled up information on the "Vermicomposting: An Eco-friendly Approach for Recycling/Management of Organic Wastes" in Chap. 8. Vermicomposting technology has been presented as a promising tool for recycling/management of organic waste for being socially acceptable, economically viable, and environmentally sustainable at global level. The information presented discusses the fact that organic wastes are broken down by different species of earthworms and a fine humus-like material known as vermicompost is produced. Authors' findings stress that vermicompost can be used as a replacement for harmful inorganic fertilizers. Chapter 9 is titled as "Bio-Fertilizers: Eco-Friendly Approach for Plant and Soil Environment," and written jointly by the scientists from Pakistan and Saudi Arabia. They mention that application of bio-fertilizers may serve as low cost and environment-friendly fertilization strategy for a sustainable crop production. Another group of workers from India have prepared Chap. 10 titled "Phytoremediation of Heavy Metals: an Eco-Friendly and Sustainable Approach." It presents information on the phytoremediation as the only alternative solution for remediation of toxic contaminants from the environment in an efficient way. As per the authors, phytoremediation has gained importance due to its green approach of remediating heavy metals. The investigators from Mexico have pooled up data on the "Credibility of In Situ Phytoremediation for Restoration of Disturbed Environments" in Chap. 11. The chapter summarizes data on *in situ* phytoremediation studies carried out in Mexico, mostly the sites polluted by trace metals following mining activities. They stress that information presented here will be useful for planning as well as remediation of contaminated sites. Authors have given data related to the trace metal hyperaccumulator plants like *Hydrocotyle ranunculoides*, *Parietaria pensylvanica*, and *Commelina diffusa* for zinc and *Rorippa nasturtium-aquaticum* for copper. They point out that native species must be studied to establish mechanisms of phytoextraction of metals and interaction with water/soil and microorganisms to improve the efficiency of *in situ* phytoremediation. Chapter 12 titled as "Role of White Willow (*Salix alba* L.) for Cleaning up the Toxic Metal Pollution" has been prepared by scientists from India. The section portrays the decision of eco-accommodating *Salix alba* species and their correct situation within urban locations to overcome the contamination issues. It discusses the potential for the phytoremediation of substantial metal-defiled land by the *Salix alba*. "Mycoremediation: A Sustainable Tool for Abating Environmental Pollution" has been reviewed by the workers from India in Chap. 13. They emphasize mycoremediation tool can be applied for different types of contaminated environs eco-friendly. The system is based on enzymes produced by the fungi like *Pleurotus ostreatus*, *Aspergillus niger*, *Tramates hirsute*, and some others. As per the authors, these are having tremendous potential for degradation and remediation of toxic pollutants. In Chap. 14, authors from Pakistan and Malaysia present information on the "Microbial Biofilm Cell Systems for Remediation of Wastewaters." The aim of authors has been to present the role of immobilized microbial cell systems in bioremediation of different pollutants and future research outlooks in this area. Chapter 15, as the closing chapter, has been written by the scientists from Brazil. It is titled as "Pollution Remediation By Way Of Using Genetically Modified Plants (GMP)." Authors mention that transgenic plants deserve to be highlighted once their capacity to contribute can surpass the one offered by microorganisms in the uptake, transforming and limiting the toxicity of different contaminants as cost-effective strategy. The chapter briefly reviews phytoremediation using GMP.

The chapters included and information presented in this book cover most important aspects. The editors need to be congratulated for their academic efforts spent on bringing out this book.

Münir Öztürk

Faculty of Forestry
Universiti Putra Malaysia
Seri Kembangan, Selangor, Malaysia
ICCBS, Karachi University
Karachi, Pakistan
Botany Department, Faculty of Science
Ege University
Izmir, Turkey
Centre for Environmental Studies
Ege University
Izmir, Turkey

Preface

The world is advancing towards modernization, and there is a price to pay for that in the form of environmental contaminants. The survival of terrestrial, aerial, and aquatic organisms including human beings has been endangered by pollutants produced as a result of development. While all the components of the environment are hugely impacted by the outcome of modernization, aquatic ecosystems are the ultimate sinks for the contaminants. Environmental contamination is the outcome of human activities like urbanization, industrialization, and agricultural activities. The overuse of pesticide, fertilizers, and the sewage discharges from residential as well as industrial areas ultimately finds its way to aquatic and terrestrial environments. As a result of pollution, aquatic invasions are a serious threat to lake ecosystems. By spreading and growing rapidly, invasive species displace the native species decreasing the efficacy of ecosystem services.

Considering the impact of pollutants on various ecosystems, this book is an attempt to address certain issues in a scientific and sustainable manner. The book aims at ecological stability which aims at replacing pollutants with the application of substitute techniques. So far as pesticides are concerned, the content of the book reflects that research is required to produce eco-friendly species-specific pesticides so that the damage to non-target organisms is avoided. Bio-pesticides have also been discussed in a detailed manner. Ample space has been provided to the knowledge of the dispersal strategies of invasive plants, time period of perpetuation, and mode of invasion which is a prerequisite for management of invasive alien species.

An attempt has been made to provide an overview of modern technologies, i.e., the microsystems of detection, control, and automation that have adapted to the needs of modern times and allow environmental analysis to provide real-time, on-site analysis information with high reliability as well as minimum costs therefore, providing the opportunity to make decisions to respond to social, scientific, or technological problems.

The book discusses the various remediation techniques devised across the world to tackle environmental pollution like remediation of contaminated soil and groundwater; ex situ remediation techniques (dig and dump, pump-and-treat, incineration, oxidation, adsorption, ion-exchange, pyrolysis remediation, physical separation, dehalogenation technique, bioremediation, solidification remediation, constructed wetlands, etc.); and in situ remediation techniques (biological treatments, physical or chemical treatments, thermal treatments).

In order to enhance the degradation potential of heterotrophic microorganisms in bioremediation, different bio-augmentation techniques have been deliberated upon vermicomposting technology which is a promising tool for recycling/management of organic waste and is socially acceptable, economically viable, and environmentally sustainable technology throughout the world has been detailed at length.

Bio-fertilizers and phytoremediation have been elaborated to gain a detailed insight into the current research and technology status in these areas. The use of mycoremediation as an eco-friendly and sustainable technique has also received its due place in the book.

Overall, the book is a complete amalgam of the problems associated with environmental contamination and the control as well as management strategies that can be and should be opted to tackle them. The book is a useful reference and resource for students, researchers, and scientists working in this field.

We are highly grateful to all our contributors for readily accepting our invitation for not only sharing their knowledge and research, but also for venerably integrating their expertise in dispersed information from diverse fields in composing the chapters and enduring editorial suggestions to finally produce this venture. We greatly appreciate their commitment. We are also thankful to Prof. Munir Ozturk for his suggestions and writing the foreword for this volume.

We thank Springer-International team for their generous cooperation at every stage of the book production.

Jeddah, Saudi Arabia — Khalid Rehman Hakeem
Srinagar, India — Rouf Ahmad Bhat
Srinagar, India — Humaira Qadri

About the Book

Being a serious concern, toxic substances threaten aquatic and terrestrial ecosystems and ultimately human health. The book is a thoughtful effort in bringing forth the role of biotechnology for bioremediation and restoration of the ecosystems degraded by toxic and heavy metal pollution. The introductory chapters of the book deal with the understanding of the issues concerned with the pollution caused by toxic elements and heavy metals and their impacts on the different ecosystems followed by the techniques involved in monitoring of the pollution. These techniques include use of bio-indicators as well as modern techniques for the assessment and monitoring of toxicants in the environment. Detailed chapters discussing the role of microbial biota, aquatic plants, and terrestrial plants to enhance the accumulation efficiency of these toxic and heavy metals are followed by remediation techniques involving mycoremediation, bio-pesticides, bio-fertilizers, phytoremediation, and rhizo-filtration. A sizeable portion of the book has been dedicated to the advanced bioremediation techniques which are finding their way from the laboratory to the field for revival of the degraded ecosystems. These involve biofilms, micro-algae, genetically modified plants, and filter feeders. Furthermore, the book is a detailed comprehensive account for the treatment technologies from unsustainable to sustainable. Academicians, researchers, and students shall find it as a complete wrap up regarding biotechnological intervention for sustainable treatment of pollution and shall suffice for the diverse needs of teaching and research.

Contents

About the Editors

Khalid Rehman Hakeem, Ph.D. is Professor at King Abdulaziz University, Jeddah, Saudi Arabia. After completing his doctorate (Botany; specialization in Plant Eco-physiology and Molecular Biology) from Jamia Hamdard, New Delhi, India, in 2011, he worked as a lecturer at the University of Kashmir, Srinagar, for a short period. Later, he joined Universiti Putra Malaysia, Selangor, Malaysia, and worked there as Postdoctorate Fellow in 2012 and Fellow Researcher (Associate Professor) from 2013 to 2016. Dr. Hakeem has more than 10 years of teaching and research experience in plant eco-physiology, biotechnology and molecular biology, medicinal plant research, plant–microbe–soil interactions as well as in environmental studies. He is the recipient of several fellowships at both national and international levels; also, he has served as the visiting scientist at Jinan University, Guangzhou, China. Currently, he is involved with a number of international research projects with different government organizations. So far, Dr. Hakeem has authored and edited more than 35 books with international publishers, including Springer Nature, Academic Press (Elsevier), and CRC Press. He also has to his credit more than 80 research publications in peer-reviewed international journals and 55 book chapters in edited volumes with international publishers. At present, Dr. Hakeem serves as an editorial board member and reviewer of several high-impact international scientific journals from Elsevier, Springer Nature, Taylor and Francis, Cambridge, and John Wiley Publishers. He is included in the advisory board of Cambridge Scholars Publishing, UK. He is also a fellow of Plantae group of the American Society of Plant Biologists, member of the World Academy of Sciences, member of the International Society for Development and Sustainability, Japan, and member of Asian Federation of Biotechnology, Korea. Dr. Hakeem has been listed in Marquis Who's Who in the World, since 2014–2019. Currently, Dr. Hakeem is engaged in studying the plant processes at eco-physiological as well as molecular levels.

Rouf Ahmad Bhat, Ph.D. is working in Cluster University Srinagar (J&K), India, in the capacity of Assistant Professor and has his specialization in Limnology, Toxicology, Phytochemistry, and Phytoremediation. Dr. Bhat has been teaching graduate and postgraduate students of Environmental Sciences for the past two years. He is an author of more than 50 scientific articles and 15 book chapters and has published 10 books with international publishers. He has presented and participated in numerous State, National, and International conferences, seminars, workshops, and symposium. Dr. Bhat has worked as an Associate Environmental Expert in World Bank-funded Flood Recovery Project and also Environmental Support Staff in Asian Development Bank (ADB)-funded development projects. He has received many awards, appreciations, and recognitions for his services to the science of water testing, air and noise analysis. He has served as editorial board member and reviewer of reputed international journals. Dr. Bhat is still writing and experimenting with diverse capacities of plants for use in aquatic pollution.

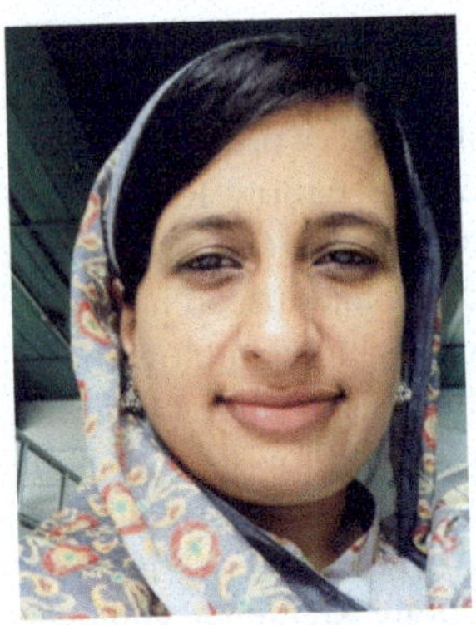

Humaira Qadri, Ph.D. is actively involved in teaching the postgraduate students of Environmental Science for the past ten years in Sri Pratap College Campus, Cluster University Srinagar, J&K, India. She is also heading the Department of Environment and Water Management. She is a gold medallist at her master's level, and she has bagged a number of awards and certificates of merit. Her specialization is in Limnology, Nutrient Dynamics, and Phytoremediation. She has published scores of papers in international journals and has more than ten books with national and international publishers. She is also the reviewer of various international journals and is the principal investigator of some major projects on phytoremediation. She is guiding a number of research students for Ph.D. program and has supervised more than 60 master's dissertations. She also has been on the scientific board of various international conferences and holds life memberships of various international organizations. With a number of national scientific events to her credit, she is an active participant in national and international scientific events and has organized a number of national conferences on science.

Chapter 1
Concerns and Threats of Contamination on Aquatic Ecosystems

Ishrat Bashir, F. A. Lone, Rouf Ahmad Bhat, Shafat A. Mir, Zubair A. Dar, and Shakeel Ahmad Dar

1.1 Introduction

Anthropogenic activities bring almost contamination and subsequent pollution to our varied ecosystems. "Pollution is defined as the production and or introduction by man, directly or indirectly of substances or energy into the environment, resulting in deleterious effects to living resources, including human beings or interfere with amenities and other uses of the environment (Don-Pedro 1990)." Pollution is one of the prime problems that humans face in the whole world particularly in the developing countries. However, produced by humans and their activities, it has harmful effects on man's environment and resources (Mendil and Uluözlu 2007). The discharge of various pollutants into the aquatic environments is the outcome of countless anthropogenic activities, threatening the health of the living beings and damaging the quality of the environment by rendering water bodies unsuitable (Abowel and Sikoki 2005; Ekubo and Abowel 2011). Aquatic environments are pickers for anthropogenic contamination and industrial wastes and leaks, whether chemicals or solid pollutants (Hampel et al. 2015; Bhat et al. 2017). These wastes can be "heavy metals, detergents, microfibers, plastic or non-plastic origin," etc., and contribute to "aquatic pollution problems" (Hampel et al. 2015). Aquatic environs are addressee for plenty of pollutants and their outrageous toxic actions (Hampel et al. 2015). "Chemicals reaching aquatic ecosystems include radioactive elements" ("strontium, cesium, radon"), metals ("cadmium, mercury, lead"), industrial solvents and "volatile organic compounds" ("tri- and tetrachloroethylene, chlorofluorocarbons, benzene, xylenes, formaldehyde"), "agrochemicals" ("fertilizers and pesticides"), household products ("detergents, cleaners, paints"), "fuel combustion" ("N and sulfur oxides," "polycyclic aromatic hydrocarbons," "carbon monoxide," and "carbon dioxide"), "nanoparticles," personal care products, "microplastics,

I. Bashir (✉) · F. A. Lone · R. A. Bhat · S. A. Mir · Z. A. Dar · S. A. Dar
Sher-e-Kashmir University of Agricultural Sciences and Technology,
Jammu, Jammu and Kashmir, India

K. R. Hakeem et al. (eds.), *Bioremediation and Biotechnology*,
https://doi.org/10.1007/978-3-030-35691-0_1

antibiotics," as well as a huge variety of prescription (Hampel et al. 2015)" and "nonprescription drugs and pharmaceuticals of human and veterinary medicine" (Hughes et al. 2013; Larsson 2014; Malaj et al. 2014; Hampel et al. 2015).

"Aquatic ecosystems, particularly the freshwater ecosystems, are exposed to supplementary contamination than other environs, as water is used in various industrial practices as well as release of discharges commencing from industry" and urban growths (Demirak et al. 2006; Fernandesa et al. 2007). "Water pollution is a worldwide task that has augmented in both advanced and emerging nations" (Mateo-Sagasta et al. 2017). "Universally, 80% of municipal wastewater is discharged into water bodies untreated, and industry is responsible for dumping millions of tons of heavy metals" (Mateo-Sagasta et al. 2017), "solvents, toxic sludge and other wastes into water bodies each year" (WWAP 2017; Mateo-Sagasta et al. 2017). Agriculture, exploits "70% of water globally and plays a key part in water contamination" (Mateo-Sagasta et al. 2017). Huge amounts of "agrochemicals," "organic matter," "drug residues," "sediments," and "saline drainage" from agricultural lands are released into water bodies (Mateo-Sagasta et al. 2017) and hence poses significant threats to "aquatic environments," "human health," and "productive activities" (UNEP 2016; Mateo-Sagasta et al. 2017). Most aquatic ecosystems have a natural tendency to dilute pollution to some extent, but severe contamination of aquatic ecosystems results in alteration in the fauna and flora of the community (Mateo-Sagasta et al. 2017). The onset of human civilization in itself discloses the history of aquatic pollution (Mateo-Sagasta et al. 2017). Moreover, aquatic pollution did not receive significant consideration until a threshold level was reached with hostile outcomes on the "ecosystems" and "living organisms" including "humans" (Halpern et al. 2008; Mateo-Sagasta et al. 2017). Therefore, "pollution and its effects are considered as one of man's greatest crimes against himself. Pollutants may cause primary damage, with direct identifiable impact on the environment, or secondary damage in the form of minor perturbations in the delicate balance of the biological food web that are detectable only over long time periods" (Sharma 2012; Al Naggar et al. 2014; Ghani 2015). Thus, "maintaining the quality of aquatic ecosystems represents one of the most formidable challenges facing global society in the twenty-first century" (Hampel et al. 2015).

1.2 Aquatic Ecosystems

Aquatic ecosystems are water-based environments in which biotic components interact with abiotic components of the aquatic ecosystem. "Aquatic ecosystems" are usually divided into two types: the "marine ecosystem" and the "freshwater ecosystem" (Barange et al. 2010). Marine ecosystem is the largest water ecosystem which covers over 70% of the Earth's surface. The marine ecosystem is subdivided into "oceans," "estuaries," "coral reefs," and "coastal ecosystems." Freshwater ecosystems cover less than 1% of the Earth's surface. The various kinds of freshwater ecosystems are lotic ecosystem, lentic ecosystem, and wetland ecosystem.

1.3 Human Activities Resulting in Contamination of Aquatic Ecosystems and Their Adverse Impacts

Anthropogenic activities such as "deforestation," "filling and construction of canals," "dams," "roads and bridges," "agricultural," and "industrial and domestic activities" result in contamination of aquatic environments. Human settlements, industries, and agriculture are the main sources of water pollution (Table 1.1).

Table 1.1 Sources and route of pollutant discharge into aquatic environs (NEST 1991; Mateo-Sagasta et al. 2017)

Contamination	Source	Route
"Oxygen-demanding wastes (organic pollutants)"	"Domestic sewage, human and animal wastes (such as wastes from canneries and wood pulp mills)"	"Thrown, dumped or released into streams and rivers or into gutters, drains from where they may get washed by run-off into water bodies"
"Infectious disease agents"	"Domestic sewage, human and animal wastes"	"Washing, swimming or working in paddy rice fields and on irrigated land"
"Plant nutrients such as nitrate, phosphate and others"	"Fertilized farm lands, ashes and detergent"	"Run-off from fertilized farmlands"
"Pesticides (insecticides and herbicides)"	"Organic and inorganic chemicals"	"Run-off from pesticides associated with farmlands"
"Industrial effluents which include DDT, dyes, mercury, cadmium"	"Textile factories, distilleries pulp and paper mills, fertilized plants, chemical and allied industry, food, beverages and tobacco industries, soap, detergents and confectionery industries"	"Human discharges"
"Eroded sediments"	"Deforestation and accelerated soil erosion"	"Soil erosion, urban storm water runoffs and dredging activities"
"Other solid wastes"	"Metals, plastics, artificial fibers etc."	"Dumping by human beings due to poor management of waste disposal"
"Petroleum products"	"Drill cuttings," "drilling mud (fluids used to stimulate the production processes)," accidental discharges of "crude petroleum," "refinery effluents" which include "oil" and "grease," "phenol," "cyanide," "sulphide," "suspended solids," "chromium," and "biologically oxygen demanding organic matter"	"Petroleum, exploration, exploitation, refining, transportation, storage, marketing, use and ruptured oil pipelines"

In most developed nations, agriculture is the major factor in the degradation of water ecosystems. In the "European Union, 38% of water ecosystems are significantly under agricultural pressure" (WWAP 2015; Mateo-Sagasta et al. 2017). In the USA, "agriculture is the leading source of pollution in rivers and streams" (Mateo-Sagasta et al. 2017), the second main source in wetlands, and the third main source in lakes (USEPA 2016; Mateo-Sagasta et al. 2017). In China, "agriculture is accountable for a huge portion of surface-water pollution and is responsible almost entirely for groundwater pollution" (Mateo-Sagasta et al. 2017) by nitrogen (FAO 2013; Mateo-Sagasta et al. 2017). In emerging nations, the unlimited amounts of raw municipal and industrial wastewater are major threats (Mateo-Sagasta et al. 2017).

1.3.1 Agrochemicals

The ever-increasing "demand for food has led to the land clearance and the expansion of agriculture" which have "contributed to the higher pollution loads in the water" (Mateo-Sagasta et al. 2017). Increase in the population growth has increased the food demand, which has resulted in the increase in the quantity of agrochemicals used to increase the production (Schwarzenbach et al. 2010). The "unsustainable use of agrochemicals" ("fertilizers, pesticides, herbicides and plant hormones") to rise the production has resulted in "greater pollution masses" in the environment, including "rivers," "lakes," "aquifers," and "coastal waters" (Mateo-Sagasta et al. 2017; Bhat et al. 2018; Mushtaq et al. 2018). More importantly, "agricultural areas gather an extensive variety of agrochemicals from nearby fields" due to "run off," "direct drift," and "leaching," and these areas are "the principal receivers of agrochemicals" (Rathore and Nollet 2016).

1.3.1.1 Nutrients

When "fertilizers are applied at a higher rate than they are fixed by the soil, or taken up by the crops or when they are taken off through surface runoff from the soil surface leads to water pollution." "Excess nitrogenous fertilizers and phosphate fertilizers can leach into groundwater or reach into surface water bodies through surface runoff" (Mateo-Sagasta et al. 2017). If "organic manure" is applied "in excess in the agricultural fields," it will lead to "diffuse water pollution." Mostly, "manure is not stored in confined areas and during heavy rainfall events it can be washed into waterways via surface runoff." The "high-nutrient concentration together with other substances results in the nutrient enrichment eutrophication" of "lakes," "reservoirs," "ponds," and "coastal waters," which leads to excessive growth of aquatic plants—"algae blooms" that destroy other aquatic plants and animals. "About 415 coastal areas have been identified worldwide which experience eutrophication" (Mateo-Sagasta et al. 2017), of which 169 are hypoxic (WRI 2008; Mateo-Sagasta et al. 2017). The "excessive buildup of nutrients may also increase the adverse health effects" (Mateo-Sagasta et al. 2017), such as "blue-baby syndrome- due to high levels of nitrate in drinking water" (Mateo-Sagasta et al. 2017). "Nitrate from agriculture leaches into the groundwater is the most common chemical contaminant in the world's groundwater aquifers" (Mateo-Sagasta et al. 2017).

1.3.1.2 Pesticides

Pesticides such as "insecticides," "herbicides, and fungicides" are applied extensively in agriculture fields in several nations (Schreinemachers and Tipraqsa 2012; Mateo-Sagasta et al. 2017; Bhat et al. 2018) and get washed into aquatic ecosystems and pollute the water resources (Mateo-Sagasta et al. 2017). They contain "carcinogens and other poisonous substances that may kill aquatic life" or may be absorbed by them (Mateo-Sagasta et al. 2017) and pass through the "food chain until they become toxic to humans" (Mateo-Sagasta et al. 2017). "Millions of tons of pesticides are used in agriculture fields" (FAO 2016a; Mateo-Sagasta et al. 2017). "Acute pesticide poisoning causes significant human morbidity" (Mateo-Sagasta et al. 2017) and "mortality worldwide, especially in low income countries, where poor farmers often use highly hazardous pesticides" (Mateo-Sagasta et al. 2017).

1.3.1.3 Salts

Through irrigation, accumulated salts in soils are transported into receiving water bodies by drainage water and cause salinization (Mateo-Sagasta et al. 2017). The "intrusion of saline seawater into groundwater aquifers as a result of excessive groundwater extractions for agriculture is another important cause of salinization in coastal areas" (Mateo-Sagasta and Burke 2010; Mateo-Sagasta et al. 2017). "Major water-salinity problems have been reported in Argentina, Australia, China, India, the Sudan, the United States of America, and many countries in Central Asia" (FAO 2011). "Highly saline waters alter the geochemical cycles of major elements such as carbon, iron, nitrogen, phosphorus, silicon and sulphur" (Herbert et al. 2015; Mateo-Sagasta et al. 2017) with overall impacts on ecosystems (Mateo-Sagasta et al. 2017). Salinization can affect freshwater biota (Mateo-Sagasta et al. 2017) by "causing changes within species and community composition" (Mateo-Sagasta et al. 2017) and results "in decline of the biodiversity of microorganisms, algae, plants and animals" (Lorenz 2014; Mateo-Sagasta et al. 2017).

1.3.1.4 Emerging Pollutants

New "agricultural pollutants such as antibiotics, vaccines, growth promoters and hormones have emerged in the last two decades" (Mateo-Sagasta et al. 2017). These pollutants can reach water via "leaching and runoff from livestock" (Mateo-Sagasta et al. 2017) and "aquaculture farms, as well as through the application of manure and slurries to agricultural land" (OECD 2012; Mateo-Sagasta et al. 2017). Today, "more than 700 emerging pollutants and their metabolites and transformation products are listed as present in European aquatic environments" (Norman 2016; Mateo-Sagasta et al. 2017). "Agriculture is not only a source of emerging pollutants, it also contributes to the spread and reintroduction of such pollutants into aquatic environments through wastewater reuse for irrigation and the application of municipal biosolids to land as fertilizers" (Mateo-Sagasta et al. 2017). "An estimated 35.9 Mha of agricultural lands are subjected to the indirect use of wastewater" (Thebo et al. 2017; Mateo-Sagasta et al. 2017). The "potential risks to human health posed by exposure to emerging pollutants via contaminated agricultural products needs attention" (Mateo-Sagasta et al. 2017).

1.3.2 Sewage

The greatest volume of "waste discharged into the aquatic ecosystems is sewage." Sewage contains "industrial wastes, municipal wastes and domestic wastes which include wastes from baths, washing machines, kitchens and faecal matter." Fresh water sources "serve as best sinks for the discharge of these wastes" (Das and Acharya 2003; Tukura et al. 2009). It is estimated that "58% of the wastewater from urban areas and 81% of industrial wastes are discharged directly into water bodies with no or inadequate treatment results in contamination of ~73% of the water bodies" (Vargas-Gonzalez et al. 2014). The release of "sewage has led to the increase in water pollution and depletion of clean water resources" (Avalon Global Research 2012). "Huge loads of such wastes are generated daily from highly populated cities and are finally washed out by the drainage systems which generally open into nearby rivers or aquatic systems" (Tukura et al. 2009). It has resulted in "extensive ecological degradation such as a decline in water quality and availability, intense flooding, loss of species, and changes in the distribution and structure of the aquatic biota" (Oberdorff et al. 2002). The "negative impact of sewage is based on the composition and concentration of the contaminants as well as the volume and frequency of wastewater effluents entering water bodies" (Akpor and Muchie 2011; Bhat et al. 2017). "Sewage is comprised of several microorganisms, heavy metals, nutrients, radionuclides, pharmaceutical, and personal care products." Sewage is primarily organic in nature; owing to the organic load of sewage, the "oxygen concentration in the receiving waters is reduced, thus sewage is said to have a high BOD." The "effect of maltreated sewage on surface water is largely determined by the oxygen balance of the aquatic ecosystem, and its presence is essential in maintaining biological life within the system" (DFID 1999; Morrison et al. 2001; Momba et al. 2006). "DO concentrations below 5 mg/L can have a negative effect on the living organisms in the aquatic ecosystem" (Momba et al. 2006). Low dissolved oxygen concentration can affect "functioning of some fish species and can eventually lead to the death of fish population" (Igbinosa and Okoh 2009; Mehmood et al. 2019).

Decaying "organic matter" and "nutrients such as nitrites, nitrates, and phosphorus" in sewage can induce "eutrophication of water courses." "Eutrophication can lead to growth of plants and algae blooms" in the "aquatic ecosystem" (Bhat et al. 2017). "Algal blooms result in toxin production." Fish species "feeding in water contaminated" by "algal toxins will absorb these toxins and are subject to mass mortality" (Hernandez et al. 1998). Due to "eutrophication turbidity of the water increases, plant and animals' biomass increases, sedimentation rate increases, species diversity decreases, and anoxic conditions may develop, and this could give rise to change in dominant species of the aquatic biota" (Edokpayi 2016).

"Sewage effluent entering into surface waters contains a variety of pathogenic organisms that could result in the transmission of waterborne diseases when such contaminated water is used for domestic and other purposes" (WHO 2006; Chigor et al. 2013) thus is "detrimental to human health and the society at large" (DWA 1999). Some pathogens contaminate water resources (Mateo-Sagasta et al. 2017), via runoff (FAO 2006a; WHO 2012; Mateo-Sagasta et al. 2017). About "25% of all deaths worldwide are the result of infectious diseases caused by pathogenic microorganisms" (UNEP 2006). Scientists have identified about "1400 species of microorganisms that can cause ill health, including bacteria, protozoa, protozoan parasites, parasitic worms, fungi, and viruses" (CSIR 2010). Some common "pathogens found in sewage" are presented in Table 1.2 (WHO 2006; Christou 2011).

1.3.3 Heavy Metals

"Heavy metals enter the aquatic ecosystem from both natural and anthropogenic sources." Entry may be as a result of "direct discharges into both fresh and marine ecosystems or through indirect routes such as atmospheric deposition and surface run-off" (Biney et al. 1994). Important "natural sources are volcanic activity and weathering of rocks." "Heavy metals are natural constituents of rocks and soils and enter the environment as a consequence of weathering and erosion" (Förstner 1987). Heavy metals in "aquatic system can be naturally produced by the slow leaching from soil/rock to water, which are usually at low levels, causing no serious lethal effects on human health" (Chang et al. 2000; Rashid et al. 2019). The "development of industry and agriculture promotes the rapid increase of heavy metal pollution. Aquatic heavy metal pollution usually represents high levels of Mercury, Chromium, Lead, Cadmium, Copper, Zinc, Nickel etc. in the water system". "Arsenic, Cadmium, Copper, Mercury and Zinc are the five metals with most potential impact that enter the environment in elevated concentrations through storm water and wastewater discharges as a consequence of agricultural and industrial activity" (Alloway 2013). They are important "group of toxic contaminants because of their high toxicity and persistence in all aquatic ecosystems." Zinc and copper are present in "fertilizers as impurities, while Arsenic, Cadmium and Mercury are constituents of some fungicides and algaecides" (Fifield and Haines 2000) (Table 1.3).

Table 1.2 Microbial diseases associated with polluted aquatic environs

Agent	Species	Disease
Bacteria	"*Campylobacter jejune*" "*Escherichia coli*" "*E. coli* O157:H7" "*Helicobacter pylori*" "*Salmonella* sp." "*Salmonella typhi*" "*Shigella* sp." "*Vibrio cholera*"	"Gastroenteritis" "Gastroenteritis" "Bloody diarrhea, hemolytic uremic syndrome" "Abdominal pain, peptic ulcers, gastric cancer" "Salmonellosis, gastroenteritis, diarrhea" "Typhoid fever" "Dysentery" "Cholera"
Helminths	"*Ascaris lumbricoides* (round worm)" "*Clonorchis sinensis* (liver fluke)" "*Fasciola* (liver fluke)" "*Fasciolopsis buski* (intestinal fluke)" "*Opisthorchis viverrini*" "*Schistosoma* (blood fluke)" "*Trichuris* (whim worm)" "*Taenia* (tape worm)"	"Ascariasis" "Clonorchiasis" "Fascioliasis" "Fascioloidiasis" "Opisthorchiasis" "Schistosomiasis (bilharzia)" "Trichuriasis" "Taeniasis"
Protozoa	"*Balantidium coli*" "*Cryptosporidium parvum*" "*Cyclospora cayetanensis*" "*Entamoeba histolytica*" "*Giardia lamblia*"	"Balantidiasis (dysentery)" "Cryptosporidiosis" "Persistent diarrhea" "Amoebiasis (amoebic dysentery)" "Giardiasis"
Viruses	"*Adenovirus*" "*Astrovirus*" "*Calicivirus*" "*Coronavirus*" "*Eneroviruses*" "*Echovirus*" "*Poliovirus*" "*Hepatitis* A and E" "*Parvovirus*" "*Norovirus*" "*Rotavirus*" "*Coxsackieviruses*"	"Respiratory disease and eye infections" "Gastroenteritis" "Gastroenteritis" "Gastroenteritis" "Gastroenteritis" "Fever, rash, respiratory and heart disease, aseptic meningitis" "Paralysis, aseptic meningitis" "Infectious hepatitis" "Gastroenteritis" "Gastroenteritis" "Gastroenteritis" "Herpangina, aseptic meningitis, respiratory illness, fever, paralysis, respiratory, heart and kidney disease"

Heavy metals have "high ecological significance because they are not removed from water, but accumulate in the water reservoirs and thus enter the food chain" (Loska and Wiechuła 2003). Under "certain environmental conditions," heavy metals may "accumulate to a highly toxic concentration and cause ecological damage" (Harguinteguy et al. 2014). Once released in aquatic environments, they are generally "bound to particulate matter, which eventually settle down and become incorporated into sediments and are released into the water under the suitable conditions

Table 1.3 Different kinds of heavy metal discharge sources in aquatic environs (Fifield and Haines 2000)

Metal	Sources
Iron	"Pigments and paints, fuel, refineries, textile"
Manganese and zinc	"Batteries and electrical, pigments and paints, alloys and solders, pesticides, glass, fertilizers, refineries, fuel"
Lead	"Batteries and electrical, pigments and paints, alloys and solders, pesticides, glass, fertilizers, refineries, fuel, plastic"
Cadmium	"Batteries and electrical, pigments and paints, alloys and solders, plastic, fertilizers, fuel"
Nickel	"Batteries and electrical, pigments and paints, alloys and solders, fertilizers, fuel, catalysts"
Copper	"Batteries and electrical, pigments and paints, alloys and solders, fertilizers, pesticides, fuel, catalysts"
Chromium	"Pigments, fertilizers, textile"

such as pH values and Eh, leading to further contamination of aquatic environment" (Xu and Yang 1996). Accordingly, sediments represent one of the "ultimate sinks for heavy metals discharged into aquatic environment" (Gibbs 1973; Bryan and Langston 1992; Harguinteguy et al. 2014). "More and more attention has been drawn due to the wide spread occurrence of metal pollution in aquatic system" (Zhou et al. 2008). Some "heavy metals" may transform into the "persistent metallic compounds with high toxicity" (Zhou et al. 2008), which can be "bioaccumulated in the organisms" (Zhou et al. 2008), "magnified in the food chain, thus threatening human health" (Jin 1992; Zhou et al. 2008). "Various harmful effects including abnormal development of fetus, procreation failure, and immune deficiency has exhibited due to aquatic metal exposure" (Chang et al. 2000; Zhou et al. 2008). Some heavy metals, including mercury, chromium, cadmium, nickel, copper, and lead, introduced into water systems may pose high toxicities on the aquatic organisms (Wu and Zhao 2006). As an example, "cadmium is a priority environmental contaminant with consequences for human health and the maintenance of biodiversity in affected ecosystems" (Zhou et al. 2008) and "the timeliness of a broader, ecosystem-based approach to cadmium research is highlighted based on the overview of recent developments in the field" (Campbell 2006; Zhou et al. 2008).

1.3.4 Eutrophication

Eutrophication is a leading cause of destruction of many freshwater and marine ecosystems in the world. It is characterized by "excessive plant and algal growth due to the increased availability of one or more limiting growth factors needed for photosynthesis" (Schindler 2006), such as sunlight, carbon dioxide, and nutrients. "Eutrophication occurs naturally over centuries as lakes age and are filled in with sediments" (Carpenter 1981). However, "human activities have accelerated the rate and extent of eutrophication through both point-source discharges and non-point loadings of limiting nutrients, such as nitrogen and phosphorus, into aquatic ecosystems (i.e. cultural eutrophication), with dramatic consequences for drinking water sources, fisheries, and recreational water bodies" (Carpenter et al. 1998; Bhat et al. 2017). However, "during 1960s and 1970s, scientists linked algal blooms to nutrient enrichment resulting from anthropogenic activities such as agriculture, industry and sewage disposal" (Schindler 1974). The known "consequences of cultural eutrophication include blooms of blue-green algae (cyanobacteria), tainted drinking water supplies, degradation of recreational opportunities and hypoxia" (Bhat et al. 2017). The most obvious effect of cultural eutrophication is the creation of dense blooms of noxious, foul "smelling phytoplankton" that reduce water clarity and "harm water quality." "Algal blooms limit light penetration, reduce growth and cause death of plants in littoral zones and also lower the success of predators that need light to catch prey" (Lehtiniemi et al. 2005). Furthermore, high rates of photosynthesis associated with eutrophication can deplete dissolved inorganic carbon and raise pH to extreme levels during the day. "Elevated pH can in turn 'blind' organisms that rely on perception of dissolved chemical cues for their survival by impairing their chemosensory abilities" (Turner and Chislock 2010). When these "dense algal blooms eventually die, microbial decomposition severely depletes DO, creating a hypoxic dead zone, lacking sufficient oxygen to support most organisms." Dead zones are found in many "freshwater lakes including the Laurentian Great Lakes" (e.g., central basin of Lake Erie; Arend et al. 2011) during the summer. Furthermore, such "hypoxic events are particularly common in marine coastal environments surrounding large, nutrient-rich rivers" (e.g., Mississippi River and the Gulf of Mexico; Susquehanna River and the Chesapeake Bay) and have been shown to affect more than 245,000 square kilometers in over 400 near-shore systems (Diaz and Rosenberg 2008). "Hypoxia and anoxia as a result of eutrophication continue to threaten profitable commercial and recreational fisheries worldwide. Some algal blooms pose an additional threat because they produce noxious toxins" (e.g., microcystin and anatoxin-a) (Chorus and Bartram 1999). Over the past century, "harmful algal blooms (HABs) have been linked with (1) degradation of water quality" (Francis 1878), (2) "destruction of economically important fisheries" (Burkholder et al. 1992), and (3) "public health risks" (Morris 1999). Within freshwater ecosystems, "cyanobacteria are the most important phytoplankton associated with HABs" (Paerl 1988). "Toxigenic cyanobacteria, including *Anabaena*, Cylindrospermopsis, Microcystis, and *Oscillatoria* (Planktothrix), tend to dominate nutrient-rich, freshwater systems due to their superior competitive abilities under high nutrient concentrations, low nitrogen-to-phosphorus ratios, low light levels, reduced mixing, and high temperatures" (Downing et al. 2001; Paerl and Huisman 2009; Paerl and Paul 2012). Poisonings of "domestic animals, wildlife and even humans by blooms of toxic cyanobacteria" have been recognized throughout the world. Francis (1878) has first observed dead livestock due to "algal bloom of cyanobacteria" (Bhat et al. 2017). Also, cyanobacteria is responsible for several off-flavor compounds (e.g., methylisoborneal and geosmin) found in municipal drinking water systems as well as in aquaculture-raised fishes, resulting in large financial losses for state and regional economies (Crews and Chappell 2007). In addition to posing "significant public health risks, cyanobacteria have been shown to be poor quality food for most zooplankton grazers in laboratory studies" (Tillmanns et al. 2008; Wilson et al. 2006), thus reducing the efficiency of energy transfer in aquatic food webs and potentially preventing zooplankton from controlling algal blooms. Eutrophication is also associated with major changes in aquatic community structure. During "cyanobacterial blooms, small-bodied zooplankton tend to dominate plankton communities, and past observational studies have attributed this pattern to anti-herbivore traits of cyanobacteria" (e.g., toxicity, morphology, and poor food quality) (Porter 1977). However, the biomass of planktivorous fish is often positively related to nutrient levels and ecosystem productivity. Piscivorous fishes (e.g., bass, pike) tend to dominate the fish community of nutrient-poor, oligotrophic lakes, while planktivorous fishes (e.g., shad, bream) become increasingly "dominant with nutrient enrichment" (Jeppesen et al. 1997). Thus, an alternative explanation for the lack of zooplankton control of cyanobacterial blooms could include consumption of zooplankton by planktivores.

1.3.5 Plastics and Microplastics

Among the several human pressures on aquatic ecosystems, the accumulation of plastic debris is one of the most apparent but least studied. "Plastics generate significant benefits to the human society" (Andrady and Neal 2009), but due to its "durability, unsustainable use and inappropriate waste management plastics accumulate extensively in the natural habitats" (Barnes et al. 2009). Because of "high mobility, plastic debris has practically permeated the global marine environment" (Cole et al. 2011; Ivar do sul and Costa 2014), including the "polar region" (Barnes et al. 2009), "mid-ocean islands" (Ivar do sul et al. 2013), and "the deep sea" (Van Cauwenberghe et al. 2013). The sources of marine plastics are not very well characterized. A rough estimation predicts that "70 to 80% of marine litter, most of it is plastics, originate from inland sources and are emitted by rivers to the oceans" (GESAMP 2010). Rivers transport considerable amounts of plastics and "thus contribute significantly to the marine plastics pollution" (Moore et al. 2005; Lechner et al. 2014). "Plastics are dumped in huge volumes in beaches, lakes, navigation channels and other forms of water masses" (Lechner et al. 2014). The volume of plastic is even bigger in low-income countries with poor waste disposal regulations. In the marine environment, "plastics of various size classes and origins are omnipresent and affect numerous species that become entangled in or ingest plastics as well as an aesthetic problem" (Gregory 1999, 2009). "Plastics have been reported as a problem in the marine environment since the 1970s, but only recently the issue of plastic pollution in marine and freshwater environments been identified as a global problem" (Carpenter and Smith 1972). It has been reported that "single-use plastics (plastic bags and micro beads) are a major source of this pollution" (Desforges et al. 2014; Perkins 2015).

Under environmental conditions, larger plastic items degrade to so-called microplastics (MPs), typically smaller than 5 mm in diameter. "MPs are considered an emerging global issue by various experts" (Sutherland et al. 2010; Depledge et al. 2013) and international institutions (GESAMP 2010; UNEP 2011). Recent studies suggest that "risks of microplastics in the marine environment may pose more threat than macroplastics" (Thompson 2015; Diamond et al. 2018).

Potential sources of "MPs include wastewater treatment plants, runoff from urban, agricultural, touristic, and industrial areas, as well as shipping activities, beach litter, fishery and harbors" (Zubris and Richards 2005; Norén 2007; GESAMP 2010; Claessens et al. 2011; Dubaish and Liebezeit 2013). Another "potential source is sewage sludge that typically contains more MPs than effluents" (Leslie et al. 2012). A "broad spectrum of aquatic organisms are prone to MP ingestion ranging from plankton and fish to birds and even mammals, and accumulate throughout the aquatic food web" (Wright et al. 2013). Due to their large "surface-to-volume ratio and chemical composition, MPs accumulate environmental chemicals from the surrounding environment including metals" (Ashton et al. 2010) and "persistent, bioaccumulative, and toxic compounds" (Koelmans et al. 2013) transferring these contaminants from water to biota. "Plastic particles are also dominated by certain human pathogens like specific members of the genus *Vibrio*". Therefore, MPs can act as a vector for waterborne "human pathogens" influencing the water quality. In addition, "plastics contain and release a multitude of chemical additives" (Rochman 2013; Dekiff et al. 2014) and adsorb organic contaminants from the surrounding media (Bakir et al. 2012). Compounds such as MPs can transfered to organisms upon ingestion (Zarfl and Matthies 2010), this may increase "the chemical exposure of the ingesting organism and thus, toxicity" (Oehlmann et al. 2009; Teuten et al. 2009).

1.3.6 Oil Spills

An "oil spill" is defined as the discharge of "liquid petroleum hydrocarbons" into the environment, mainly in the "marine ecosystem" caused by human activity. "Environmental pollution caused by an oil spill is detrimental" (Broekema 2016). This is because "petroleum hydrocarbons are toxic to all forms of life and harm both aquatic and terrestrial organisms." The pollution of marine environments has caught the attention of researchers and environmentalists. This is due to the severe impact of oil spills on marine life. "A 1% increase in spill size has been estimated to increase the damage by some US$0.718 million" (Alló and Loureiro 2013). "Oil spills, which result from damage, transportation accidents and various other industrial and mining activities, are classified as hazardous waste" (Bartha and Bossert 1984). They are considered to be the most "frequent organic pollutants of aquatic ecosystems" (Bossert et al. 1984; Margesin and Schinnur 1997). Oil spills can occur from multiple sources including "oil tankers" (35.7%), facilities (27.6%), "non-tank vessels" (19.9%), "pipelines" (9.3%), and other sources (7.4%) (Benko and Drewes 2008). "Marine ship-source oil spill can occur as a result of ship accidents or operations, or the intentional discharge of oily wastes into oceans" (Knapp and Van de Velden 2011).

1.3.6.1 Major Oil Spills in the History

It is estimated that "3.2 million tons of oil is released per year from all sources into the environment. The majority of this oil is due to general shipping and industrial activities" (ITOPF 1990). During the Iran–Iraq war (1980–1988), approximately 2 million barrels of oil were discharged into the Arabian Gulf sea water. These included "1.5 million barrels from the Nawruz blow-out in 1983" (Watt 1994). Following the "Gulf War in 1991, 4 to 8 million barrels of oil were released into the Gulf and the Kuwaiti Desert and, making this the largest oil spill in the history at that time" (Purvis 1999). Previous observations indicated that the number of large "oil spills (>700 tons) has decreased significantly over the last 30 years" (ITOPF 1990). During the 1990s, the average number of large oil spills per year was about a third of the amount that was witnessed during the 1970s. It should be noted that "1,133,000 tons of oil was lost in the 1990s and 2000s, during 2010–2013, about 22,000 tons of oil was lost" (Levy and Gopalakrishnan 2010; Carriger and Barron 2011). The BP Deepwater Horizon oil spill on April 20, 2010, caused the discharge of more than 2.6 million gallons of oil into the Gulf of Mexico over just about 3 months. This "oil spill was the second largest in human history" (Levy and Gopalakrishnan 2010; Carriger and Barron 2011). During the 1991 Gulf War, the deliberate release of "over 6 million barrels of oil" (Randolph et al. 1998) into the marine environment was considered as the largest in history.

1.3.6.2 Impact on Human Health

Oil spills pose a great danger to humans. Direct "contact with crude oil or indirect contact through inhalation of vapors or consumption of contaminated seafood can cause deleterious health effects ranging from dizziness and nausea to certain types of cancers and issues with the central nervous system" (Aguilera et al. 2010; Major and Wang 2012). Toxic chemicals contained in the oil such as benzenes, toluene, poly-aromatic hydrocarbon, and oxygenated polycyclic aromatic hydrocarbons can harm the air quality (Tidwell et al. 2015). As witnessed in the "Kuwait Oil Fires, between January 16, 1991 and November 6, 1991, produced air pollution which caused respiratory distress" (Petruccelli et al. 1999). Oil-related disasters cause water contamination when the oil spillage comes in contact with any drinking water supply, for example, the 2013 incident in Miri, Malaysia, contaminated the water supply for 300,000 natives.

1.3.6.3 Impact on Coral Reefs

Coral reefs are considered to be important components of marine ecosystems. This is because "coral polyps are important nurseries for shrimp, fish and other animals" (Perkol-Finkel and Benayahu 2007). The aquatic organisms that live within and around the coral reefs are at risk of exposure to the toxic substances within oil. They are rapidly degrading because of a variety of environmental and anthropogenic pressures. Thus, they are suffering significant changes in "species diversity," "species abundance," "species evenness," and "habitat structure" worldwide (Hughes et al. 2007). "Oil dispersants are potentially harmful to marine life including coral reefs" (Shafir et al. 2007). A study using coral nubbins in coral reef eco-toxicology testing (Shafir et al. 2003) found that dispersed oil and oil dispersants are harmful to soft and hard coral species at early life stages.

1.3.6.4 Impact on Marine Mammals

Marine mammals include "bottlenose dolphins, fins, humpbacks, rights, sei whales, sperm whales, manatees, cetaceans, seals, sea otters and pinnipeds." The physical contact of oil with furred mammals affects these animals because they rely on their outer coats for buoyancy and warmth. Consequently, "these animals often succumb to hypothermia, drowning and smothering when oil flattens and adheres to the outer layer" (Lin and Tjeerdema 2008).

1.3.6.5 Impact on Seabirds

Physical contact is one of the major routes of exposure, and it usually affects seabirds (Table 1.4). For example, thousands of African penguins (*Spheniscus demerus*) were oiled following the 2000 treasure oil spill in South Africa.

1.3.7 Aquaculture Activities

Aquaculture is the farming of aquatic organisms. The "rapidly growing human population is creating an increase in the demand for fish worldwide" (Tidwell and Allan 2001). The amount of "fish captured in fisheries is no longer meeting this demand because the annual production of captured fish has not changed significantly since 2011" (FAO 2016b). "Aquaculture is becoming a more popular fish production method as it has an annual increase of 6% and is projected to produce over half of the fish consumed by 2025" (FAO 2016b). "Aquaculture has tremendous benefits for the humans like seafood production by fisheries and contributes with 15 to 20% of average animal protein consumption to 2.9 billion people worldwide" (Smith et al. 2010). The nutritional quality of aquatic products has "high standard and

Table 1.4 Mass motility of seabirds collected at "Exxon Valdez and Braer oil spills" (Dauvin 1998)

"Species group"	"Alaskan spill"	"Shetland spill"
"Sea ducks (eiders, etc.)"	1435–1445	168
"Mergansers"	120–125	1–2
"Loons"	390–400	12–15
"Grebes"	460–462	0
"Heron"	1	2–4
"Geese/swans"	8–10	0
"Gulls"	694–698	72–76
"Kittiwakes"	1220–1230	130–135
"Cormorants/shags"	835–837	862–865
"Shearwaters"	3400	0
"Fulmars"	868–872	30–32
"Guillemots/murres"	20,560–20,562	218–222
"Other auks"	2172–2176	228–230
"Bald eagles"	124–126	NA
"Other birds"	3150–3153	0
Total	35,466–35,468	1535–1538

represents an important source of macro and micronutrients for the people from developing countries" (Roos et al. 2007). Despite the undeniable benefits of aquaculture such as the provision of good quality and accessible food for population and the generation of millions of jobs and billion dollars in budget for the developing countries, the activity is one of the most criticized worldwide, mainly because of the environmental impacts (FAO 2016c). The most common "negative environmental impacts that are associated with aquaculture is water eutrophication, water quality, alteration or destruction of natural habitats, introduction and transmission of diseases" (FAO 2006b).

1.3.8 Harmful Impacts Related to the Aquaculture Activities Are as Follows

1.3.8.1 Eutrophication of Receiving Waters

Aquaculture can be "a major contributor to eutrophication or organic loads in the receiving waters" (Mateo-Sagasta et al. 2017). It is mainly produced by "uneaten feed (especially due to overfeeding), lixiviation of aquaculture feedstuffs" (Focardi et al. 2005; Crab et al. 2007), "decomposition of died organisms and over fertilization" (Feng et al. 2004; Gyllenhammer and Hakanson 2005). In Scotland, for example, "the discharge of untreated organic waste from salmon production is equivalent to 75% of the pollution discharged by the human population" (Mateo-Sagasta et al. 2017). "Shrimp aquaculture in Bangladesh generates 600 tons of waste per day"

(SACEP 2014; Mateo-Sagasta et al. 2017). It is well documented that from "the total nitrogen supplemented to the cultured organisms, only 20–50% is retained as biomass by the farmed organisms, while the rest is included into the water column or sediment" (Jackson et al. 2003; Schneider et al. 2005) and "eventually discharged into the receiving ecosystems, increases the risk of eutrophication and algal blooms (like toxic microalgae-red tides) in lakes" (Mateo-Sagasta et al. 2017), reservoirs, and coastal areas (Alonso-Rodriguez and Paez-Osuna 2003; Mirto et al. 2009; Mateo-Sagasta et al. 2017). "Organic pollutants consume dissolved oxygen (DO) in the water as it degrades quality characteristics of fresh water, with the result DO drops, fish and other aquatic life are exposed to extreme conditions or killed due to hypoxia in water bodies" (Mateo-Sagasta et al. 2017).

1.3.8.2 Introduction of Exotic Species

Aquaculture comes in multiple versions, two of which are open systems and closed systems. "Open systems are found offshore in coastal areas, exposed to natural environments" (Lawson 1995). These systems are high-risk because they allow unchecked interactions between the farmed fish and surrounding environment, which leads to "free exchange of diseases, parasites and fecal matter" (Ali 2006). The recent study has revealed "a parasite transmission of sea lice from captive to wild salmon" (Krkosek et al. 2007). The only barrier between the harvested fish and the wild population is a rigid cage or netting system. When these netting systems are damaged during inclement weather such as snowstorms or hurricanes, it allows "fish to escape from the open systems" (Centre for Food Safety 2012). There were "25 million reported fish escapes worldwide and the majority occurred when netting was damaged during severe weather conditions" (Centre for Food Safety 2012). The escaping of "exotic aquaculture species into the natural ecosystem causes the displacement of native populations, competition for food, space, mates and prey" (Naylor et al. 2005).

1.3.8.3 Destruction of Mangrove Forests

"Aquaculture farms" are constructed in "mangrove forests" (Dewalt et al. 2002; Stickney and McVey 2002; Rajitha et al. 2007). "Mangrove forests" are important ecosystems as they act as nurseries for many "aquatic species" as well as nesting areas for "birds, reptiles, crustaceans and other taxonomic groups" (Paez-Osuna 2005). The cover of mangrove forest has decreased worldwide from "19.8 million hectares in 1980 to less than 15 millions hectares in 2000." The annual "deforestation rate was 1.7% from 1980 to 1990 and 1% from 1990 to 2000" (FAO 2007), and the "problem of deforestation still continue today." "Aquaculture has been responsible for the deforestation of millions of hectares of mangrove forest in Thailand, Indonesia, Ecuador, Madagascar and other countries" (Harper et al. 2007).

1.3.8.4 Contamination of Water for Human Consumption

"Inland aquaculture" has been responsible for the "degradation of water bodies used for human consumption" (Paez-Osuna 2001). Aquaculture activities cause death of benthic organisms as well as undesirable odors and the presence of pathogens in the discharge sites (Martinez-Cordova and Enriquez-Ocana 2007). The spread and the "outbreaks of diseases are negative consequences of the expansion and diversification of the aquaculture sector" (Crisafi et al. 2011; Mancuso 2013; Mancuso et al. 2013).

1.3.9 Preventive Measures and some Humanistic Solutions

"Water contamination" can be reduced from a "personal level" to "national and international level." Every individual has a duty to prevent pollution of water resources. "Water is a basic need for our survival," and hence it should be our first priority to keep all "water resources" free from contamination. There are various "sources of water contamination." Thus, the control of water contamination needs a range of preventive measures. "Measures of prevention and control are essential in improving the quality of water" and reducing the "costly treatment measures that are taken to treat water." Preventive measures and possible solutions to "control water contamination" are given as follows (Xiong et al. 2015; Lan et al. 2015; Xanthos and Walker 2017; Barmentlo et al. 2018):

- "Do not throw rubbish away in places like the beach, riverside and water bodies rather put it in trash can."
- "Use water wisely. Do not keep the tap running when not in use."
- "Do not throw chemicals, oils, plastics, paints and medicines down the sink drain, or the toilet."
- "Buy more environmentally safe cleaning liquids for use at home and other public places."
- "Not to overuse pesticides and fertilizers in farms. This will reduce runoffs of the chemical into nearby water sources."
- "Natural fertilizers such as peat, compost, manure should be preferred while gardening and farming."
- "Implementing water quality laws they can help in protecting aquatic ecosystems by imposing acceptable concentrations of pollutants and prevents the release of pollutants into water resources."
- "Proper use and disposeal of chemicals prevent the contamination of aquatic environments."
- "Use detergents with low or no phosphate because high phosphate content causes eutrophication of lakes."
- "Control storm water runoff. As the storm water runoff flows over impervious surfaces, it collects debris, sediments, chemicals and other pollutants which can have negative effects on the quality of water if the runoff is left untreated."

- "Decrease water resistant surfaces such as cement around homes to reduce surface runoff. Vegetation, porous materials, gravel, wood decking etc. can be used instead of cement."
- "Avoid throwing garbage into lakes, rivers and streams and help in cleaning litter around water resources."
- "Wash your automobiles at carwashes instead of washing it yourself. The wastewater from these carwashes is drained into the sewer and treated which reduce the amount of pollutants in the water."
- "Speak up against industries that dump waste into local streams, rivers, and beach fronts to reduce water pollution in your community."
- "Implement existing environmental laws. There are very strict laws that help minimize water pollution. These laws are usually directed at industries, hospitals, schools and market areas on how to dispose, treat and manage sewage."
- "Do not dispose non-degradable products such as plastic bags or plastic wrappers down the drain."

1.4 Conclusion

The degradation of aquatic ecosystems is largely due to human activities. Increased urbanization and industrialization are greatly responsible for water pollution. Human contribution to water pollution is enormous, such as dumping of solid wastes, industrial wastes, and domestic wastes. Water pollution is a major concern to the world. Environmental education is very important to reduce the pollution of aquatic ecosystems.

References

Abowel JFN, Sikoki FD (2005) Water pollution management and control. Double Trust, Port Harcourt

Aguilera F, Méndez J, Pásaro E, Laffon B (2010) Review on the effects of exposure to spilled oils on human health. J Appl Toxicol 30(4):291–301

Akpor OB, Muchie M (2011) Environmental and public health implications of wastewater quality. Afr J Biotechnol 10(13):2379–2387

Al Naggar Y, Naiem E, Mona M, Giesy J, Seif A (2014) Metals in agricultural soils and plants in Egypt. Toxicol Environ Chem 96(5):730–742

Ali AMS (2006) Rice to shrimp: land use/land cover changes and soil degradation in southwestern Bangladesh. Land Use Policy 23:421–435

Alló M, Loureiro ML (2013) Estimating a meta-damage regression model for large accidental oil spills. Ecol Econ 86:167–175

Alloway BJ (2013) Introduction: in heavy metals in soils. Springer, Dordrecht, pp 3–9

Alonso-Rodriguez R, Paez-Osuna F (2003) Nutrients, phytoplankton and harmful algal blooms in shrimp ponds: a review with special reference of the situation in the Gulf of California. Aquaculture 219(1–4):317–336

Andrady AL, Neal MA (2009) Applications and societal benefits of plastics. Philos Trans R Soc Lond Ser B Biol Sci 364:1977–1984
Arend KK, Beletsky D, DePinto J, Ludsin SA, Roberts JJ (2011) Seasonal and inter annual effects of hypoxia on fish habitat quality in Central Lake Erie. Freshwat Biol 56:366–383
Ashton K, Holmes L, Turner A (2010) Association of metals with plastic production pellets in the marine environment. Mar Pollut Bull 60:2050–2055
Avalon Global Research (2012) Water and waste water treatment opportunity in India: an overview
Bakir A, Rowland SJ, Thompson RC (2012) Competitive sorption of persistent organic pollutants onto microplastics in the marine environment. Mar Pollut Bull 64:2782–2789
Barange M, Field JG, Harris RP, Eileen E, Hofmann EE, Perry RI, Werner F (2010) Marine ecosystems and global change. Oxford University Press, Oxford, p 464
Barmentlo SH, Schrama M, Hunting ER, Heutink R, van Bodegom PM, de Snoo GR, Vijver MG (2018) Assessing combined impacts of agrochemicals: aquatic macroinvertebrate population responses in outdoor mesocosms. Sci Total Environ 631:341–347
Barnes DK, Galgani F, Thompson RC, Barlaz M (2009) Accumulation and fragmentation of plastic debris in global environments. Philos Trans R Soc Lond Ser B Biol Sci 364:1985–1998
Bartha R, Bossert I (1984) The treatment and disposal of petroleum refinery wastes. In: Atlas RM (ed) Petroleum microbiology. Macmillan, New York, pp 1–61
Benko KL, Drewes JE (2008) Produced water in the Western United States: geographical distribution, occurrence, and composition. Environ Eng Sci 25:239–246
Bhat RA, Shafiq-ur-Rehman MMA, Dervash MA, Mushtaq N, Bhat JIA, Dar GH (2017) Current status of nutrient load in Dal Lake of Kashmir Himalaya. J Pharmacog Phytochem 6(6):165–169
Bhat RA, Beigh BA, Mir SA, Dar SA, Dervash MA, Rashid A, Lone R (2018) Biopesticide techniques to remediate pesticides in polluted ecosystems. In: Wani KA, Mamta (eds) Handbook of research on the adverse effects of pesticide pollution in aquatic ecosystems. IGI Global, Hershey, PA, pp 387–407
Biney C, Amuzu AT, Calamari D, Kaba N, Mbome IL, Naeve H, Ochumba PB, Osibanjo O, Radegonde V, Saad MA (1994) Review of heavy metals in the African aquatic environment. Ecotoxicol Environ Saf 28(2):134–159
Bossert ID, Kachel WM, Bartha R (1984) Fate of hydrocarbons during oily sludge disposal in soil. Appl Environ Microbiol 47:763–767
Broekema W (2016) Crisis-induced learning and issue politicization in the Eu: the Braer, sea empress, Erika and prestige oil spill disasters. Public Adm 94:381–398
Bryan GW, Langston WJ (1992) Bioavailability, accumulation and effects of heavy metals in sediments with special reference to United Kingdom estuaries: a review. Environ Pollut 76(2):89–131
Burkholder JM, Noga EJ, Hobbs CH Jr, Glasgrow HB (1992) New 'phantom' dinoflagellate is the causative agent of major estuarine fish kills. Nature 358:407–410
Campbell PGC (2006) Cadmium-a priority pollutant. Environ Chem 3:387–388
Carpenter SR (1981) Submerged vegetation: an internal factor in lake ecosystem succession. Am Nat 118:372–383
Carpenter EJ, Smith KL (1972) Plastics on the Sargasson Sea surface. Science 175(4027):1240–1241
Carpenter SR, Caraco NF, Correl DL, Howarth RW, Sharpley AN, Smith VH (1998) Nonpoint pollution of surface waters with phosphorus and nitrogen. Ecol Appl 8:559–568
Carriger J, Barron MG (2011) Minimizing risks from spilled oil to ecosystem services using influence diagrams: the Deepwater horizon response. Environ Sci Technol 45:7631–7639
Centre for Food Safety (2012) Reported escapes from fish farm
Chang XX, Wen CH, Wang HJ (2000) Effect of heavy metal pollution on human health and sustainable development. Yunnan Environ Sci 19:59–61
Chigor VN, Sibanda T, Okoh AI (2013) Studies on the bacteriological qualities of the Buffalo River and three source water dams along its course in the eastern Cape Province of South Africa. Environ Sci Pollut Res 20:4125–4136
Chorus I, Bartram J (1999) Toxic cyanobacteria in water: a guide to their public health consequences, monitoring, and management. E & FN Spon, London

Christou L (2011) The global burden of bacterial and viral zoonotic infections. Clin Microbiol Infect 17(3):326–330

Claessens M, De Meester S, Van Landuyt L, De Clerck K, Janssen CR (2011) Occurrence and distribution of microplastics in marine sediments along the Belgian coast. Mar Pollut Bull 62:2199–2204

Cole M, Lindeque P, Halsband C, Galloway TS (2011) Microplastics as contaminants in the marine environment: a review. Mar Pollut Bull 62:2588–2597

Crab R, Avnimelech Y, Defoirdt T, Bossier P, Verstraete W (2007) Nitrogen removal techniques in aquaculture for a sustainable production. Aquaculture 270(1–4):1–14

Crews JR, Chappell JA (2007) Agriculture and natural resources U.S. catfish industry outlook. Auburn University, Auburn, AL

Crisafi F, Denar R, Genovese M, Cappello S, Mancuso M, Genovese L (2011) Comparison between 16SrDNA and ToxR genes as targets for the detection of vibrio anguillarum in Dicentrarchus labrax kidney and liver. Res Microbiol 162:223–230

CSIR (2010) A CSIR perspective on water in South Africa. CSIR report no. CSIR/NRE/PW/IR/2011/0012/A

Das J, Acharya BC (2003) Hydrology and assessment of lotic water quality in Cuttack City, India. Water Air Soil Pollut 150:163–175

Dauvin JC (1998) The fine sand abra alba community of the Bay of Morlaix twenty years after the Amoco Cadiz oil spill. Mar Pollut Bull 36:669–676

Dekiff JH, Remy D, Klasmeier J, Fries E (2014) Occurrence and spatial distribution of microplastics in sediments from Norderney. Environ Pollut 186:248–256

Demirak A, Yilmaz F, Levent Tuna A, Ozdemir N (2006) Heavy metals in water, sediment and tissues of Leuciscus cephalus from a stream in southwestern Turkey. Chemosphere 63(9):1451–1458

Depledge MH, Galgani F, Panti C, Caliani I, Casini S, Fossi MC (2013) Plastic litter in the sea. Mar Environ Res 92:279–281

Desforges JP, Galbraith M, Dangerfield N, Ross PS (2014) Widespread distribution of microplastics in subsurface seawater in the NE Pacific Ocean. Mar Pollut Bull 79(1–2):94–99

Dewalt BR, Ramirez-Zavala JR, Noriega L, Gonzalez E (2002) Shrimp aquaculture, the people and the environment in coastal Mexico, Tech. Rep., World Bank, NACA, WWF y FAO consortium program on shrimp farming and the environment

DFID (1999) In: Pearce GR, Chaudhry MR, Ghulum S (eds) A simple methodology for water quality monitoring. Department for International Development, Wallingford, p 100

Diamond J, Altenburger R, Coors A, Dyer SD, Focazio M, Kidd K, Tolls J (2018) Use of prospective and retrospective risk assessment methods that simplify chemical mixtures associated with treated domestic wastewater discharges. Environ Toxicol Chem 37(3):690–702

Diaz RJ, Rosenberg R (2008) Spreading dead zones and consequences for marine ecosystems. Science 321:926–929

Don-pedro KN (1990) Pesticide pollution-biological resources for control and management. In: Proceedings of the conference on pesticide pollution detection and management at the University of Agriculture, Abeokuta, Nigeria

Downing JA, Watson SB, McCauley E (2001) Predicting cyanobacteria dominance in lakes. Can J Fish Aquat Sci 58:1905–1908

Dubaish F, Liebezeit G (2013) Suspended microplastics and black carbon particles in the jade system, southern North Sea. Water Air Soil Pollut 224:1352

DWA (1999) Wastewater limit values applicable to the discharge of wastewater into a river resource. National Water Act, Government Gazette No. 20528

Edokpayi JN (2016) Assessment of the efficiency of wastewater treatment facilities and the impact of their effluent on surface water and sediments in Vhembe District, South Africa, PhD thesis submitted to the University of Venda, South Africa

Ekubo AJ, Abowel JFN (2011) Aspects of aquatic pollution in Nigeria. Res J Environ Earth Sci 3:673–693

FAO (2006a) Food and Agriculture Organization of the United Nations, Livestock's Long Shadow, Rome
FAO (2006b) State of world aquaculture, FAO fisheries technical paper no. 500, Rome, p 134
FAO (2007) El estadomundial de la pesca y acuicultura, Rome, Italy
FAO (2011) Food and Agriculture Organization of the United Nations (FAO) and London, the state of the world's land and water resources for food and agriculture, Earthscan, Rome
FAO (2013) Guidelines to control water pollution from agriculture in China, Water report 40
FAO (2016a) Food and Agriculture Organization of the United Nations (FAO). FAOSTAT, Rome
FAO (2016b) The state of world fisheries and aquaculture, contributing to food security and nutrition for all. Food and Agriculture Organization of United Nations, Rome, p 200
FAO (2016c) The state of world fisheries and aquaculture, contributing to food security and nutrition for all. Food and Agriculture Organization, UN, Rome
Feng YY, Hou LC, Ping NX, Ling TD, Kyo CI (2004) Development of mariculture and its impacts in Chinese coastal waters. Rev Fish Biol Fish 14(1):1–10
Fernandesa C, Fontainhas FA, Peixotoc F, Salgadod MA (2007) Bioaccumulation of heavy metals in Liza saliens from the Esmoriz –Paramos coastal lagoon, Portugal. Ecotoxicol Environ Saf 66(3):426–431
Fifield FW, Haines PJ (2000) Environmental analytical chemistry. Wiley-Blackwell, Hoboken, NJ
Focardi S, Corsi I, Franchi E (2005) Safety issues and sustainable development of European aquaculture: new tools for environmentally sound aquaculture. Aquacult Int 13(1–2):3–17
Förstner U (1987) Metal speciation in solid wastes-factors affecting mobility, speciation of metals in water, sediment and soil systems. Springer, Berlin, Heidelberg, pp 11–41
Francis G (1878) Poisonous Australian lake. Nature 18:11–12
GESAMP (2010) IMO/FAO/UNESCO-IOC/UNIDO/WMO/IAEA/UN/UNEP Joint Group of Experts on the Scientific Aspects of Marine Environmental Protection. In: Proceedings of the GESAMP international workshop on micro-plastic particles as a vector in transporting persistent, bio-accumulating and toxic substances in the oceans
Ghani SAA (2015) Trace metals in seawater, sediments and some fish species from MarsaMatrouh beaches in North-Western Mediterranean coast, Egypt. Egypt J Aquat Res 41(2):145–154
Gibbs RJ (1973) Mechanisms of trace metal transport in rivers. Science 180(4081):71–73
Gregory MR (1999) Plastics and South Pacific Island shores: environmental implications. Ocean Coast Manag 42:603–615
Gregory MR (2009) Environmental implications of plastic debris in marine settings: entanglement, ingestion, smothering, hangers-on, hitch-hiking and alien invasions. Philos Trans R Soc Lond B Biol Sci 364:2013–2025
Gyllenhammer A, Hakanson L (2005) Environmental consequence analyses of fish farm emissions related o different scales and exemplified by data from the Baltic—a review. Mar Environ Res 60(2):211–243
Halpern BS, Walbridge S, Selkoe KA, Kappel CV, Micheli F, D'agrosa C, Fujita R (2008) A global map of human impact on marine ecosystems. Science 319(5865):948–952
Hampel M, Blasco J, Segner H (2015) Molecular and cellular effects of contamination in aquatic ecosystems. Environ Sci Pollut Res 22:17261–17266
Harguinteguy CA, Cirelli AF, Pignata MI (2014) Heavy metal accumulation in leaves of aquatic plant Stuckenia filiformis and its relationship with sediment and water in the Suquía river (Argentina). Microchem J 114:111–118
Harper GJ, Steininger MK, Tucker CJ, Juhn D, Hawkins F (2007) Fifty years of deforestation and forest fragmentation in Madagascar. Environ Conserv 34(4):325–333
Herbert ER, Boon P, Burgin AJ, Neubauer SC, Franklin RB, Ardón M, Hopfensperger KN, Lamers LPM, Gell P (2015) A global perspective on wetland salinization: ecological consequences of a growing threat to freshwater wetlands. Ecosphere 6(10):1–43
Hernandez M, Robinson I, Aguilar A, Gonzalez LM, Lopez- Jurado LF, Reyero MI, Cacho E, Franco J, Lopez-Rodas V, Costas E (1998) Did algal toxins cause monk seal mortality. Nature 393:28–29

Hughes TP, Rodrigues MJ, Bellwood DR, Ceccarelli D, Hoegh-Guldberg O, McCook L, Moltschaniwskyj N, Pratchett MS, Steneck RS, Willis B (2007) Phase shifts, herbivory and the resilience of coral reefs to climate change. Curr Biol 17:360–365
Hughes SR, Kay P, Brown LE (2013) Global synthesis and critical evaluation of pharmaceutical data sets collected form river systems. Environ Sci Technol 47:661–677
Igbinosa EO, Okoh AI (2009) The impact of discharge wastewater effluents on the physiochemical qualities of a receiving watershed in a typical rural community. Int J Environ Sci Technol 6(2):175–182
ITOPF (1990) Response to marine oil spills. Wither by, International Tanker Owners Pollution Federation, London
Ivar do Sul JA, Costa MF (2014) The present and future of microplastic pollution in the marine environment. Environ Pollut 185:352–364
Ivar do Sul JA, Costa MF, Barletta M, Cysneiros FJ (2013) Pelagic microplastics around an archipelago of the equatorial Atlantic. Mar Pollut Bull 75:305–309
Jackson C, Preston N, Thompson PJ, Burford M (2003) Nitrogen budget and effluent nitrogen components at an intensive shrimp farm. Aquaculture 218(1–4):397–411
Jeppesen E, Jensen JP, Sondergaard M, Lauridsen T, Pedersen LJ, Jensen L (1997) Top-down control in freshwater lakes: the role of nutrient state, submerged macrophytes and water depth. Hydrobiologia 342(343):151–164
Jin L (1992) Environmental bionomy, 1st edn. High Education Press, Beijing
Knapp S, Van de Velden M (2011) Global ship risk profiles: safety and the marine environment. Trans Res D Trans Environ 16(8):595–603
Koelmans AA, Besseling E, Wegner A, Foekema EM (2013) Plastic as a carrier of POPs to aquatic organisms: a model analysis. Environ Sci Technol 47:7812–7820
Krkosek M, Ford JS, Morton A, Lele S, Myers RA, Lewis A (2007) Declining wild salmon populations in relation to parasites from farm salmon. Science 318(5857):1772–1775
Lan D, Liang B, Bao C, Ma M, Xu Y, Yu C (2015) Marine oil spill risk mapping for accidental pollution and its application in a coastal city. Mar Pollut Bull 96(1–2):220–225
Larsson DGJ (2014) Pollution from drug manufacturing: review and perspectives. Phil Trans R Soc Lond B Biol Sci 369(1656):20130571
Lawson TB (1995) Aquaculture in open systems. In: Fundamentals of aquacultural engineering. Springer, Boston, MA, pp 58–83
Lechner A, Keckeis H, Lumesberger-Loisl F, Zens B, Krusch R, Tritthart M, Glas M, Schludermann E (2014) The Danube so colourful: a potpourri of plastic litter outnumbers fish larvae in Europe's second largest river. Environ Pollut 188:177–181
Lehtiniemi M, Engstrom-Ost J, Viitasalo M (2005) Turbidity decreases anti-predator behaviour in pike larvae, Esox Lucius. Environ Biol Fish 73(1):1–8
Leslie HA, Moester M, de Kreuk M, Vethaak AD (2012) Pilot study on emissions of microplastics from wastewater treatment plants. H2O 14(15):45–47
Levy J, Gopalakrishnan C (2010) Promoting ecological sustainability and community resilience in the US Gulf Coast after the 2010 Deep Ocean horizon oil spill. J Nat Res Poll Res 2:297–315
Lin C, Tjeerdema RS (2008) Crude oil, oil, gasoline and petrol. In: Jorgensen SE, Fath BD (eds) Encyclopedia of ecology, Ecotoxicology, vol 1. Elsevier, Oxford, pp 797–805
Lorenz JJ (2014) A review of the effects of altered hydrology and salinity on vertebrate fauna and their habitats in northeastern Florida bay. Wetlands 34:189–200
Loska K, Wiechuła D (2003) Application of principal component analysis for the estimation of source of heavy metal contamination in surface sediments from the Rybnik reservoir. Chemosphere 51(8):723–733
Major DN, Wang H (2012) How public health impact is addressed: a retrospective view on three different oil spills. Toxicol Environ Chem 94:442–467
Malaj E, von der Ohe PC, Grote M, Kühne R, Mondy CP, Usseglio-Polatera P, Brack W, Schäfer RB (2014) Organic chemicals jeopardize the health of freshwater ecosystems on the continental scale. Proc Natl Acad Sci U S A 111:9549–9554

Mancuso M (2013) Fish welfare in aquaculture-editorial. J Aquacult Res Develop 3:4–6
Mancuso M, Caruso G, Adone R, Genovese L, Crisafi E, Zaccone R (2013) Detection of Photobacterium damselae sub sp. piscida in sea waters by fluorescent antibody. J Appl Aquac 25:337–345
Margesin R, Schinnur F (1997) Efficiency of endogenous and inoculated cold-adapted soil microorganisms for biodegradation of diesel oil in alpine soils. Appl Environ Microbiol 63:2660–2664
Martinez-Cordova LR, Enriquez-Ocana F (2007) Study of the benthic fauna in a discharge lagoon of the shrimp faro with special emphasis on polychaeta. J Biol Sci 7:12–17
Mateo-Sagasta J, Burke J (2010) Agriculture and water quality interactions: a global overview. SOLAW background thematic report-TR08. Food and Agriculture Organization of the United Nations (FAO), Rome
Mateo-Sagasta J, Zadeh SM, Turral H, Burke J (2017) Water pollution from agriculture: a global review. Food and Agriculture Organization of the United Nations, Rome and the International Water Management Institute on behalf of the Water Land and Ecosystems Research Program, Colombo
Mehmood MA, Qadri H, Bhat RA, Rashid A, Ganie SA, Dar GH, Shafiq-ur-Rehman (2019) Heavy metal contamination in two commercial fish species of a trans-Himalayan freshwater ecosystem. Environ Monit Assess Environ 191:104. https://doi.org/10.1007/s10661-019-7245-2
Mendil D, Uluözlu ÖD (2007) Determination of trace metal levels in sediment and five fish species from lakes in Tokat, Turkey. Food Chem 101(2):739–745
Mirto S, Bianchelli S, Gambi C, Krzelj M, Pusceddu A, Scopa M (2009) Fish-farm impact on metazoan meiofauna in the Mediterranean Sea: analysis of regional vs. habital effects. Mar Environ Res 69:38–47
Momba MNB, Osode AN, Sibewu M (2006) The impact of inadequate wastewater treatment on the receiving water bodies-case study: Buffalo City and Nkokonbe Municipalities of the Eastern Cape Province. Water SA 32:5
Moore CJ, Lattin GL, Zellers AF (2005) A snapshot of land-based contributions of plastic and other trash to coastal waters and beaches of Southern California, working our way upstream. Algalita Marine Research Foundation, Long Beach, CA
Morris JG (1999) Harmful algal blooms: an emerging public health problem with possible links to human stress on the environment. Annu Rev Energy Environ 24:367–390
Morrison G, Fatoki OS, Persson L, Ekberg A (2001) Assessment of the impact of point source pollution from the Keiskammahoek sewage treatment plant on the Keiskamma river—pH, electrical conductivity, oxygen demanding substance (COD) and nutrients. Water SA 27(4):475–480
Mushtaq N, Bhat RA, Dervash MA, Qadri H, Dar GH (2018) Biopesticides: the key component to remediate pesticide contamination in an ecosystem. In: Environmental contamination and remediation. Cambridge Scholars Publishing, Cambridge, pp 152–178
Naylor R, Hindar K, Fleming IA, Goldburg S, Williams S, Volpe J, Whoriskey F, Eagle J, Kelso D, Mangel M (2005) Fugative salmon: assessing the risks of escaped fish from net-pen aquaculture. Bioscience 55(5):427–437
NEST (1991) Nigeria's threatened environment: a national profile. Nigerian Environmental Study/Action Teams, Ibadan, p 288, ISBN-13: 9789783120303
Norén F (2007) Small plastic particles in coastal Swedish waters. KIMO, Sweden
NORMAN (2016) List of emerging substances, Network of Reference Laboratories, Research Centres and related Organisations for Monitoring of Emerging Environmental Substances (NORMAN)
Oberdorff T, Pont D, Hugueny B, Porcher J (2002) Development and validation of a fish based index for the assessment of 'river health' in France. Fresh Water Biol 47(9):1720–1734
OECD (2012) Organisation for Economic Co-operation and Development. New and emerging water pollutants arising from agriculture, prepared by Alistair B.A. Boxall, Paris
Oehlmann J, Schulte-Oehlmann U, Kloas W, Jagnytsch O, Lutz I, Kusk KO, Wollenberger L, Santos EM, Paull GC, Van Look KJ, Tyler CR (2009) A critical analysis of the biological impacts of plasticizers on wildlife. Philos Trans R Soc Lond B Biol Sci 364:2047–2062

Paerl HW (1988) Nuisance phytoplankton blooms in coastal, estuarine and inland waters. Limnol Oceanogr 33:823–847
Paerl HW, Huisman J (2009) Climate change: a catalyst for global expansion of harmful cyanobacterial blooms. Environ Microbiol Rep 1:27–37
Paerl HW, Paul VJ (2012) Climate change: links to global expansion of harmful cyanobacteria. Water Res 46:1349–1363
Paez-Osuna F (2001) The environmental impact of shrimp aquaculture: causes, effects and mitigating alternatives. Environ Manag 28(1):131–140
Paez-Osuna F (2005) Restos y perspectivas de la camaroniculturaen la zona costera. Revistalatinoamericana de Recursosnaturales 1:21–31
Perkins S (2015) Nearly every seabird may be eating plastic by 2050. Science. https://doi.org/10.1126/science.add1694
Perkol-Finkel S, Benayahu Y (2007) Differential recruitment of benthic communities on neighboring artificial and natural reefs. J Exp Mar Biol Ecol 340:25–39
Petruccelli BP, Goldenbaum M, Scott B, Lachiver R, Kanjarpane D, Elliott E, Francis M, McDiarmid MA, Deeter D (1999) Health effects of the 1991 Kuwait oil fires: a survey of US army troops. J Occup Environ Med 41(6):433–439
Porter KG (1977) The plant-animal interface in freshwater ecosystems. Am Sci 65:159–170
Purvis A (1999) Ten largest oil spills in history. Planet Watch Time Int 153:12
Rajitha K, Mukherjee CK, VinuChandran R (2007) Applications of remote sensing and GIS for sustainable management of shrimp culture in India. Aquacult Eng 36(1):1–17
Randolph RC, Hardy JT, Fowler SW, Price ARG, Pearson WH (1998) Toxicity and persistence of nearshore sediment contamination following the 1991 Gulf War. Environ Int 24:33–42
Rashid A, Bhat RA, Qadri H, Mehmood MA (2019) Environmental and socioeconomic factors induced blood lead in children: an investigation from Kashmir, India. Environ Monit Assess 191(2):76. https://doi.org/10.1007/s10661-019-7220-y
Rathore HS, Nollet LM (2016) Handbook of pesticides: methods of pesticide residues analysis. CRC Press, Boca Raton, FL
Rochman CM (2013) Plastics and priority pollutants: a multiple stressor in aquatic habitats. Environ Sci Technol 47:2439–2440
Roos N, Wahab MA, Chamnan C, Thilsted SH (2007) The role of fish in food-based strategies to combat vitamin A and mineral deficiencies in developing countries. J Nutr 137(4):1106–1109
SACEP (2014) South Asian Co-operative Environmental Programme, Nutrient loading and eutrophication of coastal waters of the South Asian Seas—a scoping study
Schindler DW (1974) Eutrophication and recovery in experimental lakes: implications for lake management. Science 174:897–899
Schindler DW (2006) Recent advances in the understanding and management of eutrophication. Limnol Oceanogr 51:356–363
Schneider O, Sereti V, Eding EH, Verreth AJ (2005) Analysis of nutrient flows in integrated intensive aquaculture systems. Aquac Eng 32(3–4):379–401
Schreinemachers P, Tipraqsa P (2012) Agricultural pesticides and land use intensification in high, middle and low income countries. Food Policy 37:616–626
Schwarzenbach RP, Egli T, Hofstetter TB, Von Gunten U, Wehrli B (2010) Global water pollution and human health. Annu Rev Environ Res 35:109–136
Shafir S, Rijn JV, Rinkevich B (2003) The use of coral nubbins in coral reef ecotoxicology testing. Biomol Eng 2:401–406
Shafir S, Rijn JV, Rinkevich B (2007) Short and long term toxicity of crude oil and oil dispersants to two representative coral species. Environ Sci Technol 41:5571–5574
Sharma YC (2012) A guide to the economic removal of metals from aqueous solutions. Wiley, Hoboken, NJ
Smith MD, Roheim CA, Crowder LB, Halpern BS, Turnipseed M, Anderson JL, Asche F, Bourillon L, Guttormsen AG, Khan A, Liguori LA, McNevin A, O'connor MI, Squires D, Tyedmers P, Brownstein C, Carden K, Klinger DH, Sagarin R, Selkoe KA (2010) Sustainability and global seafood. Science 327(5967):784–786

Stickney RR, McVey JP (2002) Responsible marine aquaculture. New York, World Aquaculture Society
Sutherland WJ, Clout M, Cote IM, Daszak P, Depledge MH, Fellman L, Fleishman E, Garthwaite R, Gibbons DW, De Lurio J, Impey AJ, Lickorish F, Lindenmayer D, Madgwick J, Margerison C, Maynard T, Peck LS, Pretty J, Prior S, Redford KH, Scharlemann JPW, Spalding M, Watkinson AR (2010) A horizon scan of global conservation issues for 2010. Trends Ecol Evol 25:1–7
Teuten EL, Saquing JM, Knappe DR, Barlaz MA, Jonsson S, Bjorn A, Rowland SJ, Thompson RC, Galloway TS, Yamashita R, Ochi D, Watanuki Y, Moore C, Viet PH, Tana TS, Prudente M, Boonyatumanond R, Zakaria MP, Akkhavong K, Ogata Y, Hirai H, Iwasa S, Mizukawa K, Hagino Y, Imamura A, Saha M, Takada H (2009) Transport and release of chemicals from plastics to the environment and to wildlife. Philos Trans R Soc Lond B Biol Sci 364:2027–2045
Thebo AL, Drechsel P, Lambin EF, Nelson KL (2017) A global, spatially explicit assessment of irrigated croplands influenced by urban wastewater flows. Environ Res Lett 12:074008
Thompson RC (2015) Mircoplastics in the marine environment: sources, consequences and solutions. In: Bergmann M, Gutow L, Klages M (eds) Marine anthropogenic litter. Springer, Heidelberg, pp 185–200
Tidwell JH, Allan GL (2001) Fish as food: aquaculture's contribution: ecological and economic impacts and contributions of fish farming and capture fisheries. EMBO Rep 2(11):958–963
Tidwell LG, Allan SE, O'Connell SG, Hobbie KA, Smith BW, Anderson KA (2015) Polycyclic aromatic hydrocarbon (PAH) and oxygenated PAH (OPAH) air–water exchange during the deepwater horizon oil spill. Environ Sci Technol 49:141–149
Tillmanns AR, Wilson AE, Pick FR, Sarnelle O (2008) Meta-analysis of cyanobacterial effects on zooplankton population growth rate: species-specific responses. Fund Appl Limnol 171:285–295
Tukura BW, Kagbu JA, Gimba CE (2009) Effects of pH and seasonal variations on dissolved and suspended heavy metals in dam surface water. Chem Class J 6:27–30
Turner AM, Chislock MF (2010) Blinded by the stink: nutrient enrichment impairs the perception of predation risk by freshwater snails. Ecol Appl 20:2089–2095
UNEP (2006) Water quality for ecosystem and human health. United Nations Environment Programme Global Environment Monitoring System (GEMS)/Water Programme
UNEP (2011) UNEP yearbook: emerging issues in our global environment. UNEP Division of Early Warning and Assessment, Nairobi
UNEP (2016) A snapshot of the world's water quality: towards a global assessment. United Nations Environment Programme, Nairobi
USEPA (2016) Water quality assessment and TMDL information. United States Environmental Protection Agency, Washington, DC
Van Cauwenberghe L, Vanreusel A, Mees J, Janssen CR (2013) Microplastic pollution in deep- sea sediments. Environ Pollut 182:495–499
Vargas-Gonzalez HH, Arreola-Lizarraga JA, Mendoza-Salgado RA, Mendez-Rodriguez LC, Lechuga-Deveze CH, Padilla-Arrendondo GP, Cordoba-Matson M (2014) Effect of sewage discharge on trophic state and water quality in a coastal ecosystem of the gulf of California. Sci World J 2014:618054
Watt I (1994) An outline for the development of a contingency plan to combat oil pollution in the Gulf sanctuary. In: Feltamp E, Krupp F (eds) Establishment of a marine habitat and wildlife sanctuary for the Gulf region. Final report for phase II. CEC/NCWCD, Jubail and Frankfurt, pp 38–80
WHO (2006) Guidelines for the safe use of wastewater. Excreta, and greywater. World Health Organization, Geneva
WHO (2012) Animal waste, water quality and human health. World Health Organization, Geneva
Wilson AE, Sarnelle O, Tillmanns AR (2006) Effects of cyanobacterial toxicity and morphology on the population growth of freshwater zooplankton: meta-analyses of laboratory experiments. Limnol Oceanogr 51:1915–1924

WRI (2008) World Resources Institute, Eutrophication and hypoxia in coastal areas: a global assessment of the state of knowledge. WRI Policy Note, Washington, DC
Wright SL, Thompson RC, Galloway TS (2013) The physical impacts of microplastics on marine organisms: a review. Environ Pollut 178:483–492
Wu SD, Zhao HF (2006) The analytical methods in the monitoring of water and wastewater. China Environmental Science Press, Beijing
WWAP (2015) United Nations world water assessment programme, the United Nations world water development report 2015: water for a sustainable world. United Nations Educational, Scientific and Cultural Organization, Paris
WWAP (2017) United Nations world water assessment programme, the United Nations world water development report 2017: wastewater, the untapped resource. United Nations Educational, Scientific and Cultural Organization, Paris
Xanthos D, Walker TR (2017) International policies to reduce plastic marine pollution from single-use plastics (plastic bags and microbeads): a review. Mar Pollut Bull 118(1–2):17–26
Xiong S, Long H, Tang G, Wan J, Li H (2015) The management in response to marine oil spill from ships in China: a systematic review. Mar Pollut Bull 96(1–2):7–17
Xu JL, Yang JR (1996) Heavy metals in terrestrial ecosystem. China Environmental Science Press, Beijing
Zarfl C, Matthies M (2010) Are marine plastic particles transport vectors for organic pollutants to the Arctic? Mar Pollut Bull 60:1810–1814
Zhou Q, Zhang J, Fu J, Shi J, Jiang G (2008) Biomonitoring: an appealing tool for assessment of metal pollution in the aquatic ecosystem. Anal Chim Acta 606(2):135–150
Zubris KA, Richards BK (2005) Synthetic fibers as an indicator of land application of sludge. Environ Pollut 138:201–211

Chapter 2
Effect of Pesticides on Fish Fauna: Threats, Challenges, and Possible Remedies

Shafat Ali, Adil Farooq Wali, Ali Mohd Yatoo, Sabhiya Majid, Saiema Rasool, Rehan Khan, Md Niamat Ali, Javaid Ahmad Wani, Sanah Farooq, Shabhat Rasool, Hilal Ahmad Wani, and Muneeb U. Rehman

2.1 Introduction

Pesticides are the chemical compounds with toxic nature purposefully employed to destroy a range of harmful organisms. Pesticides not only encompass insecticides but also fungicides, herbicides, and those substances that are toxic to pests (Matthews 2006). The major proportion of pesticides synthesized all over the world is utilized

---

S. Ali
Centre of Research for Development, University of Kashmir,
Srinagar, Jammu and Kashmir, India

Department of Biochemistry, Government Medical College (GMC),
Srinagar, Jammu and Kashmir, India

A. F. Wali
RAK College of Pharmaceutical Sciences, RAK Medical and Health Sciences University,
Ras Al Khaimah, United Arab Emirates

A. M. Yatoo
Department of Environmental Science, University of Kashmir,
Srinagar, Jammu and Kashmir, India

S. Majid · J. A. Wani · S. Farooq · S. Rasool · H. A. Wani
Department of Biochemistry, Government Medical College (GMC),
Srinagar, Jammu and Kashmir, India

S. Rasool
Forest Biotech Laboratory, Department of Forest Management, Faculty of Forestry,
University of Putra Malaysia, Serdang, Selangor, Malaysia

R. Khan
Department of Nano-therapeutics, Institute of Nano-science and Technology (DST-INST),
Mohali, Punjab, India

M. N. Ali
Centre of Research for Development, University of Kashmir,
Srinagar, Jammu and Kashmir, India

K. R. Hakeem et al. (eds.), *Bioremediation and Biotechnology*,
https://doi.org/10.1007/978-3-030-35691-0_2

in the agriculture sector with a purpose of keeping check over the diverse populations of pests. Furthermore, they are applicable in thwarting diseases such as malaria and dengue and growth of redundant plants in lawns, parks, etc., as well as that of pests and microorganism in various items and gadgets (Gilden et al. 2010). However, there is enormous threat to all living beings including man even on accidental contact with these pesticides of toxic nature (Sarwar 2015). The exposure to pesticides during their application, consumption of pesticide-contaminated foodstuffs or fluids, or breathing in of air infested with pesticides is injurious to public health (Pimentel et al. 2013). Even the contact with minute quantities of pesticides during the period of early growth and development has detrimental consequence on health (Damalas and Eleftherohorinos 2011). Pesticides have been reported as a pollutant of various water sources, for instance, groundwaters (Gilliom 2007), oceans (Day 1990), rivers (Gilliom 2007; Malaj et al. 2014), and lakes (Hull et al. 2015). The water bodies are exposed to the pesticides via diverse pathways (Schulz 2004; Hageman et al. 2006). The basic pathways that carry the pesticides to little nonirrigating surface water bodies during heavy rains include the water that runs off the ground surface and tile drainage system (Rabiet et al. 2010; Taghavi et al. 2010; Bereswill et al. 2012; Stehle and Schulz 2015). Several geological and climatic factors including the inclination of cultivated land, hydrology, rainfall amount and intensity, and the dampness of soil determine the penetration of pesticides into water bodies (Schulz 2004). The improvement in the production of crop seems as an indispensable element of the contemporary agricultural sector in view of the growing demand of the rising population (Omer et al. 2010; Sabir et al. 2013; Hakeem 2015; Pierart et al. 2015). There is a tremendous increase in the utilization of pesticides during the last few decades. It has been estimated that the amount of pesticides utilized annually all over the globe figures round to 5.2 billion pounds. The practice of pesticide application in order to lessen the pests has been globally widespread. They are applied not only to protect the crops but also to put a check on the growth of household organisms such as ticks, cockroaches, fleas, mosquitoes, and rats. As a result, the foodstuffs we eat and the air we breathe are often infested with the pesticides (Pesticides n.d.). The widespread and indiscriminate application of toxic substances has contaminated the water bodies by leaching, runoff, drift, and drainage (Cerejeira et al. 2003). The aquatic pollution has become a global problem. The main insecticides that are generally used include organophosphate, chlorinated hydrocarbons, pyrethroids, carbamate, and nicotinoids. The exploitation of insecticides intimidates the long-lasting survival of key bionetworks, disarrays the ecological relationships among living beings, and causes the loss of biological diversity (Banaee 2013). Among a range of noxious pesticides, organophosphates have been

M. U. Rehman (✉)
Department of Biochemistry, Government Medical College (GMC), Srinagar, Jammu and Kashmir, India

Department of Clinical Pharmacy, College of Pharmacy, King Saud University, Riyadh, Saudi Arabia extensively used throughout the world, replacing the importunate and problematical organochloride pesticides as a result of their low-persistence in the atmosphere (Oruc et al. 2006) and quick biodegradability (Ye et al. 2010). Dichlorvos (dimethyl-2,2-dichlorovinyl phosphate) is one among the organophosphate insecticide category that is extensively used to combat a variety of pests infecting domesticated animals and stored grains. This pesticide is also used to thwart ectoparasitic infection in tropical aquaculture and control mosquito vectors of various tropical diseases (Assis et al. 2007). Fish species are mostly sensitive to water pollution (Banaee 2013). Organophosphate pesticides such as dichlorvos are extremely noxious to fish and other nontarget aquatic life forms. These pesticides are potent nerve toxins as they slow down the acetyl cholinesterase (AChE) activity in the nervous system via blocking nerve transmission at synapses in cholinergic neurons. This interruption of the role of nerves leads to disorders in parasympathetic nerve system and ultimately the death of organism (Nguyen et al. 2008). Fishes are the main aquatic dwellers that are often exposed to and distressed by these lethal pesticides (Scott and Sloman 2004) since it is assumed that despite the place of pollution, it will finally finish up in the aquatic ecosystem (Firat et al. 2011). Pesticides may accumulate in the fish body and influence the health of human beings as well through ecological cycling and biomagnifications (Chebbi and David 2011). The pesticide toxicity of fish induces biochemical transformations that result in disturbances in metabolism, enzyme activity inhibition, growth retardation, decrease in the fertility, and prolonged existence of the living being. The organs of fish that are mostly at risk include the liver, brain, gills, and kidneys when it becomes exposed to the medium polluted by toxins (Malik and Maurya 2015). Eco-toxicological studies are wanted in order to find out the toxic nature and possible threat posed by these noxious chemical substances via several biomarkers in fish so that the quality of aquatic ecosystems and health of life forms living in them may be monitored. Therefore, the presence of different types of chemicals with pesticidal properties in the aquatic ecological system drastically affects a range of biochemical and physiological processes and causes grave harm to the fish health (Banaee 2013).

2.2 Historical Background of Pesticide Use

The use of pesticides began with the onset of agriculture and became prominent gradually as a result of rise in the pest population along with declining fertility of soil (Muir 2002). Human beings have been using pesticides since 2000 BC to protect the crops from the attack of pests, insects, and diseases. The chemicals used in the beginning were simple elements and phyto-derivatives (Tierney et al. 2014). The people of ancient Roman used to destroy pests by burning sulfur and bitter substances, ashes, and salts in weed management (History of pesticide use 1998). Pesticides containing sulfur were used by Sumerians more than 4500 years back. Sulfur was used in china as antifungal and antibacterial chemical almost 3000 years ago. Arsenic was employed in the Central East countries more than 2000 years before for various management purposes (Bentley and Chasteen 2002). A range of elements including lead, arsenic, and mercury were usually applied for the protection of crops close to the fifteenth century. In the distant past, human beings explored that the chemical derivatives of plants have the ability to control pest including insects (Niazi et al. 2012). Ants were controlled by using an amalgam of arsenic and honey during the 1600s (Delaplane 2000). Early on during the seventeenth century, nicotine sulfate extracted from tobacco was employed for the insect control. Subsequently, rotenone was brought in for the protection of crops. During the late 1800s, in the USA, the crop growers began the application of several chemical substances including sulfur, calcium arsenate, and nicotine sulfate for agricultural purposes, but the utilization of obsolete application techniques rendered the use of these chemicals unproductive (Delaplane 2000). In the USA, during the year 1867, unrefined copper and arsenic elements were employed for controlling the outburst of Colorado potato beetle (History of pesticide use 1998). Until the 1950s, pesticides that contain arsenic were leading; for instance, arsenical pesticides were employed for the management of ticks in livestock, and pesticides containing chromium–copper–arsenate were used for the preservation of wood timber in several countries that include China, New Zealand, Australia, and the USA (Niazi et al. 2012).

A revolution took place in the development of pesticides around World War II during which a number of potent and cheap pesticides were manufactured. During this period, a range of chemicals were explored to have pesticide properties that include aldrin, dieldrin, 2,4-dichlorophenoxyacetic acid, endrin, dichlorodiphenyltrichloroethane (DDT), β-benzene hexachloride (BHC), and endrin (Delaplane 2000). Fungicides such as glyodin and captan and organophosphate insecticides such as Malathion were launched during the period 1950–1955, and subsequently the triazine herbicides were discovered between 1955 and 1960 (Jabbar and Mallick 1994). Monsanto developed Agent Orange herbicide for war purposes between 1961 and 1971 and was experimented in Vietnam War (History of pesticide use 1998). The application of pesticides attained climax during the year 1961 followed by a decline in the production of further pesticides due to the ecological hazards caused by their unsystematic use. Rachel Carson in 1962 reported the sudden loss of nontargeted life forms from the crop fields as a result of toxicity caused by DDT spraying (Delaplane 2000). The integrated pest management (IPM) that involves the utilization of bio-predators or parasites for pest control was started toward the end of 1960s as a replacement technique for chemical pesticides. IPM was effective in regulating the pests but could not serve as an alternative for the pesticides of chemical nature (Delaplane 2000).

During the late years of the 1970s, several chemical pesticides were used commonly that include organochlorides, carbamates, and organophosphates. DDT and pyrethrin were the predominant pesticides all over the globe during the period of 1970–1980 (Shahid et al. 2016). At present, bio-rational pesticides are preferred for pest regulation as they do not affect nontarget life forms (Delaplane 2000).

2.3 Pesticide Classification

Pesticides have been classified on the basis of several criteria that include chemistry, toxicity, way of action, and functional groups (Garcia et al. 2012). The description of pesticide classification based on toxicity, target pest, chemistry, and mode of action is presented in Tables 2.1, 2.2, 2.3, and 2.4, respectively. Most pesticides contain organic, i.e., carbon containing, or inorganic active components. The inorganic components include copper sulfate, copper, sulfur, ferrous sulphate, lime, and so on (Gunnell et al. 2007). Chemical substances present in pesticides of organic nature have greater complexity and poor solubility compared to those present in inorganic group of pesticides (Debost-Legrand et al. 2016). Organic pesticides may also be subcategorized into natural and synthetic pesticides. Natural pesticides include those anti-pest chemicals that are obtained from natural sources, while synthetic ones are produced by humans via chemical synthesis. Pesticides regulate target pests in diverse ways. Additionally, some pesticides may trigger the task of growth regulators in plants while as certain others can efficiently regulate the process of plant photosynthesis. Similarly, a fungicide might influence cell cleavage while some others may inhibit the synthesis of some chemical substances in fungi. Occasionally, pesticides are categorized on the basis of their application against the

Table 2.1 Pesticide classification based on toxicity (WHO 2009)

Type of pesticide		Ia	Ib	II	U
Level of toxicity		Very dangerous	Highly unsafe	Moderately harmful	Unlikely to present acute risk
[a]LD50 for rat (mg/kg body weight)	Oral	<5	5–50	50–2000	5000 or higher
	Dermal	<50	50–200	200–2000	

[a]LD50 indicates the quantity of the chemicals needed to destroy half of the population of test organism

Table 2.2 Pesticide classification based on the criterion of target pest (Aktar et al. 2009; Zacharia 2011)

Pesticide category	Target organism
Bactericides	Bacteria
Insecticides	Insects
Fungicides	Fungi
Herbicides	Weeds
Miticides/acaricides	Mites
Nematicides	Nematodes
Rodenticides	Rodents
Algaecides	Algae
Piscicides	Fish
Avicides	Birds
Molluscicides	Snails, slugs
Virucides	Virus

Table 2.3 Classification of pesticides on the basis of chemistry (Kim et al. 2016)

Type of pesticide	Example	Structure	Activity
Organochlorines (insecticides)	DDT	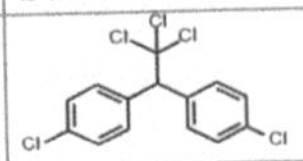	Paralysis, convulsions, and finally death of the insect occurs as these damage nervous system
	Lindane	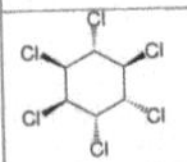	
	Endosulfan	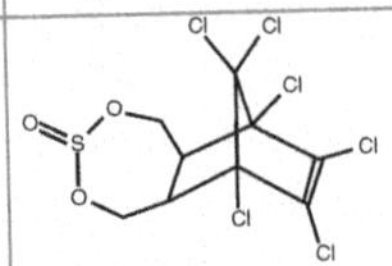	
	Aldrin	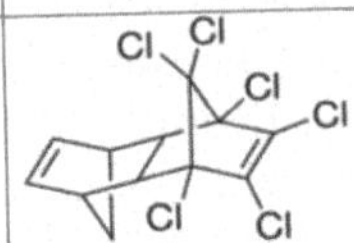	
Organophosphorus (insecticides)	Parathion	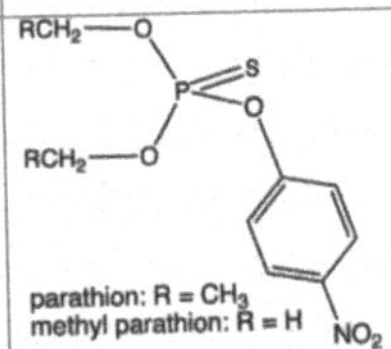	Failure of neurotransmission at synapses leading to rapid convulsion of voluntary muscles that paralyzes the organism and causes death
	Malathion	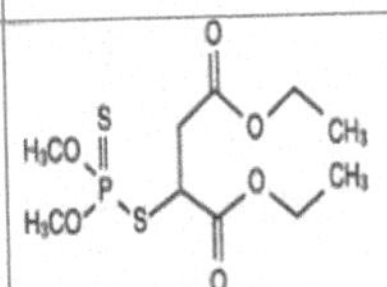	
	Diazinon	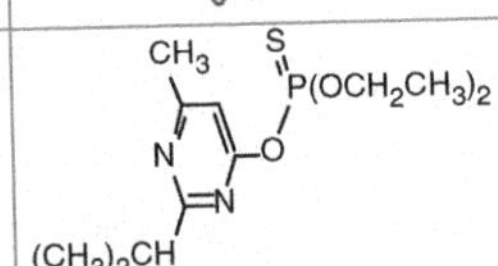	
Inorganic (fungicides)	Benomyl	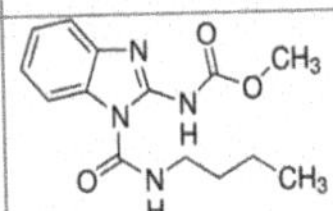	These chemicals are predominant gastric toxins
	Oxine copper	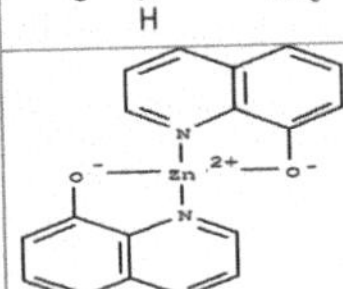	

Table 2.4 Classification of pesticides on the basis of mode of action (Buchel 1983; Zacharia 2011)

Class of pesticide	Examples	Activity
Systemic pesticides	2,4-D and glyphosate	Produce the desired effect via entering the tissues of plant and moving through the vascular system
Nonsystemic/contact pesticides	Diquat dibromide, Paraquat	Produce the desired effects through contact of the pest organism with pesticides without penetrating into the body
Stomach poisons	Fungicides, rodenticides	Produce effects after ingestion

kind of target pest; for instance, the pesticides used to check the growth of mites, fungi, weeds, and insects are categorized as miticides, fungicides, herbicides, and insecticides, respectively. Insecticides enter the body of insect through dermis, oral, or respiratory routes and kill them. Herbicides kill weeds either by way of direct exposure or after their absorption into the plant body through root, leaf, and shoot system. Many pesticides toxic to pest neuroendocrine system have been manufactured for the management of pests (Mnif et al. 2011).

2.4 Pesticide Toxicity in Fish

Pesticides are the chemical compounds applied for controlling insects, weeds, and diseases in plants. The application of pesticides on crops to destroy pests tremendously poisons the nontarget life forms especially fish and deteriorates the health of fish by impairing their metabolic process that occasionally causes death of fish (Shankar et al. 2013). Increase in human population along with rapid industrialization lead to problems in disposing waste water. The household wastes and untreated or partly treated waste matter of industries additionally contaminated with various organic substance, heavy metals, and pesticides have killed fish in huge numbers in water ecosystems (Dhasarathan et al. 2000; Pazhanisamy and Indra 2007) (Fig. 2.1). Pesticides toxicity can be divided into two categories such as acute and chronic toxicity. The potential of a pesticide to cause detrimental consequences due to single exposure is termed acute toxicity. A little quantity of pesticides that have higher acute toxicity may be lethal. It can be estimated as acute oral, dermal, and inhalation. Conversely, chronic toxicity means the capability of a pesticide to produce damaging consequence because of long time exposure and may produce a range of harmful effects including carcinogenesis, teratogenesis, mutagenesis, disorders in blood, endocrinological disorders, and reproductive toxicity (Maurya and Malik 2016). The chronic pesticide toxicity of fish leads to damages due to oxidation, mutagenesis, acetyl cholinesterase activity inhibition, carcinogenicity, and histopathological and developmental modifications (Wasim et al. 2009). The existence of various pesticides including organophosphates in the atmosphere may well induce

Fig. 2.1 Mass killing of fish species due to contamination of water bodies with toxic pesticides

fatal or sub fatal consequences in fish (Mathur 1999). The capability of pesticide to produce harmful effect on fish and other aquatic dwellers greatly depends on its toxicity, dosage rate, duration of contact, and persistence power in surroundings. Lethal dosage describes the pesticide concentration required for bringing about mortality since all individuals of a species does not die at similar dosage; thus, the estimation of a standard toxicity dosage known as lethal concentration 50 (LC50) has been employed which refers to the pesticide concentration that kills half number of individuals of a test fish population in a fixed time duration generally verified after 24–96 h. Table 2.5 shows the toxicity acuteness of diverse classes of chemical pesticides based on the criteria of species of fish and time span of exposure.

The risk rate ranges associated with commonly employed herbicides, fungicides, and insecticides along with their LC50 are shown in Table 2.6.

The exposure of aquatic organisms including fish fauna to pesticides is determined by the bioavailability, biological concentration, biological magnification, and persistence of the pesticides in the aquatic surroundings. Bioavailability is the magnitude of pesticide available in the surroundings for fish and wildlife forms

Table 2.5 The acute toxicity (LC50) of several pesticides against various species of fish

S.No.	Pesticide	Target fish species	Exposure duration	LC50 value	References
1	Alachlor	Rainbow trout	96 h	2.4 μg/L	Johnson and Finley (1980)
2	Acephate	Feathed M.	96 h	> 1000 μg/L	Johnson and Finley (1980)
3	Malathion	Labeo rohita	97 h	15 μg/L	
4	Akton	Channel catfish	96 h	400 μg/L	Johnson and Finley (1980)
5	DDT	Rainbow trout	96 h	8.7 μg/L	Johnson and Finley (1980)
6	Endosulfan	Channel catfish	96 h	1.5 μg/L	Johnson and Finley (1980)
7	Cypermethrin	Labeo rohita	96 h	4.0 μ/L	Marigoudar et al. (2009)
8	Permethrin	Cyprinus carpio	24 h	35 μg/L	Sial et al. (2009)
9	Methyl parathion	Catla catla	96 h	4.8 ppm	Ilavazhahan et al. (2010)
10	Rogar	Puntius stigma	96 h	7.1 and 7.8 ppm	Bhandare et al. (2011)
11	Endosulfan	Channa striatus	96 h	0.0035 ppm	Ganeshwad et al. (2012)
12	Malathion	Heteropneustes fossilis	96 h	0.98 ppm	Sanjoy and Rita (2012)
13	Endosulfan	Cirrhinus mrigala	96 h	1.06 μg/L	Ilyas and Jave (2013)
14	Termifos	Clarias gariepinus	96 h	0.86 mg/L	Nwani et al. (2013)
15	λ cyhalothrin	Labeo rohita	96 h	0.7 μg/L	Dey and Saha (2014)
16	Karate	Cyprinus carpio	96 h	0.160 μg/L	Bibi et al. (2014)
17	Dimethoate	Labeo rohita	96 h	24.55 μg/L	Dey and Saha (2014)

Table 2.6 Risk rates of pesticides

Risk rate	Toxicity	Minimum	Minor	Moderate	Higher	Extreme	Super
	LC50 (mg/L)	>100	10–100	1–10	0.1–1.0	0.01–0.1	>0.01

(Wikipedia 2013). Several pesticides quickly degrade subsequent to their use. However, the bioavailability of various pesticides is reduced as they become firmly attached to stream floor or the suspended particles of soil in water column. Some pesticides are rapidly watered down in aquatic medium or quickly mixed with the air by undergoing volatilization and as result are available to aquatic life forms in minute amounts. Some pesticides present in the aquatic ecosystems enter the food chain and get accumulated at each consecutive trophic level. This refers to biological magnification. As a result, greater pesticide quantities become accumulated in fishes such as trout due to recurring consumption of the contaminated animals that feed on plants that have absorbed the pesticides present in water. These toxic substances are passed on to human beings who consume fish. The time span during which a pesticide exists in the surroundings is known as pesticide persistence and is generally represented as "half-life" of a pesticide. The pesticide persistence is determined by the rate of its degradation that mainly depends on its chemical constituents and conditions of the milieu. Pesticides may breakdown via photo decomposition, thermal decay, and microbial decay, and their degradation also depends on the moisture and soil conditions such as pH. The pesticides that decay slowly exist in the environment for long time and become accessible to aquatic animals (Seyler et al. 1994).

2.5 Routes of Fish Exposure to Pesticides

Pesticides enter the body of fish and other aquatic organisms via three routes: dermal, respiratory, and oral (Helfrich et al. 2009) (Fig. 2.2). Dermal route involves the direct absorption of pesticides via skin from water polluted with pesticides. Respiratory route involves the direct pesticides uptake via gills during breathing. Orally, the pesticides enter into the body of fish via intake of pesticide-polluted water or eating of prey that has pesticides accumulated in the body. A number of secondary factors also bring fish and aquatic fauna in contact with pesticides and

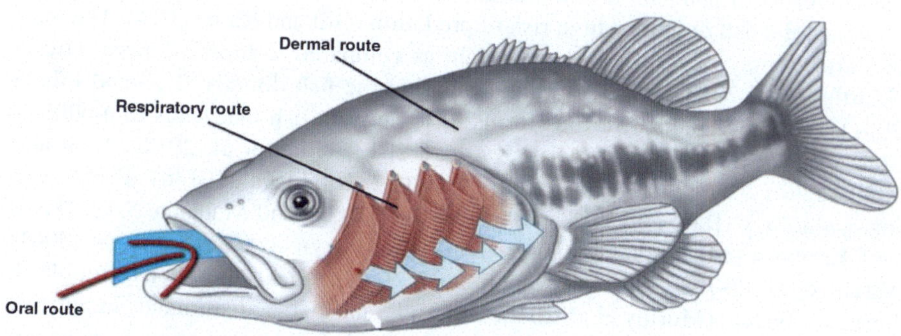

Fig. 2.2 Exposure routes of fish to pesticides

finally cause toxicity (Kingsbury and Kreutzweiser 1980; Schnick et al. 1980; Spradley 1985). The exposure of aquatic life forms including fish could be a more common problem than realized. Majority of fish killed due to pesticide poisoning remain unreported; as a result, the death toll of fish is over and over again miscalculated. Several factors such as water clarity, depth, camo coloring, and diminutive size make it difficult to enumerate the exact numbers of fish causalities. Besides, the dead bodies of fish are removed by scavengers from the death spot, and the fish that are stressed and dying may hide themselves inside a dense shelter or go away entirely from the spot. The factors that determine the possible effects as a result of pesticides on fish fauna and other water dwellers include the kind of pesticide products, usage rate, conditions of weather, species involved, magnitude of fish causalities (problem extent), site, and size of affected water body (Shankar et al. 2013).

2.6 Pesticide Effects and Threats to Fish

Pesticides penetrate into water bodies by means of drifting, running off, leaching via ground or directly by spraying these chemicals on water surface to inhibit the growth of mosquitoes. The water polluted with pesticides causes an immense threat to water dwellers. These pesticides distress aquatic vegetation, lessen the amount of dissolved oxygen in water, and alter the physiology and behavior of fish species. Different studies have reported the presence of those pesticides in streams, ponds, and lakes that are used in lawns (How Pesticides Affect the Environment n.d.). When pesticides are used on soil, they drift into water bodies and result in toxic consequences in fish species and other nontarget life forms. Additionally, these toxic pesticides interact with stressors including detrimental algal blooms. The excessive use of these pesticides has reduced the populations of various species of fish (Scholz et al. 2012). Pesticides significantly reduce the number of organisms on which fish feed in aquatic ecosystem (Helfrich et al. 2009) that indirectly leads to interruption in the food availability to fish and also alter the habitation of aquatic bodies (Maskaoui et al. 2005). Additionally, the pesticides reduce the suitability of fish habitat and produce changes in their behavior, and as a result of which, the fish remain at high risk of predation (Gill and Raine 2014). The indirect effects may be vital to a large extent as compared to those of direct effects (Murthy et al. 2013). Diverse pesticides affect the fish directly (Rao and Pillala 2001). Pesticides induce various forms of toxicity in fish that leads to modifications in the behavior of fish (Satyavardhan 2013; Ullah et al. 2014c; Rani and Kumaraguru 2014), histology (Saeedi et al. 2012; Ullah et al. 2014d), disorders in histopathology (Rani and Venkataramana 2012); Deka and Mahanta 2012; David and Kartheek 2014), genotoxicity, changes in enzymes (Gartiser et al. 2001; Vargas et al. 2001; Çavas and Könen 2007), biochemical alterations, trouble in hormone system (Murthy et al. 2013; Dey and Saha 2014), disturbance in nutrient profiles (Muthukumaravel et al. 2013; Bibi et al. 2014), deviation in nutrition

(Bhandare et al. 2011; Ravindran et al. 2012), alterations in antioxidant defense system (Nwani et al. 2010; Muthukumaravel et al. 2013), and changes in acetylcholinesterase action (Joseph and Raj 2011; Bibi et al. 2014).

Pesticides induce toxicity in different species of fish at varied concentrations. The changes produced in different body parts show difference among each other and in reaction to diverse pesticides. The effects produced by pesticide contamination have been reported almost in each part and of the system of fish body.

Several alterations and threats induced in fish due to pesticide pollution are discussed below under various subheadings.

2.7 Lethal Effects of Pesticides

The exposure of Catla catla to small concentrations of methyl parathion such as 4.8 ppm, 8 ppm, and 10 ppm brings about death in 50%, 80%, and 100% individuals, respectively (Ilavazhahan et al. 2010). λ cyhalothrin and dimethoate are reported fatal for Labeo rohita (Dey and Saha 2014). These pesticides in sublethal amounts subtly change the life of different species of fish and as a result threaten their continued existence (Scott and Sloman 2004; Rani and Kumaraguru 2014). The pesticide pollution of fish at sublethal levels disturbs the normal life of fish via changing their behavior, histopathology, hematology, protein content, immunity, biochemistry, and reproductive biology and inducing neurotoxicity and genotoxicity (Fig. 2.3).

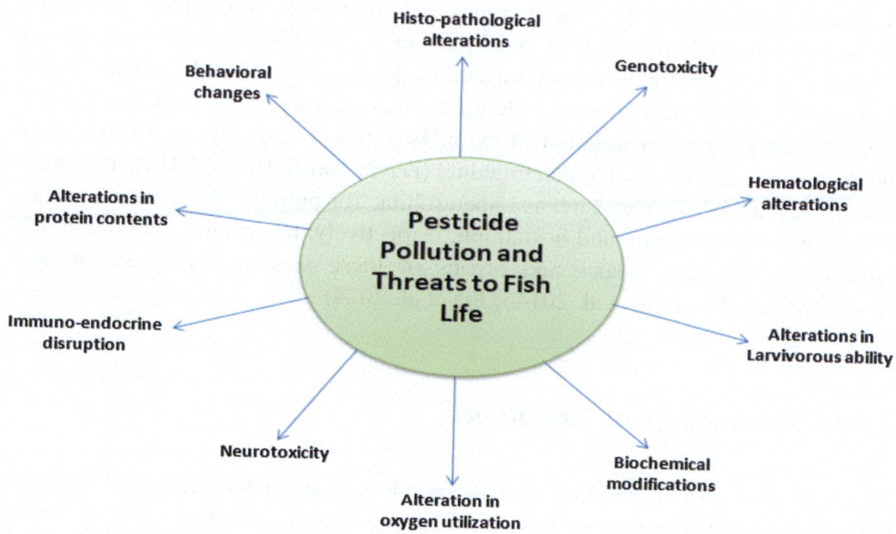

Fig. 2.3 Pesticide pollution threats to fish life

2.7.1 Behavioral Changes

The exposure of different fish to various pesticides makes them lethargic and brings changes in their capability of swimming due to which they are at high risk of predation. Their feeding, position maintaining, and territory defending abilities are also reduced (Prashanth et al. 2011). Pesticides interrupt the behavior of schooling (Gill and Raine 2014) in fish species as they show trouble in swimming and aberrant motions such as dangling, irregular, and erratic (Nagaraju et al. 2011) which make them more easily available to prey. The exposure of diverse fish species such as Labeo rohita, Catla catla, Cyprinus carpio, Oreochromis mossambicus, and Cirrhinus mrigala to sodium cyanide alters their behavior by inducing hyper-excitability, irregular, and darting motions in addition to disproportion to their swimming ability (David et al. 2010). The pesticide contamination of migratory fish causes changes in their migratory behavior (Nagaraju et al. 2011) and hence upset their life cycle.

2.7.2 Histopathological Alterations

Pesticides such as insecticides have been found to induce histopathological injury (Fanta et al. 2003). The contact of Cirrhinus mrigala and Cyprinus carpio with pesticides results in the development of liver lesions (Velmurugan et al. 2009). The changes in histopathology have also been observed in fish treated with dichlorvos and diazinon at sublethal doses (Banaee et al. 2013). Heteropneustes fossilis show modifications in tissues of different organs including the kidneys, liver, and ovaries on exposure to malathion (Deka and Mahanta 2012). The exposure of Cyprinus carpio to sublethal amounts of sodium cyanide leads to a range of histopathological alterations in the kidneys that include glomerular disintegration, necrosis, infiltration of lymphocytes, vacuolation of cytoplasm, blood clogging, size variation in tubular lumen, and injury in collecting duct (David and Kartheek 2014). The exposure of various fish species such as Labeo rohita, Tor putitora, and Channa gachua to atrazine, cypermethrin, and hostathion, respectively, has been reported to cause damages in different tissues and organs of these species (Jayachandran and Pugazhendy 2009; Ullah et al. 2014d; Jha et al. 2014).

2.7.3 Hematological Alterations

Fatal effects on fish hematology such as blood features, histological modifications in leucocytes and erythrocytes, content of hemoglobin, and packed volume of cells have been reported in Puntius ticto and *Cyprinus carpio* (Satyanarayan et al. 2004) and Tor putitora (Ullah et al. 2014d) because of their exposure to various pesticides such as DDT, sulfone, deltamethrin, aldrin, BHC, and cypermethrin. These hematological changes have also been observed in *Cyprinus carpio* when it comes in contact with diazinon (Svoboda et al. 2001), Labeo rohita on exposure to cypermethrin (Adhikari et al. 2004), and Oreochromis mossambicus caused by potassium dichromate and potassium chlorate (Sivanatarajan and Sivaramakishnan 2013).

2.7.4 *Neurotoxicity*

Acetylcholinesterase (AChE) is extremely responsive to carbamate and organophosphate pesticides than other classes of pesticides and pollutants (Murthy et al. 2013). Cypermethrin has been observed to drastically affect the brain and inhibit the activity of AChE in Labeo rohita as well as subsequently alter gills, muscle, and liver tissues (Marigoudar et al. 2010). The inhibition of AChE results in acetylcholine accumulation in cholinergic synapses that causes hyperactivation in Labeo rohita (Marigoudar et al. 2009, 2010). The neurotoxication of Colisa fasciatus due to cypermethrin brings about changes in the activities of AChE, lactic dehydrogenase, and succinic dehydrogenase (Singh et al. 2010). The changes in AChE activity has also been reported in Leporinus obtusidens (Glusczak et al. 2006) and Rhamdia quelen (Glusczak et al. 2007) in response to Glyphosate. However, decline in the activity of AChE has been observed in Cyprinus carpio due to Karate (Bibi et al. 2014). The interruption in the activity of AChE in fish eventually reduces and disturbs the ability of swimming and performance in fish that may lead to additional detrimental effects (Rao 2006; Rao et al. 2007).

2.7.5 *Biochemical Modifications*

Pesticides have been shown to exert destructive effects in different biochemical events (Ullah et al. 2014c). The application of sublethal levels of organophosphates interrupts the activities of different metabolic enzymes especially glutaminases present in the tissue of brain of Labeo rohita (Mastan and Shaffi 2010). Cypermethrin toxication affects the tissues of different organs of Tor putitora such as the brain, gills, muscle, and liver and alters the enzymatic activities of catalase, peroxidase, lipid peroxidase, and glutathione reductase of these organs (Ullah et al. 2014c). Cypermethrin also produces change in the liver enzymes of Labeo rohita (Marigoudar et al. 2012). Due to sodium cyanide contamination, the gill, muscle, and liver tissue enzymes such as phosphorylase, glucose-6-phosphate dehydrogenase, succinate dehydrogenase, alkaline phosphatase, lactate dehydrogenase, and acid phosphates show gradual and steady decline in Labeo rohita (Dube et al. 2013) and that of catalase activity in Cyprinus carpio (David et al. 2008). Pesticide-induced abnormal biochemical changes in different fish species have been reported in several other studies (Nwani et al. 2010; Muthukumaravel et al. 2013).

2.7.6 Genotoxicity

Different toxic chemicals on contact with fish induce chromosomal abnormality. Fenvalerate induces chromatid separation, gaps, deletions, breaks, and fragments as well as ringlike chromosomes (Saxena and Chaudhari 2010). Alteration in DNA replication and DNA abnormality that results in mutation has been shown to be associated with the toxicity of various pesticides (Gilot-Delhalle et al. 1983). The nucleic acids in the tissues of sex organs of Colisa fasciatus are altered due to exposure to cypermethrin (Singh et al. 2010). Scientific reports have shown that pollutants induce carcinogenesis (El Adlouni et al. 1995; Erickson and Larsson 2000), teratogenesis, mutation, and clastogenesis in different fish species (Obiakor et al. 2012) that eventually result in anomalous development, reduction of fish growth and survival during early and adult stages of life, and development of malignancies and various imperfections in organs of the body (Akpoilih 2012).

2.7.7 Alterations in Protein Contents

Different fish species have been showing changes in their protein contents of various tissues including the intestines, liver, muscles, blood, and gills due to high pesticide effects. Heteropneustes fossilis shows declined protein quantity as a result of nickel contamination (Nanda et al. 2000). The exposure of Tor putitora to cypermethrin reduces protein content, causing considerable damage in its body (Ullah et al. 2014c). Colisa fasciatus also shows significant decline in the protein level when it becomes exposed to Cypermethrin (Singh et al. 2010). Malathion lessens the protein amount in Labeo rohita (Thenmozhi et al. 2011) as well as in Clarias batrachus (Khare and Singh 2002). The total liver protein in Oreochromis niloticus is influenced by thiamethoxan (Bose et al. 2011). Similarly, thiodon leads to a drastic change in the total liver protein of Clarias gariepinus (Aguigwo 2002). Dichlorvos considerably affects the total protein, glycogen content of tissues, muscle albumen content, liver albumen content, and albumen content of kidney in Oreochromis mossambicus (Lakshmanan et al. 2013). Karate lessens the content of protein in Cyprinus carpio (Bibi et al. 2014), whereas monocrotrophos causes decline in the content of proteins, carbohydrates, and lipids in different tissues of Labeo rohita (Muthukumaravel et al. 2013).

2.7.8 Alteration in Oxygen Utilization

Different species of fish when become exposed to various pesticides show increase or decrease in the consumption of oxygen such as dimethoate (Shereena et al. 2009), and lead (James et al. 1993) induces lethal alteration in oxygen utilization in

Oreochromis mossambicus. Similarly, Labeo rohita shows variation in the oxygen use due to toxic effects of dimethyl parathion (Bengeri et al. 1984). DDT contamination alters the consumption of oxygen in Lepidocephalichthys thermalis (Gurusamy and Ramadoss 2000), and various pesticides induce toxic changes in the use of oxygen in Puntius ticto (Magare and Patil 2000). The oxygen concentration is drastically declined in water as a result of death of aquatic plant species caused by herbicides that results in the choking of fish and ultimately declined production of fish (Helfrich et al. 2009).

2.7.9 *Alterations in Larvivorous Ability*

Some pesticides such as organo-phosphorus are commonly used due to their properties such as higher insecticidal capability, lower persistence, quick biodegradability, and minor toxic effects on mammals (Bhandare et al. 2011); however, they directly and indirectly pose threat to nontarget organisms including the fish species with larvivorous ability (Roger and Bhuiyan 1990). Oryzias carnaticus shows decreased larvivorous potential on exposure to hostathion and kitazin pesticides (Ravindran et al. 2012).

2.7.10 *Alteration in Immune System and Endocrine Disruptors*

Fish immune system has been shown to be interrupted once it gets exposed to pesticides (Bols et al. 2001; Maskaoui et al. 2005). Pesticides at small amounts serve as imitators or sex hormone blockers resulting into anomalous sexual development, male feminization, aberrant sex ratios, and atypical mating behavior (Satyavardhan 2013). Furthermore, it may change other fish hormonal processes including bone development and appropriate thyroid activity (Murthy et al. 2013). λ-cyhalothrin and dimethoate have been observed to induce toxic effects on thyroid hormone in Labeo rohita (Dey and Saha 2014).

2.8 Challenges in Monitoring Pesticides in Small Freshwater Bodies

Pesticide pollution monitoring of small water bodies is an exigent task since different time- and space-related factors have an effect on the utmost climax concentrations of a pesticide in these aquatic systems (Lorenz et al. 2017). These water bodies are the essential part of freshwater systems since they comparatively support a huge percentage of biological diversity than bigger freshwater ecosystems (Biggs et al. 2014) and represent a significant inland water–carbon flux (Holgerson and Raymond 2016). The programs that intend to estimate the concentrations of different pesticides in freshwater systems require to properly take into consideration various temporal and spatial factors. Several monitoring approaches have been put forth in order to assess the pesticide exposure of freshwater bodies. One of the simplest techniques is manual grab sampling that can be carried out intermittently or at preset occasions (Day 1990; Laabs et al. 2002). However, the inadequacy of this method is that it likely fails to notice the upper limit of exposure to pesticides (Richards and Baker 1993; Leu et al. 2004) or in other words identify minor pesticide concentration (Xing et al. 2013). A substitute technique in the form of continuous water sampling method that uses passive or automatic water samplers was developed with the aim to overcome the abovementioned limitations. The automatic samplers quantify time-integrated concentrations of different pesticides neglecting the pertinent events of entry (Kreuger 1998). The water samples can be incorporated for more than a week (Kreuger 1998; Bischoff et al. 2003) to two-week period (Stenrød 2015). The episodes of storm flow (Liess and von der Ohe 2005; Xing et al. 2013) as well as the events of running off of rainwater (Bereswill et al. 2012) may also be mechanically or manually sampled subsequent to the spray or run-off episodes (Bischoff et al. 2003). Additionally, many passive samplers with the purpose of monitoring freshwater bodies for various polar and nonpolar contaminants have been designed (Vrana et al. 2005; Stuer-Lauridsen 2005; Mills et al. 2014). Different studies have comparatively analyzed the practical performance of various sampling approaches (Schafer et al. 2008; Xing et al. 2013) and found that majority of the pollutant concentration pertinent to eco-toxicology could not be traced by fixed interval sampling approach (Stehle et al. 2013; Xing et al. 2013). Therefore, the monitoring programs that depend on fixed interval approach constantly miscalculate the eco-toxicological threshold exceedances such as field-derived effect thresholds and regulatory acceptable concentrations (RACs); as a result of which, the utility of this method becomes restricted for monitoring highly short-lived and very noxious chemical substances (Stehle et al. 2013). It has been documented in some latest studies that grab sampling strategy (Rasmussen et al. 2015) as well as time integrated sampling approach with 1 week pooling frequency (Bundschuh et al. 2014) undervalued the upper limit of pesticide levels and the toxicity estimation by a factor of as a minimum 10 in contrast with water samplers triggered by runoff.

The records obtained from manual event driven monitoring of pesticides in lotic small water bodies (Bischoff et al. 2003; Su et al. 2006) have shown that manual sampling performs better than weekly integrated sampling done by automatic water samplers. Some studies reported that passive and event driven sampling approaches may be equally effectual for depicting rapport between exposure to pesticides and effects on ecology (Schafer et al. 2008; Fernandez et al. 2014). Nevertheless, the protection of small water bodies necessitates the application of those monitoring approaches that are capable of detecting the threshold exceedances stimulated through each pertinent entry pathways, mainly when the spray drifting is a key concern.

2.9 Remedies/Alternatives

Alternative measures for pesticide application may be effective for pest management and avoiding toxic effects on nontarget life forms particularly fish and the whole biosphere. Some of the remedies in the form of alternative measures include the following (Fig. 2.4).

2.9.1 Integrated Pest Management

Integrated pest management (IPM) is a strategy that recognizes and decreases the hazards due to pests. It integrates the environmental biology of pest and various existing technology via the most inexpensive ways in order to thwart undesirable intensity of damage caused by pests as well as reduce the hazard to humans, resources, and the surroundings. It offers an efficient approach for the management of pests in urbanized farming fields, inhabited and public areas, to natural areas and wilds. IPM presents an efficient low-risk strategy to save humans and resources from pests (USDA 2013). IPM coordinates many management approaches in such a manner that permits the systems of production to retreat from the usual management which is based on chemicals to ecologically healthy tactics (MacHardy 2000; Prokopy 2003). The application of chemicals is directed by financial and remedial thresholds depending on monitoring pests, useful life forms, and conditions of the environment (Cooley and Coli 2009). IPM has the capability to deal with every pest complex such as insects, weeds, diseases, vertebrates, and others and may be adapted to any production goals such as conventional, organic, and sustainable (Bidddinger and Rajotte 2015). IPM concentrates on long-lasting pest deterrence or the harm they cause, via combining many techniques including biocontrol, manipulating habitation, modifying cultural practices, and utilizing resistant varieties. Substances for the control of pests are chosen and used in such a way that lessen hazards to the health of human beings, useful and nontarget life forms, as well as the

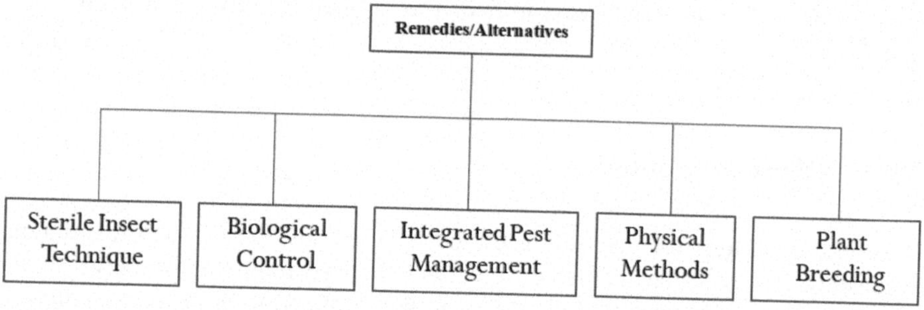

Fig. 2.4 Remedial techniques that may be used as alternative approaches for pest management

surroundings. Additionally, it has been emphasized that there should be complementation between the alternative procedures and the existing chemical control practices (Gurusubramanian et al. 2005; Rahman et al. 2005a, b; Gurusubramanian and Borthakur 2005).

2.9.2 *Biological Control*

The utilization of natural enemies to regulate the population of a species is known as biological control, e.g., the augmentation of local herbivores for weed control, introduction of predators or parasites for insect control (Atalah et al. 2013), or application of biopesticides (Jackson 2007). All over the globe, weeds have been controlled very successfully by means of biological control. The use of biocontrol agents including insects and pathogens has effectively regulated nearly 41 weed species. Additionally, local fungi have been used as myco-herbicides for the management of three weed species (Mcfadyen 2000). During the last decade, in Australia 43 arthropod and pathogen species were utilized in 19 separate projects with the aim to successfully various exotic weeds. Efficient biocontrol was obtained in various projects; however, exceptional achievement was attained in regulating Cryptostegia grandiflora (rubber vine) and Asparagus asparagoides (Bridal creeper) (Palmer et al. 2010). In certain cases, the use of biological control against target weeds has been able to decrease the trouble by 50–83% (Fowler et al. 2010; Paynter et al. 2010).

2.9.3 *Sterile Insect Technique*

Globally, sterile insect technique (SIT) plays an important part in the programs meant for the suppression and eradication of some pests particularly fruit flies (Klassen and Curtis 2005). The application of SIT as an approach for controlling pests is advantageous as it is species specific, compatible with the application of other regulatory approaches, and effective in the less dense populations of pests. SIT was parameterized on light-brown apple moth for employing it in California (Kean et al. 2011), and its success in field was verified in New Zealand (Stringer et al. 2013).

2.9.4 *Physical Methods*

In several situations for controlling the pests belonging to vertebrates, different physical interferences are available which include mass trapping (Warburton et al. 2008), shooting (Choquenot et al. 1999), or the elimination of pests manually in order to lessen the populations of pests (Yamanaka 2007).

2.9.5 Plant Breeding

Plant breeding is an important nonpesticide technique to control pests. The use of obligate endophyte, Epichloe, in the pastures of New Zealand has been extremely successful in controlling ryegrass (Lolium spp.) and tall fescue (Schedonorus arundinaceus) pests (Johnson et al. 2013). Lucerne (M. sativa) has been successfully bred to develop resistance against aphids, and cereals resistant to striped rust Puccinia striiformis Westend have been developed (Cromey 1992).

2.10 Conclusion

Pesticides are the chemical compounds applied for controlling insects, weeds, and diseases in plants. The pesticides have been used by human beings since 2000 BC. The major proportion of pesticides synthesized throughout the globe is utilized for the management of pests in the agriculture sector. There is a tremendous increase in the utilization of pesticides during the last few decades. The widespread and unsystematic pesticide application has contaminated the water bodies via drifting, running off, leaching, or directly by spraying these chemicals on water surface to inhibit the growth of mosquitoes. The water polluted with pesticides cause an immense threat to water dwellers. The main pesticides that are generally used include organophosphate, chlorinated hydrocarbons, pyrethroids, carbamate, and nicotinoids. The pesticides threaten the long-lasting survival of key bionetworks, disarray the ecological relationships among living beings, and cause the loss of biological diversity. Among a range of noxious pesticides, organophosphates have been extensively used throughout the world, replacing the importunate and problematical organochloride pesticides as a result of their low persistence in the atmosphere and quick biodegradability. Fish species are mostly sensitive to water pollution. Organophosphate pesticides such as dichlorvos are extremely noxious to fish and other nontarget aquatic life forms. These pesticides are potent nerve toxins as they slow down AChE activity in the nervous system via blocking the transmission at synapses in cholinergic neurons. This interruption of the role of nerves leads to disorders in parasympathetic nerve system and ultimately the death of organism. Fishes are the main aquatic dwellers that are often exposed to and distressed by these lethal pesticides. The pesticide toxicity of fish induces biochemical transformations that result in disturbances in metabolism, enzyme activity inhibition, growth retardation, decrease in the fertility, and prolonged existence of the living being. Eco-toxicological studies are wanted in order to find out the toxic nature and possible threat posed by these noxious chemical substances via several biomarkers in fish so that quality of aquatic ecosystems and health of life forms living in them may be monitored. Additionally, many techniques such integrated pest management, biological control, genetic control, physical interventions, and sterile insect technique may be used as substitutes for pesticides. The use of selected pesticides may also reduce the pollution of water ecosystems. Further research is required to produce eco-friendly species-specific pesticides so that the damage to nontarget organisms is avoided.

Acknowledgments Special thanks are due to Research Centre, College of Pharmacy, King Saud University, Riyadh, and Deanship of Scientific Research, King Saud University, Kingdom of Saudi Arabia.

References

Adhikari S, Sarkar B, Chatterjee A, Mahapatra CT, Ayyappan S (2004) Effects of cypermethrin and carbofuran on certain hematological parameters and prediction of their recovery in a freshwater teleost, Labeo rohita (Hamilton). Ecotoxicol Environ Saf 58:220–226

Aguigwo JN (2002) The toxic effect of cymbush pesticide on growth and survival of the African cat fish Clarias gariepinus. J Aqua Sci 17:81–84

Akpoilih BU (2012) Fish Ecogenotoxicology: an emerging science, an emerging tool for environmental monitoring and risk assessment. Glob J Bio Sci Biotechnol 1:141–151

Aktar MW, Sengupta D, Chowdhury A (2009) Impact of pesticides use in agriculture: their benefits and hazards. Interdiscip Toxicol 2:1–12

Assis CRD, Amaral IPG, Castro PCF, Carvalho JLB, Bezerra RS (2007) Effect of Dichlorvos on the acetylcholinesterase from Tambaqui (Colossoma macropomum) brain. Environ Toxicol Chem 26:1451–1453

Atalah J, Bennett H, Hopkins G, Forrest BM (2013) Evaluation of the sea anemone Anthothoe albocincta as an augmentative biocontrol agent for biofouling on artificial structures. Biofouling 29:559–571

Banaee M (2013) Physiological dysfunction in fish after insecticides exposure. In: Insecticides-development of safer and more effective technologies. Intech Open, pp 103–143

Banaee M, Sureda A, Mirvagefei AR, Ahmadi K (2013) Histopathological alterations induced by Diazinon in rainbow trout (Oncorhynchus mykiss). Int J Environ Res 7:735–744

Bengeri KV, Shivaraj KM, Patil HS (1984) Toxicity of dimethyl parathion to freshwater fish, Labeo rohita and oxygen uptake rate of exposed fish. Environ Ecol 2:1–4

Bentley R, Chasteen TG (2002) Microbial methylation of metalloids: arsenic, antimony, and bismuth. Microbiol Mol Biol Rev 66:250–271

Bereswill R, Golla B, Streloke M, Schulz R (2012) Entry and toxicity of organic pesticides and copper in vineyard streams: erosion rills jeopardize the efficiency of riparian buffer strips. Agri Ecos Environ 146:81–92

Bhandare RY, Pathan TS, Shinde SE, More PR, Sonawane DL (2011) Toxicity and behavioural changes in fresh water fish Puntius stigma exposed to pesticide (Rogor). Am Euras J Toxicol Sci 3:149–152

Bibi N, Zuberi A, Naeem M, Ullah I, Sarwar H, Atika B (2014) Evaluation of acute toxicity of karate and its sub-lethal effects on protein and acetylcholinesterase activity in Cyprinus carpio. Int J Agri Biol 16:731–737

Bidddinger DJ, Rajotte EG (2015) Integrated pest and pollinator management-adding a new dimension to an accepted paradigm. Curr Opin Insect Sci 10:204–209

Biggs J, Nicolet P, Mlinaric M, Lalanne T (2014) Report of the Workshop on the protection and management of small water bodies, Brussels, 14th November 2013

Bischoff G, Stahler M, Ehlers K, Pestemer W (2003) Biological-chemical monitoring in drainage ditches in the 'Altes Land' orchading region. Part 1: Application of plant protection products and residues of a.i. in surface water. In Del Re, A, Capri E, Padovani L, Trevisan M (eds), Pesticide in air, plant, soil & water system. Proceedings of the 12. Symposium pesticide chemistry, Piacenza, Italia, pp 831–840

Bols NC, Brubacher JL, Ganassin RC, Lee LEJ (2001) Ecotoxicology and innate immunity in fish. Develop Comp Immunol 25:853–873
Bose S, Nath S, Sahana SS (2011) Toxic impact of thiamethoxam on the growth performance and liver protein concentration of a freshwater fish Oreochromis niloticus (TREWAVAS). Ind J Fund App Life Sci 1:274–280
Buchel KH (1983) Chemistry of pesticides. Wiley, New York
Bundschuh M, Goedkoop W, Kreuger J (2014) Evaluation of pesticide monitoring strategies in agricultural streams based on the toxic-unit concept—experiences from long term measurements. Sci Total Environ 484:84–91
Çavas T, Könen S (2007) Detection of cytogenetic and DNA damage in peripheral erythrocytes of goldfish (Carassius auratus) exposed to a glyphosate formulation using the micronucleus test and the comet assay. Mutagenesis 22:263–268
Cerejeira MJ, Viana P, Batista S, Pereira T, Silva E, Valerio MJ et al (2003) Pesticides in Portugal surface and ground waters. Water Res 37:1055–1063
Chebbi SG, David M (2011) Modulation in the protein metabolism under sublethal concentration of Quinalphos intoxication in the freshwater common carp, Cyprinus carpio (Linnaeus, 1758). Int J Pharma Biol Arch 2:1183–1189
Choquenot D, Hone J, Saunders G (1999) Using aspects of predator-prey theory to evaluate helicopter shooting for feral pig control. Wildlife Res 26:251–261
Cooley DR, Coli WM (2009) Implementation of Apple IPM: The Massachusetts Experience. In: Biorational tree fruit pest management, pp. 145–170
Cromey MG (1992) Adult plant resistance to stripe rust (Puccinia striiformis) in some New Zealand wheat cultivars. New Zeal J Crop Hort Sci 20:413–419
Damalas CA, Eleftherohorinos GE (2011) Pesticide exposure, safety issues, and risk assessment indicators. Int J Environ Res Public Health 8:1402–1419
David M, Kartheek RM (2014) Sodium cyanide induced Histopathological changes in kidney of fresh water fish Cyprinus carpio under sublethal exposure. Int J Pharma Chem Biol Sci 4:634–639
David M, Munaswamy V, Halappa R, Marigoudar SR (2008) Impact of sodium cyanide on catalase activity in the freshwater exotic carp, Cyprinus carpio (Linnaeus). Pest Biochem Physiol 92:1518
David M, Haragi SB, Chebbi SG, Patil VK, Halappa R, Chittaragi JB et al (2010) Assessment of sodium cyanide toxicity on freshwater teleosts. Rec Res Sci Tech 2:1–5
Day KE (1990) Pesticide residues in freshwater and marine zooplankton: a review. Environ Pollu 67:205–222
Debost-Legrand A, Warembourg C, Massart C, Chevrier C, Bonvallot N, Monfort C et al (2016) Prenatal exposure to persistent organic pollutants and organophosphate pesticides, and markers of glucose metabolism at birth. Environ Res 146:207–217
Deka S, Mahanta R (2012) A study on the effect of organophosphorus pesticide malathion on hepatorenal and reproductive organs of heteropneustes fossilis (Bloch). Sci Probe 1:1–13
Delaplane KS (2000) Pesticide usage in the United States: history, benefits, risks, and trends. Cooperative extension service. The University of Georgia, Athens
Dey C, Saha SK (2014) A comparative study on the acute toxicity bioassay of Dimethoate and lambda-cyhalothrin and effects on thyroid hormones of freshwater teleost fish Labeo rohita (Hamilton). Int J Environ Res 8:1085–1092
Dhasarathan P, Palaniappen R, Ranjit SJA (2000) Effect of endosulfan and butachloron on the digestive enzyme and proximate composition of the fish, Cyprinus carpio. Indian J Environ Ecoplan 3:611–614
Dube PN, Alavandi S, Hosetti BB (2013) Effect of exposure to sublethal concentrations of sodium cyanide on the carbohydrate metabolism of the Indian major carp Labeo rohita (Hamilton, 1822). Pesquisa Veterinária Brasileira 33:914–919
El Adlouni C, Tremblay J, Walsh P, Lagueux J, Bureau J, Laliberte D et al (1995) Comparative study of DNA adducts levels in white sucker fish (Catostomus commersoni) from the basin of the St. Lawrence River (Canada). Mol Cell Biochem 148:133–138

Erickson G, Larsson A (2000) DNA adducts in perch (Perca fluviatilis) living in coastal water polluted with bleached pulp mill effluents. Ecotoxicol Environ Saf 46:167–173

Fanta E, SantAnna R, Romao F, Vianna ACC, Freiberger S (2003) Histopathology of the fish corydoras paleatus contaminated with sublethal levels of organophosphorus in water and food. Ecotoxol Environ Saf 54:119–130

Fernandez D, ELM V, Bandow N, Munoz K, Schafer RB (2014) Calibration and field application of passive sampling for episodic exposure to polar organic pesticides in streams. Environ Pollut 194:196–202

Firat O, Cogun HY, Yüzereroglu TA, Gök G, Firat O, Kargin F et al (2011) A comparative study on the effects of a pesticide (cypermethrin) and two metals (copper, lead) to serum biochemistry of Nile tilapia, Oreochromis niloticus. Fish Physiol Biochem 37:657–666

Fowler SV, Paynter Q, Hayes L, Dodd S, Groenteman R (2010) Biocontrol of weeds in New Zealand: an overview of nearly 85 years. In: Zydenbos SM (ed) 17th Australasian weeds conference papers and proceedings: new frontiers in New Zealand, together we can beat the weeds, 26–30 September, Christchurch, New Zealand. New Zealand Plant Protection Society, Christchurch, pp 211–214

Ganeshwad RM, Dama LB, Deshmukh DR, Ghanbahadur AG, Sonawane SR (2012) Toxicity of endosulfan on freshwater fish Channa striatus. Trends Fish Res 1:29–31

Garcia FP, Ascencio SYC, Oyarzun JCG, Hernandez AC, Patricia Vazquez Alavarado PV (2012) Pesticides: classification, uses and toxicity. Measures of exposure and genotoxic risks. J Res Environ Sci Toxicol 1:279–293

Gartiser S, Stiene G, Hartmann A, Zipperle J (2001) Einsatz von Desinfektionsmitteln im Krankenhausbereich Ursache fur okotoxische and gentoxische Effekte im Krankakenhausabwasser? (German). Vom Wasser 96:71–88

Gilden RC, Huffling K, Sattler B (2010) Pesticides and health risks. J Obstes Gynecol Neonatal Nurs 39:103–110

Gill RJ, Raine NE (2014) Chronic impairment of bumble bee natural foraging behaviour induced by sublethal pesticide exposure. Funct Ecol 28:1459–1471

Gilliom RJ (2007) Pesticides in U.S. streams and groundwater. Environ Sci Technol 41:3408–3414

Gilot-Delhalle J, Colizzi A, Moutshen J, MoutshenDahmen M (1983) Mutagenicity of some organophosphorus compounds at the ade6 locus of Schizosaccharomyces pombe. Mutat Res 117:139–148

Glusczak L, dos Santos MD, Crestani M, da Fonseca MB, de Araújo PF, Duarte MF et al (2006) Effect of glyphosate herbicide on acetylcholinesterase activity and metabolic and hematological parameters in piava (Leporinus obtusidens). Ecotoxicol Environ Saf 65:237–241

Glusczak L, Miron DS, Morch BS, Simoes RR, Schetinger MRC, Morsch VM et al (2007) Acute effects of glyphosate herbicide on metabolic and enzymatic parameters of silver catfish (Rhamdia quelen). Comp Biochem Physiol 146:519–524

Gunnell D, Eddleston M, Phillips MR, Konradsen F (2007) The global distribution of fatal pesticide self-poisoning: systematic review. BMC Public Health 7:357–371

Gurusamy K, Ramadoss V (2000) Impact of DDT on oxygen consumption and opercular activity of Lepidocephalichthys thermalis. J Ecotoxicol Environ Monit 10:239–248

Gurusubramanian G, Borthakur M (2005) Integrated management of tea pests. In: Dutta AK, Baruah SK, Ahmed N, Sarma AK, Burugohain D (eds) Field management in tea. Assam Printing Works Private Limited, Tocklai Experimental Station, TRA, Jorhat, Assam, pp 159–172

Gurusubramanian G, Borthakur M, Sarmah M, Rahman A (2005) Pesticide selection, precautions, regulatory measures and usage. In: Dutta AK, Gurusubramanian G, Barthakur BK (eds) Plant protection in tea. Assam Printing Works Private Limited, Tocklai Experimental Station, TRA, Jorhat, pp 81–91

Hageman KJ, Simonich SL, Campbell DH, Wilson GR, Landers DH (2006) Atmospheric deposition of current use and historic-use pesticides in snow at national parks in the Western United States. Environ Sci Technol 40:3174–3180

Hakeem KR (2015) Crop production and global environmental issues. Springer International Publishing AG, Cham

Helfrich LA, Weigmann DL, Hipkins P, Stinson ER (2009) Pesticides and aquatic animals: a guide to reducing impacts on aquatic systems. Virginia Polytechnic Institute and State University, Blacksburg

History of pesticide use (1998). http://www2.mcdaniel.edu/Biology/eh01/pesticides/history of pesticides use.html. Accessed 7 May 2019

Holgerson MA, Raymond PA (2016) Large contribution to inland water CO_2 and CH_4 emissions from very small ponds. Nat Geosci 9:222–226

How pesticides affect the environment (n.d.) Peel Public Health. http://www.peelregion.ca/health/topics/pesticides/why-reduce/why-reduce4.htm

Hull RN, Kleywegt S, Schroeder J (2015) Risk-based screening of selected contaminants in the Great Lakes Basin. J Great Lakes Res 41:238–245

Ilavazhahan M, Tamil SR, Jayaraj SS (2010) Determination of LC50 of the bacterial pathogen, pesticide and heavy metal for the fingerling of freshwater fish Catla catla. Glob J Environ Res 4:76–82

Ilyas R, Jave M (2013) Acute toxicity of endosulfan to the fish species Catla catla, Cirrhina mrigala and Labeo rohita. Int J Agri Biol 15:149–152

Jabbar A, Mallick S (1994) Pesticides and environment situation in Pakistan (Working Paper Series No. 19). Available from Sustainable Development Policy Institute (SDPI)

Jackson TA (2007) A novel bacterium for control of grass grub. Biol Cont A Global Perspec 2007:160–168

James R, Alagurathinam S, Sampath K (1993) Heametological changes and oxygen consumption in Oreochromis mosambicus exposed to sublethal concentration of lead. Ind J Fish 40:193–196

Jayachandran K, Pugazhendy K (2009) Histopathological changes in the Gill of Labeo rohita (Hamilton) fingerlings exposed to atrazine. Amer-Eruasian. J Sci Res 4:219–221

Jha JK, Ranjana KP, Mishra AP (2014) Histopathological changes in the gills of Channa gachua, an air breathing teleost after short term exposure of hostathion. Bioscane 9:925–929

Johnson WW, Finley MT (1980) Handbook of acute toxicity of chemicals to fish and aquatic invertebrates. US Department of the Interior, fish and wildlife service. Res Pub 137:98

Johnson LJ, de Bonth AC, Briggs LR, Caradus JR, Finch SC, Fleetwood DJ et al (2013) The exploitation of epichloae endophytes for agricultural benefit. Fungal Divers 60:171–188

Joseph B, Raj JS (2011) Impact of pesticide toxicity on selected biomarkers in fishes. Int J Zool Res 7:212–222

Kean JM, Suckling DM, Stringer LD, Woods B (2011) Modeling the sterile insect technique for suppression of light brown apple moth (Lepidoptera: Tortricidae). J Economic Entomol 104:1462–1475

Khare A, Singh S (2002) Impact of Malathion on protein content in freshwater fish Clarias batrachus. J Ecotoxicol Environ Monit 12:129–132

Kim KH et al (2016) Exposure to pesticides and the associated human health effects. Sci Total Environ 575:525–535

Kingsbury PD, Kreutzweiser DP (1980) Environmental impact assessment of a semioperational permethrin application, foresty service report FMP-X-30. Environment Canada, Ottauh, pp 1–47

Klassen W, Curtis CF (2005) History of the sterile insect technique. In: Dyck VA, Hendrichs J, Robinson AS (eds) Sterile insect technique: principles and practice in area-wide integrated pest management. Springer, Dordrecht, pp 3–36

Kreuger J (1998) Pesticides in stream water within an agricultural catchment in southern Sweden, 1990–1996. Sci Total Environ 216:227–251

Laabs V, Amelung W, Pinto AA, Wantzen M, da Silva CJ, Zech W (2002) Pesticides in surface water, sediment, and rainfall of the northeastern Pantanal Basin, Brazil. J Environ Qual 31:1636–1648

Lakshmanan SA, Rajendran C, Sivasubramaniyan (2013) Impact of Dichlorvos on tissue glycogen and protein content in freshwater fingerlings, Oreochromis mossambicus (Peters). Int J Res Environ Sci Technol 3:19–25

Leu C, Singer H, Stamm C, Muller SR, Schwarzenbach RP (2004) Simultaneous assessment of sources, processes and factors influencing herbicide losses to surface waters in a small agricultural catchment. Environ Sci Technol 38:3827–3834
Liess M, von der Ohe P (2005) Analyzing effects of pesticides on invertebrate communities in streams. Environ Toxicol Chem 24:954–965
Lorenz S, Rasmussen JJ, Süß A, Kalettka T, Golla B, Horney P et al (2017) Specifics and challenges of assessing exposure and effects of pesticides in small water bodies. Hydrobiology 793:213–224
MacHardy WE (2000) Current status of IPM in apple orchards. Crop Prot 19:801–806
Magare SR, Patil HT (2000) Effect of pesticides on oxygen consumption, red blood cell count and metabolites of a fish. Puntius ticto Environ Ecol 18:891–894
Malaj E, Peter C, Grote M, Kuhne R, Mondy CP, Usseglio-Polatera P et al (2014) Organic chemicals jeopardize the health of freshwater ecosystems on the continental scale. Proc Nat Acad Sci 111:9549–9554
Malik DS, Maurya PK (2015) Heavy metal concentration in water, sediment, and tissues of fish species (Heteropneustis fossilis and Puntius ticto) from Kali River. Toxicol Environ Chem 96:1195–1206
Marigoudar SR, Ahmed RN, David M (2009) Cypermethrin induced respiratory and behavioural responses in Labeo rohita. Vet Arhiv 79:583–590
Marigoudar SR, Ahmed RN, David M (2010) Cypermethrin induced: in vivo inhibition of the acetylcholinesterase activity in functionally different tissues of the freshwater teleost, Labeo rohita (Hamilton). Toxicol Environ Chem 91:1175–1182
Marigoudar SR, Ahmed RN, David M (2012) Ultrastructural responses and oxidate stress induced by cypermethrin in liver of Labeo rohita. Chem Ecol 29:296–308
Maskaoui K, Zhou JL, Zhen TL, Hong H, Yu Z (2005) Organochlorine micropollutants in the Jiulong River estuary and Western Xiamen Sea, China. Mar Pollut Bull 51:950–959
Mastan S, Shaffi S (2010) Sub-lethal effect of pesticides on the distribution of Glutaminases in the brain of Labeo rohita (Ham.). Int J Toxicol 7(2):1–6
Mathur SC (1999) Future of Indian pesticides industry in next millennium. Pesticide Inform 22:9–23
Matthews GA (2006) Pesticides: health, safety and the environment. Blackwell, Oxford
Maurya PK, Malik DS (2016) Bisoaccumulation of xenobiotics compound of pesticides in riverine system and its control technique: a critical review. J Indus Poll Cont 32:580–594
Mcfadyen RC (2000) Successes in biological control of weeds. In: Neal RS (ed) Proceedings of the X international symposium on biological control of weeds 3, vol 3. Montana State University, Bozeman, Montana, pp 3–14
Mills GA, Gravell A, Vrana B, Harman C, Budzinski H, Mazella N et al (2014) Measurement of environmental pollutants using passive sampling devices—an updated commentary on the current state of the art. Environ Sci: Processes Impacts 16:369–373
Mnif W, Hassine AI, Bouaziz A, Bartegi A, Thomas O, Roig B (2011) Effect of endocrine disruptor pesticides: a review. Int J Environ Res Public Health 8:2265–2303
Muir P (2002) The history of pesticides use. Oregon State University Press, USA
Murthy KS, Kiran BR, Venkateshwarlu M (2013) A review on toxicity of pesticides in fish. Int J Open Sci Res 1:15–36
Muthukumaravel K, Sivakumar B, Kumarasamy P, Govindarajan M (2013) Studies on the toxicity of pesticide monocrotophos on the biochemical constituents of the freshwater fish Labeo rohita. Int J Curr Biochem Biotechnol 2:20–26
Nagaraju B, Sudhakar P, Anitha A, Haribabu G, Rathnamma VV (2011) Toxicity evaluation and behavioural studies of fresh water fish Labeo rohita exposed to Rimon. Int J Res Pharma Biomed Sci 2:722–727
Nanda P, Panda BN, Behera MK (2000) Nickel induced alterations in protein level of some tissues of Heteropneustes fossilis. J Environ Bio 21:117–119
Nguyen VC, Nguyen TP, Mark B (2008) Brain cholinesterase response in the snakehead fish (Channa striata) after field exposure to diazinon. Ecotoxicol Environ Safe 71:314–318

Niazi NK, Singh B, Zwieten LV, Kachenko AG (2012) Phytoremediation of an arseniccontaminated site using Pteris vittata L. and Pityrogramma calomelanos var. austroamericana:a long-term study. Environ Sci Pollut Res 19:3506–3515
Nwani CD, Lakra WS, Nagpure NS, Kumar R, Kushwaha B, Srivastra SK (2010) Toxicity of the herbicide atrazine: effects on lipid peroxidation and activities of antioxidant enzymes in the fresh water fish Channa punctatus (block). Int J Environ Res Public Health 7:3298–3312
Nwani CD, Ivoke N, Ugwu DO, Atama C, Onyishi GC, Echi PC, Ogbonna A (2013) Investigation on acute toxicity and behavioral changes in a freshwater African catfish, Clarias gariepinus (Burchell, 1822), exposed to organophosphorous pesticide, Termifos®. Pak J Zool 45:959–965
Obiakor MO, Okonkwo JC, Nnabude PC, Ezeonyejiaku CD (2012) Eco- genotoxicology: micronucleus assay in fish erythrocytes as in situ aquatic pollution biomarker: a review J Anim. Sci Adv 2:123–133
Omer A, Pascual U, Russell N (2010) A theoretical model of agrobiodiversity as a supporting service for sustainable agricultural intensifi cation. Ecol Econ 69:1926–1933
Oruc EÖ, Üner N, Sevgiler Y, Usta D, Drumaz H (2006) Sublethal effects of organophosphate diazinon on the brain of Cyprinus carpio. Drug Chem Toxicol 29:57–67
Palmer WA, Heard TA, Sheppard AW (2010) A review of Australian classical biological control of weeds programs and research activities over the past 12 years. Biol Control 52:271–287
Paynter Q, Fowler SV, Gourlay AH, Groenteman R, Peterson PG, Smith L et al (2010) Predicting parasitoid accumulation on biological control agents of weeds. J Appl Ecol 47:575–582
Pazhanisamy K, Indra N (2007) Toxic effects of arsenic on protein content in the fish, labeorhita (Hamilton). Nat Environ Poll Technol 6:113–116
Pesticides (n.d.). GRACE Communications Foundation. http://www.sustainabletable.org/263/pesticides. Accessed 7 May 2019
Pierart A, Shahid M, Séjalon-Delmas N, Dumat C (2015) Antimony bioavailability: knowledge and research perspectives for sustainable agricultures. J Hazard Mater 289:219–234
Pimentel D, Culliney TW, Bashore T (2013) Public Health Risks Associated with Pesticides and Natural Toxins in Foods. IPM World Textbook. Regents of the University of Minnesota
Prashanth MS, Sayeswara HA, Goudar MA (2011) Free cyanide induced physiological changes in the freshwater fish, Poecilia reticulate. J Exp Sci 2:27–31
Prokopy RJ (2003) Two decades of bottom-up, ecologically based pest management in a small commercial apple orchard in Massachusetts. Agric Ecosyst Environ 94:299–309
Rabiet M, Margoum C, Gouy V, Carluer N, Coquery M (2010) Assessing pesticide concentrations and fluxes in the stream of a small vineyard catchment—effect of sampling frequency. Environ Poll 158:737–748
Rahman A, Sarmah M, Phukan AK, Borthakur M, Gurusubramanian G (2005a) A Plant having insecticidal property for the management of tea pests. In: Proceedings of 2005 International symposium on innovation in tea science and sustainable development in tea industry. 11–15 November, 2005, Hangzhou, China, pp. 731–748
Rahman A, Sarmah M, Phukan AK, Roy S, Sannigrahi S, Borthakur M, Gurusubramanian G (2005b) Approaches for the management of tea mosquito bug, Helopeltis theivora Waterhouse (Miridae : Heteroptera). In: Barooah AK, Borthakur M, Kalita JN (eds) Proceedings of 34th Tocklai conference—strategies for quality. Tocklai Experimental Station, TRA, Jorhat, Assam, pp 146–161
Rani GI, Kumaraguru AK (2014) Behavioural responses and acute toxicity of Clarias batrachus to synthetic pyrethroid insecticide, ƛ-cyhalothrin. J Environ App Biores 2:19–24
Rani S, Venkataramana GV (2012) Effects of the organophosphorous Malathion on the branchial gills of a freshwater fish Glossogobius giuris (ham). Int J Sci Nat 3:324–330
Rao JV (2006) Sublethal effects of an organophosphorus insecticide (RPR-II) on biochemical parameters of tilapia, Oreochromis mossambicus. Comp Biochem Physiol 143:492–498
Rao AS, Pillala RR (2001) The concentration of pesticides in sediments from Kolleru lake in India. Pest Manag Sci 57:620–624
Rao JV, Kavitha P, Jakka NM, Sridhar V, Usman P (2007) Toxicity of organophosphates on morphology and locomotor behavior in brine shrimp, Artemia salina. Arch Environ Contam Toxicol 53:227–232

Rasmussen JJ, Wiberg-Larsen P, Baattrup-Pedersen A, Cedergreen N, McKnight US et al (2015) The legacy of pesticide pollution: an overlooked factor in current risk assessments of freshwater systems. Water Res 84:25–32
Ravindran JK, Daniel R, Kumari S, George S, Eapen A (2012) Effect of agricultural pesticides, Hostathion and Kitazin on the Larvivorsity of the Carnatic Rice fish, Oryzias carnaticus (Jerdon, 1849). Am-Euras. J Toxicol Sci 4:56–59
Richards RP, Baker DB (1993) Pesticide concentration patterns in agricultural drainage networks in the Lake Erie basin. Environ Toxicol Chem 12:13–26
Roger PA, Bhuiyan SI (1990) Rice field ecosystem management and its impact on disease vectors. Int J Water Resour Dev 6:2–18
Sabir M, Hanafi MM, Malik MT, Aziz T, Zia-ur-Rehman M, Ahmad HR et al (2013) Differential effect of nitrogen forms on physiological parameters and micronutrient concentration in maize (Zea mays L.). Aust J Crop Sci 7:1836–1842
Saeedi FM, Roodsari HV, Zamini A, Mirrasooli E, Kazemi R (2012) The effects of Diazinon on behavior and some Hematological parameters of fry rainbow trout (Oncorhynchus mykiss). World J Fish Mari Sci 4:369–375
Sanjoy D, Rita M (2012) A study on the effect of organophosphorus pesticide Malathion on hepato-renal and reproductive organs of Heteropneustes foddilis (block). Sci Probe 1:1–13
Sarwar M (2015) The dangers of pesticides associated with public health and preventing of the risks. Int J Bioinfor Biomed Eng 1:130–136
Satyanarayan S, Bejankiwar RS, Chaudhari PR, Kotangale JP, Satyanarayan A (2004) Impact of some chlorinated pesticides on the haematology of the fish Cyprinus carpio and Puntius ticto. J Environ Sci 16:631–634
Satyavardhan K (2013) A comparative toxicity evaluation and behavioral observations of fresh water fishes to Fenvalerate. Middle-East J Sci Res 13:133–136
Saxena KK, Chaudhari R (2010) Study of chromosomal abnormalities in Channa punctatus expose to fenvalerate. J App Nat Sci 2:70–73
Schafer RB, Paschke A, Vrana v MR, Liess M (2008) Performance of the Chemcatcher passive sampler when used to monitor 10 polar and semi-polar pesticides in 16 central European streams, and comparison with two other sampling methods. Water Res 42:2707–2717
Schnick R, Fred M, Leroy GA (1980) Guide to approved chemicals in fish production and fisheries resource management, Arkansas cooperative extension service publication MP241. University of Arkansa, Little Rock, Arkansa
Scholz NL, Fleishman E, Brown L, Werner I, Johnson ML, Brooks ML et al (2012) A perspective on modern pesticides, pelagic fish declines, and unknown ecological resilience in highly managed ecosystems. Bioscience 62:428–434
Schulz R (2004) Field studies on exposure, effects, and risk mitigation of aquatic nonpoint-source insecticide pollution: a review. J Environl Qual 33:419–448
Scott GR, Sloman KA (2004) The effects of environmental pollutants on complex fish behaviour: integrating behavioural and physiological indicators of toxicity. Aqua Toxicol 68:369–392
Seyler LD, Rutz D, Allen J, Kamrin M (1994) A pesticide information project of the cooperative extension offices of Cornell University", Extoxnet: Extension Toxicology Network. Oregon State University
Shahid M, Ahmad A, Khalid S, Siddique HF, Saeed MF, Ashraf MR et al (2016) Pesticides pollution in agricultural soils of Pakistan. In: Soil Science: Agri Environ Prospec. Springer, Berlin, pp 199–229
Shankar KM, Kiran BR, Venkateshwarlu M (2013) A review on toxicity of pesticides in fish. Int J Open Sci Res 1:15–36
Shereena KM, Logaswamy S, Sunitha P (2009) Effect of an organophosphorous pesticide (Dimethoate) on oxygen consumption of the fish Tilapia mossambica. Recent Res Sci Technol 1:4–7
Sial IM, Kazmi MA, Kazmi QB, Naqvi SN (2009) Toxicity of Biosal (Phytopesticide) and Permethrin (Pyrethroid) against common carp, Cyprinus carpio. Pak J Zool 41:235–238

Singh SK, Singh SK, Yadav RP (2010) Toxicological and biochemical alterations of cypermethrin (synthetic pyrethroids) against freshwater teleost fish colisa fasciatus at different season. World J Zool 5:25–32
Sivanatarajan P, Sivaramakishnan T (2013) Studies on some hematologic values of Oreochromis mossambicus (Peters) following its sudden transfer to various concentrations of potassium chlorate and potassium dichromate. Eur J Appl Sci 5:19–24
Spradley JP (1985) Toxicity of pesticides to fish, Arkansas cooperative extension service. Publication MP330, University of Arkansa little Rock, Arkansas
Stehle S, Schulz R (2015) Agricultural insecticides threaten surface waters at the global scale. Proc Nat Acad Sci 112:5570–5575
Stehle S, Knabel A, Schulz R (2013) Probabilistic risk assessment of insecticide concentrations in agricultural surface waters: a critical appraisal. Environ Mon Asses 185:6295–6310
Stenrød M (2015) Long-term trends of pesticides in Norwegian agricultural streams and potential future challenges in northern climate. Acta Agri Scand B Soil Plant Sci 65:199–216
Stringer LD, Sullivan NJ, Sullivan TES, Mitchell VJ, Manning LM, Mas F et al (2013) Attractiveness and competitiveness of irradiated light brown apple moths. Entomol Exp Appl 148:203–212
Stuer-Lauridsen F (2005) Review of passive accumulation devices for monitoring organic micro-pollutants in the aquatic environment. Environ Pollu 136:503–524
Su ¨ßA, Bischoff G, Mueller ACW, Buhr L (2006) Chemical and biological monitoring of the load of plant protection products and of zoocoenoses in ditches of the orchard region "Altes land". Nachrichtenblatt des Deutschen Pflanzen Schutz Dienstes 58:28–42
Svoboda M, Luskova V, Drastichova J, Zlabek V (2001) The effect of diazinon on haematological indices of common carp (Cyprinus carpio L.). Acta Vet Brno 70:457–465
Taghavi L, Probs J, Merlina G, Marchand A, Durbe G, Probst A (2010) Flood event impact on pesticide transfer in a small agricultural catchment (Montousse at Aurade, south West France). Int J Environ Ana Chem 90:390–405
Thenmozhi C, Vignesh V, Thirumurugan R, Arun S (2011) Impacts of malathion on mortality and biochemical changes of freshwater fish Labeo rohita. Iran J Environ Health Sci Eng 8:387–394
Tierney KB, Kennedy CJ, Gobas F, Gledhill M, Sekela M (2014) Organic contaminants and fish. Org Chem Toxicol Fish 33:1–52
Ullah R, Zuberi A, Ullah S, Ullah I, Dawar FU (2014c) Cypermethrin induced behavioral and biochemical changes in mahseer, Tor putitora. J Toxicol Sci 39:829–836
Ullah R, Zuberi A, Tariq M, Ullah S (2014d) Acute toxic effects of cypermethrin on hematology and morphology of liver, brain and gills of mahseer (Tor putitora). Int J Agri Biol 17(1):199–204
USDA RMA (2013) Organic farming practices. Programs aid 1912. USDA Risk Management Agency, Washington, DC
Vargas VM, Migliavacca SB, de Melo AO, Horn RC, Guidobono RR, de Sa F et al (2001) Genotoxicity assessments in aquatic environments under the influence of heavy metals and organic contaminants. Mutat Res 490:141–158
Velmurugan B, Selvanayagam M, Cengiz E, Unlu E (2009) Histopathological changes in the gill and liver tissues of freshwater fish (Cirrhinus mrigala) exposed to Dichlorvos. An Inter J Braz Arch Bio Technol 52:1291–1296
Vrana B, Allan IJ, Greenwood R, Mills GA, Dominiak E, Svensson K et al (2005) Passive sampling techniques for monitoring pollutants in water. Trends Analy Chem 24:845–868
Warburton B, Poutu N, Peters D (2008) Traps for killing stoats (Mustela erminea): improving welfare performance. Anim Welf 17:111–116
Wasim MD, Dwaipayan S, Ashim C (2009) Impact of pesticides use in agriculture: their benefits and hazards. Interdiscip Toxicol 2:1–12
WHO (2009) The WHO Recommended classification of pesticides by hazard
Wikipedia, the free encyclopedia (2013) Insecticide Dropdata, pp. 1–5
Xing Z, Chow L, Rees H, Meng F, Li S, Ernst B et al (2013) Influences of sampling methodologies on pesticide-residue detection in stream water. Arch Environ Contam Toxicol 64:208–218

Yamanaka T (2007) Mating disruption or mass trapping? Numerical simulation analysis of a control strategy for lepidopteran pests. Popul Ecol 49:75–86

Ye J, Zhao M, Liu J, Liu W (2010) Enantioselectivity in environmental risk assessment of modern chiral pesticides. Environ Poll 158:2371–2383

Zacharia JT (2011) Identity, physical and chemical properties of pesticides. In: Pesticides in the modern world—trends in pesticides analysis. InTech, London, pp 1–18

Chapter 3
Impact of Invasive Plants in Aquatic Ecosystems

Afrozah Hassan and Irshad A. Nawchoo

3.1 Introduction

Water, an important natural resource, is a universal component of life, and without water, life would cease to exist (Duran-Sánchez et al. 2018). More than 90% of the earth's freshwater resources are held in lakes (Rast 2014). As per the Millennium Ecosystem Assessment, lakes provide regulating, provisioning, cultural, and supporting services developed within the ecosystem services framework (MEA 2005). Indeed, lake ecosystems characterize valuable environmental resource consequently with high preservation, conservation, and utilization value (Marothia 2004). The ecosystem services provided by lakes are significantly important. Annually, the global value of ecosystem services provided by lakes is several trillion dollars (Postel and Carpenter 1997). Lakes constitute important bioresources and have the potential for fishery and high conservation values (Ganai and Parveen 2014). As sensitive ecosystems, lakes can undergo rapid environmental changes, often leading to variations in function and structure. Since the recent few years, freshwater ecosystems are considered as the ecosystems mostly impacted by species invasions (Burks et al. 2006).

As a predominant global change (NCR 2000; MEA 2005), the conservation of natural resources and conservation of biodiversity is challenged by biological invasions (TEEB 2010). Principally invasive species have earned the peculiarity as the second extreme cause of species extinction (Drake et al. 1989). In the current period of global environmental change, modern research has clearly identified the role of invasive species in loss of biodiversity. Besides habitat degradation, certain studies have categorized species invasions as the second major cause of biodiversity loss (Wilcove et al. 1998). As a considerable economic issue, biological invasions cause

A. Hassan (✉) · I. A. Nawchoo
Plant Reproductive Biology Genetic Diversity and Phytochemistry Research Laboratory, Department of Botany, University of Kashmir, Srinagar, Jammu and Kashmir, India

K. R. Hakeem et al. (eds.), *Bioremediation and Biotechnology*,
https://doi.org/10.1007/978-3-030-35691-0_3

major losses to tourism, fisheries, and forestry sectors all over the world (Pimentel 2002). The estimated costs of the annual damage caused by biological invasions for Australia, the USA, and Brazil are US$13 billion, US$143 billion, and US$30 billion, respectively (Shine 2006).

Due to their evolutionary isolation, lakes are ingenuous to the effects caused by a broad range of invaders (Cox and Lima 2006). Lakes that represent model systems for studying the effects of invasions (Sharma et al. 2009) are gradually threatened by invasions, pollution, hydrological changes, and habitat degradation (Dudgeon et al. 2006; Shah and Reshi 2014). Considered as a rich resource, the lakes and wetlands are mostly vulnerable to invasions (Zelder and Kercher 2004). Invasive aquatic plants alter the habitat structure of freshwater ecosystems (Valley and Bremigan 2002), impact the quality of water (Rommens et al. 2003), and hence transform them completely. In comparison to marine ecosystems, impacts of invasions were found to be more frequent in freshwater ecosystems (Ricciardi and Kipp 2008).

Biological invasion in freshwater ecosystems emerges as the most detrimental change, causing significant damage and hence cascading effects on functional integrity and structural organization of the ecosystem. Apparently in freshwaters, generally there are more losses of the species as compared to terrestrial and marine environments (Olden et al. 2006). Usually, the freshwater ecosystems are characteristically invasive owed to the great potential of increasing their spatial distribution (Richardson et al. 2000). The plant species present in waterbodies called macrophytes are of great ecological significance and hence increase the complexity of the freshwater ecosystems (Esteves 1988). Macrophytes play an important role for the functioning of freshwater ecosystems (Jeppesen et al. 1998). Along with the native macrophytes, nonnative plant species called invasive alien plants are also available. In freshwater ecosystems, invasive species are of different growth forms (Tables 3.1 and 3.2). Once established, the species displace the native species completely. Invasive macrophytes may threaten freshwater ecosystems and change them exclusively (Getsinger et al. 2014; Brundu 2015).

Table 3.1 Emergent invasive plants (Ricciardi & Mac Isaac 2005)

S. no.	Scientific name	Common name	Family
1	*Typha latifolia*	Cattail common	Typhaceae
2	*Typha angustata*	Cattail narrow leaved	Typhaceae
3	*Typha orientalis*	Cat tail	Typhaceae
4	*Phragmites communis*	Common reed	Poaceae
5	*Commelina benghalensis*	Watergrass	Commelinaceae
6	*Alisma plantago*	Water cat tail	Alismataceae
7	*Cyperus difformis*	Umbrella plant	Cyperaceae
8	*Ipomea aquatica*	Floating morning glory	Convolvulaceae
9	*Trapa bispinosa*	Water chestnut	Trapaceae
10	*Hydrocotyle umbrella*	Water pennywort	Hydrocolylaceae
11	*Jussiaea repens*	Water primrose	Onagraceae
12	*Ludwigia parviflora*	Water purslane	Onagraceae

Table 3.2 Free-floating invasive plants (Subhendu et al. 2009)

S. no.	Scientific name	Common name	Family
1	*Eichhornia crassipes*	Water hyacinth	Pontederiaceae
2	*Salvinia auriculata*	Water fern	Salviniaceae
3	*Salvinia molesta*	Water fern	Salviniaceae
4	*Salvinia natans*	Water fern	Salviniaceae
5	*Pistia stratiotes*	Water lettuce	Araceae
6	*Lemna minor*	Duck weed	Lemnaceae
7	*Spirodela polyrhiza*	Giant duckweed	Lemnaceae
8	*Azolla imbricate*	Water velvet	Salviniaceae
9	*Polygonum amphibium*	Water smart weed	Polygonaceae

Table 3.3 Extremely invasive macrophytes (Scheffer et al. 2001)

S. no.	Scientific name	Family
1	*Egeria densa*	Hydrocharitaceae
2	*Eichhornia crassipes*	Pontederiaceae
3	*Hygrophila polysperma*	Acanthaceae
4	*Lagarosiphon major*	Hydrocharitaceae
5	*Limnophila sessiliflora*	Plantaginaceae
6	*Lythrum salicaria*	Lythraceae
7	*Myriophyllum aquaticum*	Haloragaceae
8	*Phragmites australis*	Poaceae
9	*Salvinia molesta*	Salviniaceae
10	*Sparganium erectum*	Typhaceae
11	*Trapa natans*	Lythraceae

Aquatic invasions change macrophyte composition (Santos et al. 2011; Hussner 2014), alter species richness and abundance (Stiers et al. 2011), change the food web structure (Villamagna and Murphy 2010), and deplete oxygen levels (Shillinglaw 1981). Vigorously, all invasive plants have the potential for clonal propagation (Kolar 2001; Liu et al. 2006; Xu et al. 2010) and spread quickly. Some invasive aquatic plants such as *Myriophyllum aquaticum*, *Alternanthera philoxeroides*, and *Eichhornia crassipes* exclude other species by forming large stands in water (Timmons and Klingman 1958; Julien and Bourne 1988). The species exhibit vigorous growth and become highly invasive in character (Table 3.3). Clonal integration is directly correlated with the invasiveness of alien plants; hence, the clonal plants propagate quickly and spread into new locations (Maurer and Zedler 2002). Due to the damaging effects, biological invasions developed as main environmental policy issue displacing native species of both terrestrial and aquatic ecosystems at an unprecedented rate (Mack et al. 2000; Simberloff et al. 2005). Some invasive species, such as *Gunnera manicata*, *Gunnera tinctoria*, *Gymnocoronis spilanthoides*, *Egeria densa*, and *Lagarosiphon* in freshwaters of New Zealand, displace culturally important species including *Phormium tenax* and edible watercress *Lepidium sativum* (Waikato 2006). Most of the European countries were invaded by the non-indigenous aquatic plants which became a major threat to the native biota (Table 3.4).

Table 3.4 List of aquatic nonindigenous species reported from European countries (Hussner 2012)

S. no.	Name of the plant species	Name of the country
1	*Alternanthera philoxeroides*	South America
2	*Ammannia senegalensis*	Africa
3	*Aponogeton distachyos*	South Africa
4	*Azolla filiculoides*	North, Central, South America
5	*Azolla caroliniana*	North, Central, South America
6	*Bacopa monnieri*	Asia, North America
7	*Baldellia ranunculoides*	Europe, North Africa
8	*Blyxa japonica*	Asia
9	*Cabomba caroliniana*	South America
10	*Callitriche brutia*	Europe
11	*Callitriche deflexa*	Central, South America
12	*Crassula helmsii*	Australia
13	*Ceratophyllum demersum*	Asia, Africa, Europe, North, Central, South America
14	*Ceratophyllum submersum*	Europe, Asia
15	*Ceratopteris thalictroides*	Asia
16	*Cryptocoryne crispatula*	Asia
17	*Egeria densa*	South America
18	*Eichhornia crassipes*	South America
19	*Eleocharis parvula*	Europe, Asia, North, Central America
20	*Elodea callitrichoides*	South America
21	*Elodea canadensis*	North America
22	*Elodea nuttallii*	North America
23	*Gymnocoronis spilanthoides*	South America
25	*Groenlandia densa*	Europe
26	*Heteranthera limosa*	North, South America
27	*Heteranthera reniformis*	North, Central, South America
28	*Heteranthera rotundifolia*	North America
29	*Heteranthera zosterifolia*	South America
30	*Hydrilla verticillata*	Asia
31	*Hydrocharis morsus-ranae*	Europe, Asia
32	*Hydrocotyle bonariensis*	North, Central, South America
33	*Hydrocotyle moschata*	New Zealand
34	*Hydrocotyle novae*	New Zealand
35	*Hydrocotyle ranunculoides*	North, Central, South America
36	*Hydrocotyle sibthorpioides*	Asia
37	*Hydrocotyle verticillata*	North, Central, South America
38	*Hygrophila polysperma*	Asia
39	*Lagarosiphon major*	South Africa
40	*Landoltia punctata*	Australia, Asia
41	*Lemna aequinoctialis*	South America
42	*Lemna gibba*	North America, Europe, Asia
43	*Lemna minor*	North America, Asia, Africa

(continued)

Table 3.4 (continued)

S. no.	Name of the plant species	Name of the country
44	*Lemna minuta*	South America, Central, and North America
45	*Lemna perpusilla*	Asia, Africa, North, Central, South America
46	*Lemna turionifera*	Asia, North America
47	*Lilaeopsis carolinensis*	North America
48	*Ludwigia grandiflora*	South America
49	*Ludwigia peploides*	South America
50	*Murdannia keisak*	Asia
51	*Myriophyllum aquaticum*	South America
52	*Myriophyllum heterophyllum*	North America
53	*Myriophyllum verrucosum*	Australia
54	*Najas graminea*	
55	*Najas guadalupensis*	Asia, North, Central, South America
56	*Najas gracillima*	North America
57	*Najas orientalis*	Asia
58	*Nelumbo nucifera*	Asia, Africa
59	*Nuphar advena*	North America
60	*Nuphar pumila*	Asia, Europe
61	*Nuphar japonica*	Asia
62	*Nymphaea alba*	Europe, Asia
63	*Nymphaea lotus*	Asia, Africa, South America
64	*Nymphaea mexicana*	North, Central America
66	*Nymphoides peltata*	Europe, Asia
67	*Orontium aquaticum*	North America
68	*Ottelia alismoides*	Asia, Australia
69	*Pistia stratiotes*	South America
70	*Pontederia cordata*	North, South America
71	*Potamogeton epihydrus*	North America
72	*Potamogeton nodosus*	North America, Europe, Asia
73	*Potamogeton trichoides*	Europe, Asia
74	*Rotala indica*	Asia
75	*Rotala macrandra*	Asia
76	*Rotala ramosior*	North America
77	*Rotala rotundifolia*	Asia
78	*Sagittaria graminea*	North America
79	*Sagittaria latifolia*	North America
80	*Sagittaria platyphylla*	North America
81	*Sagittaria rigida*	North America
82	*Sagittaria subulata*	North South America
83	*Sagittaria sagittifolia*	Europe, Asia
84	*Salvinia auriculata*	Central, South America
85	*Salvinia adnata*	South America
86	*Salvinia natans*	Europe, Asia
87	*Saururus cernuus*	North America

(continued)

Table 3.4 (continued)

S. no.	Name of the plant species	Name of the country
88	*Shinnersia rivularis*	Central America
89	*Spirodela polyrhiza*	Europe, Asia, North, Central America
90	*Stratiotes aloides*	Europe
91	*Trapa natans*	Asia, Europe
92	*Utricularia gibba*	Asia, North, Central America
93	*Vallisneria nana*	Australia
94	*Vallisneria spiralis*	Asia, Europe, North America
95	*Wolffia arrhiza*	Europe, Asia, Africa
96	*Zannichellia repens*	Europe, North America

3.2 Spread of Invasive Plants

Invasive species are key concern in freshwater ecosystems. They spread into new sites through numerous pathways such as through commercial boating, recreational boats and intentional introductions (Darke and Mandrak Drake and Mandrak 2010; Keller and Lodge 2007; Strecker et al. 2011). Most of the recent invasions occur due to human activities linked with international trade which is accelerating the spread of organisms into new locations (Gherardi & Holdich 1999; Levine and Antonio 2003). Invasive species have been introduced through various vectors (Gido and Brown 1999) and caused a devastating effect to the freshwater ecosystems (Sala et al. 2000; Leprieur et al. 2008) even though invasive plants spread through trade in ornamentals, species for ponds and aquarium (Keller and Lodge 2007), and aquaculture (Holdich et al. 1999).

In Great Lakes Basin, the worlds most invaded Lake Basin, invasions have been occurred through several vectors that operate on multiple spatial scales and are facilitated by environmental and socioeconomic factors (Ricciardi 2006). The spread of invasives cause many changes to the habitat of the species. Habitat alteration is one of the most important threat to global biodiversity and hence to the ecosystem structure and function (Mack et al. 2000). For instance, in North America and Middle Atlantic, the invasion of *Phragmites* has been clearly understood both in freshwater and saline wetlands where it spread at the expense of displacing other plant species (Chambers et al. 1999). During the 1950s, the introduction of an exotic genotype has moderately hastened the growth and spread of *Phragmites* throughout its natural range (Saltonstall 2002) and alter habitat structure.

In New Zealand and Australia where eradication programs have become well established, public surveys were carried out to understand the perceptions of the people regarding the spread of invasive species and their attitude toward management (Johnston and Marks 1997; Fraser 2001, 2006). Most of the studies carried on biological invasions reported that hydrological changes are associated with the spread of the invasive species (Baldina et al. 1999) such as hydrological changes and anthropogenic factors in riparian wetlands of river Murray in Australia have resulted in exotic plant invasions (Catford et al. 2011). Emergent plants in river Mukwonago produce higher biomass and form stabilized water levels. Any change in the water levels resulted in the increased chances of alien invasions (Boers and Zedler 2008). The shift from clear water to turbid conditions resulted in the incidence of the invasive species *Egeria densa*. Probably the invasion of the species was related to the change in the water quality of the lake (Schallenberg and Sorell 2009). In the entire process of invasion, humans also play an important role, by changing ecosystem structure and functioning (McNeely 2001; Hulme 2009; Kueffer and Kull 2017).

The role of humans has been found in the process of invasions by serving as vectors for spread of invasions, both intentionally and accidently. Certain disease occurs in humans due to invasive species; hence, the management of the species is of prime importance, and the problem of the invasion is a social ecological problem (Perrings et al. 2000, 2002). Humans are the prime cause for the spread of thousands of freshwater, estuarine, and marine species (Cohen and Carlton 1998). Worldwide human-aided spread of nonnative species such as plants, fishes, aquatic invertebrates, and microbes has a strong ecological impact especially in lakes and rivers (Nesler and Bergersen 1991; Witte et al. 1992; Flecker and Townsend 1994; Hall and Mills 2000; Latini and Petrere 2004).

The rate of invasion is rising rapidly. It has been well studied that in the coming years, the alterations in water temperature result in the establishment of many invasive species at the expense of the native species (Stachowicz 2002). Currently, in San Francisco, new species become established in 14 weeks, whereas in the 1960s, new species have been found to invade the ecosystem after 55 weeks (Cohen and Carlton 1998). For some species, warmer temperatures in specific habitats cause longer growing seasons, earlier reproduction, migration, and increase growth and dispersal rates, hence increasing the rate of invasions (Garcia-Ramos and Rodriguez 2002). Biological invasions have extremely transformed freshwater ecosystems (Vitousek et al. 1997) by fast dispersal and growth rates (Fig. 3.1).

(a) (b)

Fig. 3.1 Invasion of *Alternanthera philoxeroides* in Dal Lake. (**a**) Selected site with clear water in (2016). (**b**) Same area occupied by invasive plant species *Alternanthera philoxeroides* in (2018)

3.3 Impacts of Invasive Plants

Nonnative species have well-recognized impacts all across the globe (Pejchar and Mooney 2009; Le Maitre et al. 2011; Schackleton et al. 2014). The impacts of invasions in freshwater ecosystems are enhanced by increased nutrient levels, loss of top predators, and altered flow regimes which occur due to increased overharvesting (Kay & Hoyle 2001). Aquatic invasions are occurring at faster pace over unprecedented spatial and temporal scales (Cohen and Carlton 1998; Leppakoski and Olenin 2000; Ruiz et al. 2000). In the USA, 5000 alien invasive plants have been established and have completely displaced native plant species (Morse et al. 1995). Around the world, there is a strong indication that invasive nonnative species have damaged the ecological systems and are threat to the economy (Pejchar and Mooney 2009) (Fig. 3.2).

Invasive species have a direct impact on economic systems. For example, the invasion of *Cabomba furcata* in Malaysian lakes that form dense monospecific stands disrupted the navigation and hence affected the ecotourism (Chew and Munirah 2009; Sharip and Jusoh 2010). In freshwaters, effects of invasions include high sedimentation rates, loss of biodiversity, alterations in water chemistry, and fluctuations in temperature (Schmitz et al. 1993). In Lake Victoria, the invasion of water hyacinth *Eichhornia crassipes* hampers boat navigation by forming thick mats on the surface of water. It provides breeding grounds for mosquitoes which resulted in the increase in transmission of vector born disease (Lodge 1993). The plant species invaded to Nigeria from Benin, clog the water channels, and hence prevents the fishing and transportation of boats. The growth of the plant species impedes light penetration required for photosynthesis (Ogunye 1988).

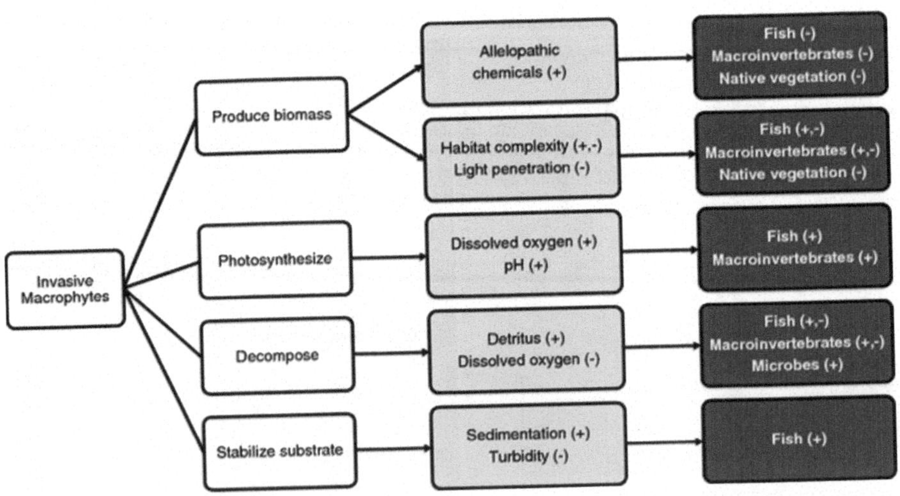

Fig. 3.2 Effects of invasive macrophytes on the ecosystem Adapted from Schultz and Dibble (2012)

Invasive species cause many socioecological problems and various diseases such as mouth foot and mouth disease in humans. The impact of invasives cause long-term effects such as changes in the composition, structure, and function of ecosystems (Lloret et al. 2004). The invasives have tremendous effects due to their excessive growth. Aquatic invasives threaten ecosystems and have both ecological and economic impacts due to their excessive growth (Brundu 2015). The average growth of macrophytes is beneficial for the lake ecosystem. Besides other benefits, plants provide shelter for fish and other organisms. Though, in some lakes, the growth of aquatic weeds can become excessive and create a serious nuisance interfering with boating, swimming, and other recreational activities. The rapid growth and spread of macrophytes is because of their adaptations (Santamaria 2002) which increase their invasive potential (Engelhardt 2011).

Usually, the impacts caused by macrophytes can spread from plant communities (Madsen et al. 1991) to various trophic levels (Theel et al. 2008) and to the entire ecosystems (Yarrow et al. 2009). The profuse macrophyte stands obstruct river flow and hence can increase the flood risks (Thouvenot et al. 2013). This results in reduced sports activities in water, decreases the economic values of lakes (Halstead et al. 2003), and hence hinders the shipping activities (Holm et al. 1969). The flora and fauna, including important medicinal and ceremonial plants, are affected by biological invasions (Williamson 1996). As soon as the invasive alien species are introduced by humans, they are established within no time, spread and hence become invasive (Latini and Petrere 2004), and add physical as well as chemical impacts to the concerned environments (Bunn and Arthington 2002; Koehn 2004). In North America and Australia, the impacts of exotic *Tamarix* spp. alter the water regime of the riparian soil and hence affect stream flows (Tickner et al. 2001).

3.4 Prevention of Invasion

"Water for life" is the main purpose of the United Nations International Decade (UNID, 2005–2015), and it is reported that the quick assessment of invasive species, well-timed prediction, effective management, and concern for conservation of biodiversity should be a priority of several research studies. In freshwaters, the management of the invasive species has been directed more toward distribution and control rather than the predictions about invasion (Wade 1997). The risk assessment of riparian plants (Pollock et al. 1998) and aquatic plants (Willby et al. 2000) has focused on biological traits such as dispersal strategies, physiological, demographic, and genetic features (Kean and Barlow 2000). Millions of dollars are spent annually by private and governmental organizations for preventing, controlling, and eradicating invasions (Lovell et al. 2006; Vilà et al. 2010).

In the USA, control and management of aquatic invasions costs millions of dollars (Pimental et al. 2005; Rockwell 2003). For mitigating the impacts of invasive species on biodiversity, eradication of the species is the best management strategy. Many invasives damage the waterbodies in the USA and affect the quality of water

(Table 3.5). The large areas which have touched an invasion threshold are challenging to control because of high management costs and much affected habitats (Byers et al. 2002). Rapid assessment tools such as remote sensing help to find and assess invasions across different regions. Remote sensing and rapid assessment tools help to find and assess invasions across different areas (Clewell and Rieger 1997).

A better understanding of remote sensing can be helpful to determine the invasion in particular sites and find effective management strategies. Identification of lakes which are susceptible to invasions helps in directing various management strategies (Vander zanden et al. 2010). Identifying the vulnerable lakes decreases the uncertainty about future invasions. This increases the probability that management strategies will be useful for eradication of both local and regional invasive species (Finnoff et al. 2007). Documentation of economic and ecological impacts of invasive species provides meaningful recommendations for preventing the spread of invasion and prioritizes management strategies (Papes et al. 2011). A complete inventory of invasive species is essential for the management of the species worldwide (Table 3.6).

Colonization, establishment, and effective control of the species is attained by early detection, quick response activities (Simberloff 2009) as well as analysis of the conditions that will allow invasive species to dominate native community (De gasperis and Motzkin 2007). Worldwide freshwater and estuarine biota are changing rapidly (Moyle and Leidy 1992). There is an urgent need to compute lost ecosystem services. Biological and economic researchers use traditional methods for quantifying the impacts of invasions (Lodge 1993; NRC 2008). Understanding the role of humans in biological invasions is one of the best researched area regarding

Table 3.5 Invasive aquatic plants in USA waterbodies

S. no.	Name of the species	Family
1	*Alternanthera philoxeroides*	Amaranthaceae
2	*Azolla pinnata*	Salviniaceae
3	*Egeria densa*	Hydrocharitaceae
4	*Eichhornia crassipes*	Pontederiaceae
5	*Eichhornia azurea*	Pontederiaceae
6	*Hydrilla verticillata*	Hydrocharitaceae
7	*Hygrophila polysperma*	Acanthaceae
8	*Lagarosiphon major*	Hydrocharitaceae
9	*Limnophila sessiliflora*	Plantaginaceae
10	*Ludwigia hexapetala*	Onagraceae
11	*Lythrum salicaria*	Lythraceae
12	*Myriophyllum aquaticum*	Haloragaceae
13	*Myriophyllum spicatum*	Haloragaceae
14	*Pistia stratiotes*	Araceae
15	*Salvinia minima*	Salviniaceae
16	*Salvinia molesta*	Salviniaceae
17	*Sparganium erectum*	Typhaceae

Table 3.6 Some of the invasive aquatic plants from Jammu (Rupinder et al. 2014)

S. no.	Name of species	Family
1	*Eichhornia crassipes*	Pontederiaceae
2	*Camboba caroliniana*	Cabombaceae
3	*Hydrilla verticillata*	Hydrocharitaceae
4	*Vallisneria americana*	Hydrocharitaceae
5	*Vallisneria spiralis*	Hydrocharitaceae
6	*Elodea canadensis*	Hydrocharitaceae
7	*Egeria densa*	Hydrocharitaceae
8	*Alternanthera philoxeroides*	Amaranthaceae
9	*Salvinia molesta*	Salviniaceae
10	*Polygonum glabrum*	Polygonaceae
11	*Polygonum barbatum*	Polygonaceae
12	*Isoetes lacustris*	Isoetaceae
13	*Marsilia quardiflora*	Marsileaceae
14	*Ipomea carnea*	Convolvulaceae
15	*Ipomea aquatic*	Convolvulaceae
16	*Potamogeton natans*	Potamogetonaceae
17	*Potamogeton crispus*	Potamogetonaceae
18	*Potamogeton lucens*	Potamogetonaceae
19	*Potamogeton pusillus*	Potamogetonaceae
20	*Scirpus acutus*	Cyperaceae
21	*Scirpus articulates*	Cyperaceae
22	*Scirpus subterminalis*	Cyperaceae
23	*Cyperus difformis*	Cyperaceae
24	*Cyperus tenuispica*	Cyperaceae
25	*Nymphoid indica*	Menyanthaceae
26	*Nymphoid aquatic*	Menyanthaceae
27	*Nasturtium officinale*	Brassicaceae
28	*Najas graminea*	Najadaceae
29	*Arundo donax*	Poaceae
30	*Typha aria*	Typhaceae
31	*Ranunculus arvensis*	Ranunculaceae
32	*Veronica anagallis-aquatica*	Plantaginaceae
33	*Ranunculus sclerata*	Ranunculaceae
34	*Azolla pinnata*	Azollaceae
35	*Caldesia parnassifolia*	Alismataceae
36	*Sagittaria subulata*	Alismataceae
37	*Nymphoid Cristata*	Menyanthaceae
38	*Hymenachne amplexicaulis*	Poaceae
39	*Paspalidium germinatum*	Poaceae
40	*Chara brauni*	Characeae
41	*Chara fragilis*	Characeae
42	*Oenanthe crocata*	Apiaceae
43	*Juncus articulates*	Juncaceae
44	*Cardamine hirsuta*	Brassicaceae
45	*Najas minor*	Najadaceae
46	*Eleocharis plantagineum*	Cyperaceae

human and social dimensions of invasion (McNeely 2001; McGeoch et al. 2010). However, much research is needed on the perceptions of people for introduction and spread of invasive species (KowarikI 2003; Kueffer and Kull 2017).

3.5 Management of Aquatic Weeds

The management of aquatic weeds is of prime importance in India, and out of the total 160 weed species, following are of important concern (Table 3.7). The losses caused by aquatic weeds are very disastrous; hence, their maintenance is very important. In India, the management of aquatic weeds can be done by various ways as follows (Narayan et al. 2017):

1. Manual and mechanical management
2. Preventive management
3. Ecological management
4. Chemical management
5. Biological management
6. Management through utilization.

3.6 Uses of Aquatic Invasive Plants

Aquatic weeds such as water hyacinth and duckweed contain 25–35% of proteins as dry matter (Taylor et al. 1971). *Myriophyllum* spp. contain rich sources of carotenes and xanthophylls (Pirie 1971). Some aquatic weeds are used for fish feed (Hazra and Tripathy 1985). Foliage of aquatic weeds is used for vegetables; roots are used as a source of carbohydrate and seeds as proteins (Gupta 1987). The *Phragmites* are used in Romania for printing, cardboards, and synthetic fibers. Some aquatic plants such as *Cyperus* and *Typha* are the source of pulp for fiber and paper (Frank 1976).

Table 3.7 Weed species which are primary concern in India (Twongo 1993; Twongo & Howard 1998)

S. no.	Name of the plant species
1	*Salvinia molesta*
2	*Eichhornia crassipes*
3	*Nelumbo nucifera*
4	*Hydrilla verticillata*
5	*Vallisneria spiralis*
6	*Nymphaea stellate*
7	*Typha angustata*
8	*Chara* spp.
9	*Nitella* spp.
10	*Ipomoea* spp.
11	*Ipomoea* spp.

3.7 Future Prospects

Preemptive strategies for weed management should be developed, because once established the species become difficult as well as expensive to control. Thinning of plants is done which reduces the chances of water blockage. Local tools should be used for the deweeding which uproots plant species completely, and hence there are less possibilities of perpetuation as compared to deweeding done by machines. Pollution of the waterbodies should be checked proactively to decrease the probabilities of weed formation. There is paucity of research in the invasion biology; hence, research in the field should be facilitated for developing effective protocols for the management of weed species. Furthermore, awareness among the locals is necessary for the timely management of the species. Harvesting of the weed plants should be done by the locals for generating economic benefits. Locals should be engaged in the weed management strategies so that good practices should be developed for complete eradication of the invasive weeds.

3.8 Conclusions

Biological invasions can devastate the ecology of the freshwater ecosystems. The perspective of locals and stakeholders should be taken for effective policy management and conservation of the ecosystems. The numerous economic benefits that the locals obtain from lakes should be incorporated in management planning. The livelihood of the locals is solely dependent on the lake resources which otherwise would not have been used, maintained, and protected through ages. For the control of invasive species, several methods should be developed which help in preventing the invasion of the species.

References

Baldina EA, Leeuw JD, Gorbunov AK, Labutina IA, Zhivogliad AF, Kooistr JF (1999) Vegetation change in the Astrakhanski Biosphere Reserve (Lower Volga Delta, Russia) in relation to Caspian Sea level fluctuation. Environ Conserv 26:169–178

Boers AM, Zedler JB (2008) Stabilized water levels and *Typha* invasiveness. Wetlands 28:676–685

Brundu G (2015) Plant invaders in European and Mediterranean inland waters: profiles, distribution, and threats. Hydrobiology 746:61–79

Bunn SE, Arthington AH (2002) Basic principles and ecological consequences of altered flow regimes for aquatic biodiversity. Environ Manag 30:492–507

Burks RL, Mulderij G, Gross EM, Jones I, Jacobsen L, Jeppesen E, Van Donk E (2006) Center stage: the crucial role of macrophytes in regulating trophic interactions in shallow lake wetlands. In: Bobbink R, Beltmann B, Verhoeven JTA, Whigham DF (eds) Wetlands: functioning, biodiversity conservation and restoration. Springer, Berlin, pp 37–59

Byers J, Reichard S, Randall J, Parker I, Smith C, Lonsdale W, Atkinson I, Seastedt T, Williamson M, Chornesky E (2002) Directing research to reduce the impacts of non-indigenous species. Conserv Biol 16:630–640

Catford JA, Downes BJ, Gippel CJ, Vesk PA (2011) Flow regulation reduces native plant cover and facilitates exotic invasion in riparian wetlands. J Appl Ecol 48:432–442

Chambers RM, Meyerson LA, Saltonstall K (1999) Expansion of Phragmites australis into tidal wetlands of North America. Aquat Bot 64:261–273

Chew MY, Munirah MY (2009) Ecological implications from the naturalization of noxious Cabomba water weeds in Malaysia. Malays Natur 63:19–21

Clewell A, Rieger J (1997) What practitioners need from restoration ecologists. Restor Ecol 5:350–354

Cohen AN, Carlton JT (1998) Accelerating invasion rate in a highly invaded estuary. Science 279:555–558

Cox JG, Lima SL (2006) Native and an aquatic-terrestrial dichotomy in the effects of introduced predators. Trend Ecol Evol 21:674–680

De gasperis BG, Motzkin G (2007) Windows of opportunity: historical and ecological 505 controls on *Berberis thunbergii* invasions. Ecology 88:3115–3125

Drake DAR, Mandrak NE (2010) Least-cost transportation networks predict spatial interaction of invasion vectors. Ecol Appl 20:2286–2299

Drake JA, Mooney HA, Castri D, Groves RH, Kruger FJ, Rejmanek M, Williamson M (1989) Biological invasions. A global perspective. Wiley, Chichester

Dudgeon D, Arthington AH, Gessner MO, Kawabata ZI, Knowler DJ, Leveque C, Sullivan CA (2006) Freshwater biodiversity: importance, threats, status and conservation challenges. Biol Rev 2:163–182

Duran-Sánchez A, Álvarez-García J, Rama M (2018) Sustainable water resources management: a bibliometric overview. Water 10:1191

Engelhardt KA (2011) Eutrophication aquatic. In: Simberloff D, Rejmánek M (eds) Encyclopedia of biological invasions. University of California Press, Berkeley, CA, pp 209–213

Esteves FA (1988) Fundamentos de Limnologia. Editora Interciencia/FINEP, Rio de Janeiro, p 574

Finnoff D, Shogren JF, Leung B, Lodge D (2007) Take a risk: preferring prevention over control of biological invaders. Ecol Econ 62:216–222

Flecker AS, Townsend CR (1994) Community—wide consequences of trout introduction in New Zealand streams. Ecol Appl 4:798–807

Frank PA (1976) Distribution and utilization research on tropical and subtropical aquatic weeds in United States. In: Aquatic weeds in S.E. Asia. Dr. W. Junk B.V., The Hague, pp 353–360

Fraser W (2001) Introduced wildlife in New Zealand: a survey of general public views. Manaaki Whenua Press, Lincoln

Fraser A (2006) Public attitudes to pest control: a literature review. Science and Technical Pub., Department of Conservation, Wellington

Ganai AH, Parveen S (2014) Effect of physico-chemical conditions on the structure and composition of the phytoplankton community in Wular Lake at Lankrishipora, Kashmir. Int J Biodiver Conserv 6:71–84

Garcia-Ramos G, Rodriguez D (2002) Evolutionary speed of species invasions. Evolution 56:661–668

Getsinger KD, Dibble E, Rodgers JH, Spencer D (2014) Benefits of controlling nuisance aquatic plants and algae in the United States. Council of Agricultural Science and Technology (CAST) Commentary, QTA, Ames, IA, pp 1–12

Gherardi F, Holdich DM (1999) Crayfish in Europe as alien species: how to make the best of a bad situation? A.A. Balkema, Rotterdam, pp 221–235

Gido KB, Brown JH (1999) Invasion of North American drainages by alien fish species. Fresh Water Biol 42:387–399

Gupta OP (1987) Aquatic weed management a text book and manual. Today & Tomorrow Printers and Publishers, New Delhi

Hall SR, Mills EL (2000) Exotic species in large lakes of the world. Aquat Ecosyst Health Manag 3:105–135

Halstead JM, Michaud J, Hallas-Burt S, Gibbs JP (2003) Hedonic analysis of effects of a nonnative invader (*Myriophyllum heterophyllum*) on New Hampshire (USA) lakefront properties. Environ Manag 32:391–398

Hazra A, Tripathy SD (1985) Nutritive value of aquatic weed *Spirodela polyrhiza* (Linn.) in grass carp. Indian J Anim Sci 55:702–705

Holdich DM, Rogers WD, Reynolds JD (1999) Native and alien crayfish in the British Isles. Crustacean Iss 11:221–236

Holm LG, Weldon LW, Blackburn RD (1969) Aquatic weeds. Science 3906:699–709

Hulme PE (2009) Trade, transport and trouble: managing invasive species pathways in an era of globalization. J Appl Ecol 46:10–18

Hussner A (2012) Alien aquatic plant species in European countries. Wed Res 52:297–306

Hussner A (2014) Long-term macrophyte mapping documents a continuously shift from native to non-native aquatic plant dominance in the thermally abnormal River Erft (North Rhine-Westphalia, Germany). Limnol Ecol Manag Inland Waters 48:39–45

Jeppesen E, Sondergaard M, Sondergaard M, Christofferson K (1998) The structuring role of submerged macrophytes in lakes. Springer, Dordrecht

Johnston MJ, Marks CA (1997) Attitudinal survey on vertebrate pest management in Victoria, vol 3. Department of Natural Resources and Environment, Agriculture Victoria, Frankston, VIC

Julien MHAS, Bourne (1988) Alligator weed is spreading in Australia. Plant Protect Quart 3:91–96

Kay SH, Hoyle ST (2001) Mail order, the internet, and invasive aquatic weeds. J Aqua Plant Manag 1:88–91

Kean JMND, Barlow (2000) Effects of dispersal on local population increase. Ecol Lett 3:479–482

Keller RP, Lodge DM (2007) Species invasions from commerce in live aquatic organisms—problems and possible solutions. Bioscience 57:428–436

Koehn JD (2004) Carp (*Cyprinus carpio*) as a powerful invader in Australian waterways. Fresh Water Biol 49:882–894

Kolar CSDM (2001) Lodge Progress in invasion biology: predicting invaders. Trend Ecol Evol 16:199–204

KowarikI (2003) Human agency in biological invasions: secondary releases foster naturalization and population expansion of alien plant species. Biol Invasions 5:293–312

Kueffer CC, Kull (2017) Non-native species and the aesthetics of nature. In: Hulme P, Vilà M (eds) Impact of biological invasions on ecosystem services. Springer, Berlin

Kumar P, TEEB (2010) The Economics of ecosystems and biodiversity: ecological and economic foundations. Earthscan, London

Latini AO, Petrere JM (2004) Reduction of a native fish fauna by alien species: an example from Brazilian freshwater tropical lakes. Fisher Manag Ecol 11:71–79

Le Maitre DC, Gaertner M, Marchante E, Ens EJ, Holmes PM, Pauchard A, Richardson DM (2011) Impacts of invasive Australian acacias: implications for management and restoration. Divers Distrib 5:1015–1029

Leppakoski E, Olenin S (2000) Nonnative species and rates of spread: lessons from the brackish Baltic Sea. Biol Invas 2:151–163

Leprieur F, Beauchard O, Hugueny B, Grenouillet G, Brosse S (2008) Null model of biotic homogenization: a test with the European freshwater fish fauna. Diver Distri 14:291–300

Levine JMD, Antonio CM (2003) Forecasting biological invasion with increasing international trade. Conserv Biol 17:322–326

Liu JM, Dong S, Miao Z, Li M, Song R, Wang (2006) Invasive alien plants in China: role of clonality and geographical origin. Biol Invasions 8:1461–1470

Lloret F, Dail F, Brundu G, Hulme PE (2004) Local and regional abundance of exotic plant species on Mediterranean islands: are species traits important? Glob Ecol Biogeogr 13:37–45

Lodge DM (1993) Biological invasions: lessons for ecology. Trends Ecol Evol 8:133–137

Lovell SJ, Stone SF, Fernandez L (2006) The economic impacts of aquatic invasive species: a review of the literature. Agric Resour Econ Rev 35:195–208
Mack RN, Simberloff D, Lonsdale WM, Evans H, Clout M, Bazzaz FA (2000) Biotic invasions: causes, epidemiology, global consequences, and control. Ecol Appl 10:689–710
Madsen JD, Sutherland JW, Bloomfield JA, Eichler LW, Boylen CW (1991) The decline of native vegetation under dense Eurasian watermill foils canopies. J Aqua Plant Manag 29:94–99
Marothia DK (2004) Restoration of lake ecosystem: an environmental economics perspective. Int Ecol Environ Sci 30:197–207
Maurer DA, Zedler JB (2002) Differential invasion of a wetland grass explained by tests of nutrients and light availability on establishment and clonal growth. Oecologia 131:279–288
McNeely JA (2001) The great reshuffling: Human dimensions of invasive alien species. IUCN, Gland
McGeoch MA, Butchart SHM, Spear D, Marais E, Kleynhans EJ, Symes A, Chanson J, Hoffmann M (2010) Global indicators of biological invasion: species numbers, biodiversity impact and policy responses. Divers Distr 16:95–108
Millennium Ecosystem Assessment (2005) Ecosystems and human well-being: current state and trends: findings of the Condition and Trends Working Group. Island Press, Washington, DC
Morse LE, Kartesz JT, Kutner LS (1995) Native vascular plants. In: La Roe ET, Farris GS, Puckett CE, Doran PD, Mac MJ (eds) Our living resources: a report to the nation on the distribution, abundance, and health of U.S. plants, animals and ecosystems. U.S. Department of the Interior, National Biological Service, Washington, DC, pp 205–209
Moyle PB, Leidy RA (1992) Loss of biodiversity in aquatic ecosystems: evidence from fish faunas. In: Conserv Biol. Springer, Boston, MA, pp 127–169
Narayan S, Nabi A, Hussain K, Khan FA (2017) Practical aspects of utilizing aquatic weeds in compost preparation
National Research Council (2000) Global change ecosystems research. National Academy Press, Washington, DC, p 2
Nesler TP, Bergersen EP (1991) Mysis in fisheries: hard lessons from headlong introductions. American Fisheries Society symposium no. 9, Bethesda
NRC (National Research Council, Transportation Research Board) (2008) Great Lakes shipping, trade, and aquatic invasive species. National Academy Press, Washington, DC, p 226
Ogunye O (1988) Water hyacinth—Nigerian experience. In: Oke OL, Imevbore AMA, Farri TA (eds) Water hyacinth menace and resource. Proceedings of international workshop. COSTED, Lagos, pp 7–12
Olden JD, JM, JT MC, Maxted WW, Fetzer JM, Zanen V (2006) The rapid spread of rusty crayfish (*Orconectes rusticus*) with observations on native crayfish declines in Wisconsin (U.S.A.) over the past 130 years. Biol Invasions 8:1621–1628
Papes M, Sallstrom M, Asplund TR, Vander Zanden MJ (2011) Invasive species research to meet the needs of resource management and. Conserv Biol 25:867–872
Pejchar L, Mooney HA (2009) Invasive species, ecosystem services and human wellbeing. Trends Ecol Evol 24:497–504
Perrings C, Williamson M, Dalmazzone S (2000) The economics of biological invasions. Edward Elgar, Cheltenham
Perrings C, Williamson M, Barbier EB, Delfino D, Dalmazzone S, Shogren J, Simmons P, Watkinson A (2002) Biological invasions risks and the public good: an economic perspective. Conserv Ecol 6:1
Pimental D, Zuniga R, Morrison D (2005) Update on the environmental and economic costs associated with alien-invasive species in the United States. Ecol Econ 52:273–288
Pimentel D (ed) (2002) Biological invasions economic and environmental costs of alien plant, animal, and microbe species. CRC Press, Boca Raton, FL
Pirie NW (1971) Leaf protein, its agronomy, preparation, quality and use, International Biological Programme Handbook No. 20. Blackwell Sci. Pub., Oxford, p 202

Pollock M, Naiman RJ, Hanley TA (1998) Plant species richness in riparian wetlands-a test of biodiversity theory. Ecology 79:94–105
Postel S, Carpenter S (1997) Freshwater ecosystem services. Nat Serv 1997:195–214
Rast W (2014) The 15th world lake conference: an overview of an informative event. Lakes Res 19:237–239
Ricciardi A (2006) Patterns of invasion in the Laurentian Great Lakes in relation to changes in vector activity. Divers Distrib 4:425–433
Ricciardi A, Kipp R (2008) Predicting the number of ecologically harmful exotic species in an aquatic system. Divers Distrib 14:374–380
Ricciardi A, Mac Isaac HJ (2005) Impacts of biological invasions on fresh water ecosystems. In: Fifty years of invasion ecology: the legacy of Charles Elton, vol 1. Blackwell, Chichester, pp 211–224
Richardson DM, Pysek P, Rejmanek M, Barbour MG, Panetta FD, West CJ (2000) Naturalization and invasion of alien plants: concepts and definitions. Divers Distrib 2:93–107
Rockwell HW (2003) Summary of a survey of the literature on the economic impact of aquatic weeds. Aquatic Ecosystem Restoration Foundation
Rommens W, Maes J, Dekeza N, Inghelbrecht P, Nhiwatiwa T, Holsters E, Ollevier F, Marshall B, Brengdonck L (2003) The impact of water hyacinth (*Eichhornia crassipes*) in a eutrophic subtropical impoundment (Lake Chivero, Zimbabwe). I. Water quality. Hydrobiologie 158:373–388
Ruiz GM, Fofonoff PW, Carlton JT, Wonham MJ, Hines AH (2000) Invasion of coastal marine communities in North America: apparent patterns, processes, and biases. Anal Rev Ecol Syst 31:481–531
Rupinder K, Bhawandeep K, Sanjay BK (2014) Sharma documentation of aquatic invasive alien flora of Jammu region, Jammu and Kashmir. Int J Interdiscip Multidiscip Stud 7:90–96
Sala OE, Chapin FS, Armesto JJ, Berlow E, Bloomfield J, Dirzo R, Huber-Sanwald E, Huenneke LF, Jackson RB, Kinzig A, Leemans R, Lodge DM, Mooney HA, Oesterheld M, Poff NL, Sykes MT, Walker BH, Walker M, Wall DH (2000) Global biodiversity scenarios for the year 2100. Science 287:1770–1774
Saltonstall K (2002) Cryptic invasion by a non-native genotype of the common reed, Phragmites australis, into North America. Proc Natl Acad Sci 99:2445–2449
Santamaria L (2002) Why are most aquatic plants widely distributed? Dispersal, clonal growth and small-scale heterogeneity in a stressful environment. Acta Oecol 23:137–154
Santos MJ, Anderson LWJ, Ustin SL (2011) Effects of invasive species on plant communities: an example using submersed aquatic plants at the regional scale. Biol Invasions 3:443–457
Schackleton RT, Le Maitre DC, Pasiecznik NM, Richardson DM (2014) *Prosopis*: a global assessment of the biogeography, benefits, impacts and management of one of the world's worst woody invasive plant taxa. AoB Plants 6:plu027
Schallenberg M, Sorell B (2009) Regime shifts between clear and turbid water in New Zealand lakes: environmental correlates and implications for management and restoration. Mar Freshw Res 43:701–712
Scheffer M, Carpenter S, Foley JA, Folke C, Walker B (2001) Catastrophic shifts in ecosystems. Nature 413:591–596
Schmitz DC, Schardt JD, Leslie AJ, Dray FA, Osborne JA, Nelson BV(1993) The ecological impact and management history of three invasive alien aquatic plant species in Florida. In: Biological pollution: the control and impact of invasive exotic species. Proceedings of a symposium held at Indianapolis, IN, USA, 25–26. Indiana Academy of Science, pp 173–194
Schultz R, Dibble E (2012) Effects of invasive macrophytes on freshwater fish and macro invertebrate communities: the role of invasive plant traits. Hydrobiologia 684:1–14
Shah MA, Reshi ZA (2014) Characterization of alien aquatic flora of Kashmir Himalaya: implications for invasion management. Trop Ecol 2:143–157
Sharip Z, Jusoh J (2010) Integrated Lake Basin management and its importance for Lake Chini and other lakes in Malaysia. Lakes Reserv 15:41–51

Sharma S, Jackson DA, Minns CK (2009) Quantifying the potential effects of climate change and the invasion of smallmouth bass on native lake trout populations across Canadian lakes. Ecography 32:517–525
Shillinglaw SN (1981) Dissolved oxygen depletion and nutrient uptake in an impoundment infested with Eichhornia crassipes (Mart.) Solms. J Limnol Soc S Afr 7:63–66
Shine C (2006) Small world means endangered world. International Herald Tribune
Simberloff D (2009) The role of propagule pressure in biological invasions. Ann Rev Ecol Evol Syst 40:81–102
Simberloff D, Parker IM, Windle PN (2005) Introduced species policy, management, and future research needs. Front Ecol Environ 1:12–20
Stachowicz JJ (2002) Linking climate change and biological invasions: ocean warming facilitates nonindigenous species invasions. Proc Natl Acad Sci U S A 99:15497–15500
Stiers I, Crohain N, Josens G, Triest L (2011) Impact of three aquatic invasive species on native plants and macroinvertebrates in temperate ponds. Biol Invas 13:2715–2726
Strecker AL, Campbell PM, Olden JD (2011) The aquarium trade as an invasion pathway in the Pacific Northwest. Fisheries 36:74–85
Taylor KG, Bates RP, Robbins RC (1971) Extraction of protein from water hyacinth. Hyacinth Contr 9:20–22
Theel HJ, Dibble ED, Madsen JD (2008) Differential influence of a monotypic and diverse native aquatic plant bed on a macro invertebrate assemblage; an experimental implication of exotic plant induced habitat. Hydrobiologia 600:77–87
Thouvenot L, Haury J, Thiebaut GA (2013) Success story: water primroses, aquatic plant pests. Aquat Conserv 23:790–803
Tickner DP, Angold PG, Gurnell AM, Mountford JO (2001) Riparian plant invasions: hydro geo morphological control and ecological impacts. Prog Phys Geog 25:22–52
Timmons FL, Klingman DL (1958) Control of aquatic and bank weeds. Soil Conserv 24:102–107
Twongo T (1993) Growing impact of water hyacinth on near shore environments of Lakes Victoria and Kyoga (East Africa). In: Johnson TC, Odada E (eds) Proceedings of a symposium: the limnology, climatology and paleoclimatology of East African Lakes, 18–22 Feb, Jinja, Uganda, 1996
Twongo T, Howard G (1998) Ways with weeds. New Sci 159:57–57
Valley RD, Bremigan MT (2002) Effect of macrophyte bed architecture on largemouth bass foraging: implications of exotic macrophyte invasions. Trans Am Fish Soc 131:234–244
Vander zanden MJ, Hansen GJ, Higgins SN, Kornis MS (2010) A pound of prevention, plus a pound of cure: early detection and eradication of invasive species in the Laurentian Great Lakes. Great Lakes Res 36:199–205
Vilà M, Basnou C, Pyšek P, Josefsson M, Genovesi P, Gollasch S, Hulme PE (2010) How well do we understand the impacts of alien species on ecosystem services? Apan-European, cross-taxa assessment. Front Ecol Environ 8:135–144
Villamagna AM, Murphy BR (2010) Ecological and socio-economic impacts of invasive water hyacinth (*Eichhornia crassipes*): a review. Freshw Biol 55:282–298
Vitousek PMD, Antonio CM, Loope LL, Rejmanek M, Westbrooks R (1997) Introduced species: a significant component of human—caused global change. N Z J Ecol 21:1–16
Wade M (1997) Predicting plant invasions: Making a start. In: Brock JH, Wade M, Pysek P, Green D (eds) Plant invasions: studies from north America and Europe. Backhuys, Leiden, pp 1–18
Waikato (2006) Pests affect our cultural heritage [www document]. http://www.ew.govt.nz/For-schools/Resources-for-teachers/Classroom-activities/Biosecurity-activities
Wilcove DS, Rothstein DJ, Phillips A, Losos E (1998) Quantifying threats to imperiled species in the United States. Bioscience 48:607–615
Willby NJVJ, Abernethy BOL, Demars (2000) Attribute based classification of European hydrophytes and its relationship to habitat utilization. Fresh Biol 43:43–74
Williamson M (1996) Biological invasions. Chapman & Hall, London

Witte F, Goldschmidt T, Wanink J (1992) The destruction of an endemic species flock: quantitative data on the decline of the haplochromine cichlids of Lake Victoria. Environ Biol Fish 34:1–28
Xu CYSS, Schooler RD, van Klinken (2010) Effects of clonal integration and light availability on the growth and physiology of two invasive herbs. J Ecol 98:833–844
Yarrow M, Marín VH, Finlayson M, Tironi A, Delgado LE, Fischer F (2009) The ecology of Egeria densa Planchon (Liliopsida: Alismatales): a wetland ecosystem engineer? Rev Chil Hist Nat 82:299–313
Zelder JB, Kercher S (2004) Causes and consequences of invasive plants in wetlands: opportunities, opportunists, and outcomes. Crit Rev Plant Sci 23:431–452

Chapter 4
Role of Modern Innovative Techniques for Assessing and Monitoring Environmental Pollution

Olga Natalia Bustos López

4.1 Introduction

In the last century, various changes were generated in the society, and industrial and technological advancement began that allowed an expansion in the chemical industry, which led to a demographic expansion, which in turn increased agricultural exploitation; all of the above generated an emission of a large number of compounds in the environment, affecting the air, water, and fertile soils (Alarcón-Herrera et al. 2012; Alloway 2013).

At the same time, there have been scientific and technological needs and demands that exist nowadays where it is necessary to design and build new equipment that allows analysis, monitoring, and field work that facilitates studies and provides reliable results and above all enable and obtain results at the moment in which the study is conducted to make decisions and measures of action (Calvo et al. 2017).

In the control of environmental contamination, it is essential to carry out an adequate sampling, with the subsequent sample treatment, to continue with the analysis of possible contaminants. To achieve a satisfactory analysis, it is essential to use the most convenient analytical technique for each type of sample, type of analysis, etc. (Da Rocha 2012).

The industries mainly and the need to optimize and control their processes in order to be more competitive have increased and evolved the use of methods of analysis of high sensitivity and accuracy, as well as those that provide information in real time (Ibañez García 2017). The tools and techniques used in analytical chemistry have reached a level of integration that they allow most analytical systems to be designed following in part or in full a sequence of unit operations that are part of

O. N. Bustos López (✉)
Reynosa-Aztlán Multidisciplinary Academic Unit, Autonomous University of Tamaulipas, Reynosa, Tamaulipas, México
e-mail: olga.bustos@uat.edu.mx

K. R. Hakeem et al. (eds.), *Bioremediation and Biotechnology*,
https://doi.org/10.1007/978-3-030-35691-0_4

an analytical process, such as sampling, sample transport, separation, reaction, measurement, transduction, signal acquisition, and processing. During this integration process, there must be connectivity in the stages of the analytical process in such a way that human intervention is minimized (Jerry 2004; Torres 2004).

4.2 Technological Advances in Analysis Systems

The process of monitoring and analysis of trace elements at the industrial level, as in other areas, follows a traditional line, which involves taking samples, transporting and analyzing them to a central laboratory, and, consequently, increasing time consumption (*off-line*), causing the control action to be delayed. By contrast, there are autonomous measurement systems, where it is possible to consider an *online* measurement, in real time, that optimizes the control of the variable to be analyzed, as well as the possibility of making measurements in situ, that is to say in the same place where the phenomenon of interest occurs (De Prada 2004; Wen 2011).

Earlier last century, polarography was discovered, marking a major step forward for the analysis of trace elements; later, construction began on the first glass electrodes, sensitive to H^+ ions, which was a breakthrough by allowing a more rapid and continuous determination of pH, with multiple applications to this day. With the need to analyze trace components and perform analyzes of industrial products in increasingly smaller times, the range of requirements in the analysis methods was broadened, including the analysis of contaminants, because the methods and techniques traditionally used did not cover these needs (gravimetric and volumetric methods) (Ortiz 2001; Lobnik 2011).

For the United States, Germany, England, among others at the beginning of the 1940s, it was common to use ultraviolet and infrared spectrophotometers, emission spectrophotometers, fluorometers, polarologists, etc. The project known as Manhattan, directly linked to the construction of the first atomic bomb, boosted considerably the analysis of trace compounds, microanalysis, and separations by ionic exchange (Kolb 1999; Rubio and Hernández 2016).

The gas chromatography and thin layer chromatography achieved a development since the middle of the last century, which, together with the achievements in infrared spectrophotometry, mass spectrometry, and nuclear magnetic resonance, gave a great advance to the diverse studies of the organic chemistry In the analysis of inorganic chemistry, atomic absorption spectroscopy was used, which helped the flame photometry technique, being of great relevance in the trace metal analysis (Faraldos 2011; Alva Díaz and Angeles 2018).

In the 1980s with the rise of computing, the instrumental analytical techniques were modified at the time of including computers and microcomputers to control the instruments and protect the information generated in the experiments. The multiple functions such as controlling sampling times, data processing, evaluating, and storing the results in a computer undoubtedly marked a great advance in the different types of analysis (Murray 2003).

The chemical analysis has played a very important role, and techniques, such as chromatography, spectrophotometry, X-ray reflection or the use of radioactive tracers, among others, for detection and quantification of pollutants have made a great guideline in environmental control. Furthermore, the monitoring methods have been used in countless analyzes. Some examples of application speaking chronologically can be mentioned (Carvalho 1998)Carvalho carried out with his collaborators a monitoring of pesticides in 18 tropical countries through the use of radioactive tracers. On the other hand, in Spain, Rosell examined the blood of agricultural workers through chromatography, detecting the presence of pesticides. At the same time in Venezuela (Bruguera 1996), using chromatography, he analyzed samples in maternal blood, finding DDT. Later (López 1993) with the use of spectrometry, he determined the cadmium content.

Some studies indicate a severe damage in the environmental area mainly in the contamination by agrochemicals; since the 1940s, the last century, the use of pesticides has increased continuously, and only in the developed countries is there a tendency to reduce their use. It has been determined that practically 0.1% of the pesticide used reaches directly to the pest, and the rest remains in the environment, theoretically contaminating the water, soil, and biota, which requires the characterization of the final destination of the pesticides, toxicity, and assessment of the associated hazards (Carvalho 1998).

Analysis carried out in New Zealand (1973–1994) showed that the levels of contamination found were linked to constant releases of organochlorine pesticides (Hendy 1996); other studies in India by Dua et al. (1996) indicated DDT levels of 2.26 ppm in soil and 0.18 ppm in water. In Thailand, it was studied that 75% of pregnant women were contaminated with organochlorine pesticides in values between 10.15 and 1.03 ppb; neonates were also affected with levels between 0.62 and 5.05 ppb, with pesticides detected and identified DDE, DDT, Lindane, HCH, and Heptachlor ; another study in Veracruz, Mexico, showed data of high levels of contamination with DDT between 9 and 20 ppm, in young people under 20 years, and also the impact and contamination in water and soil was detected (Torres et al. 2004).

Most measurement processes based on a physical principle can usually be automated much better in continuous processes, because the chemical methods are suitable for applications with discontinuous measurements. The excitation of molecules by energy is one of the main factors that must be considered in the continuous measurement of gaseous pollutants. This condition can be caused by the exposure of radiation at different wavelengths, generating high temperatures through combustion or chemical reactions. There are other methods, for example, excitation by electrical, magnetic, or nuclear forces. This type of methods can be used for laboratory analysis of contaminated air samples. Passive sampling methods, so called because the device does not involve any pumping, provide a truthful and cost-effective analysis of air quality, which indicates a good reading at average concentrations of pollution, over a considerable period of time ranging from weeks or even months (Hernandez Lucas 2002; Hernandez-Hernandez 2002; Cazes 2011).

Active sampling methods use physicochemical methods to collect contaminated air, and the analysis is done later in the laboratory. In general, a known volume of air is pumped through a collector (filter or chemical solution) for a set period of time. Then the collector is removed for further analysis. Some of the methods used to measure environmental pollution are photometry, fluorescence, chemiluminescence, flame photometry, and ionization (Hernandez Lucas 2002).

Many of the techniques mentioned above are costly due to the use of standards, which are often imported, in addition to the use of solvents such as acetone, hexane, methanol, among others, which generally involve higher costs. For this reason, some researches make proposals to optimize the techniques used in order to minimize costs and obtain optimal results (Torres 2004).

At present, it has been necessary to develop autonomous and miniaturized detection systems (mostly electrochemical and optical systems). The advantages of miniaturization in the field of analytical systems describe the evolution of μ-Tas, the materials used, trends, and applications (microfluidics integration pretreatment steps, and detection systems, among others) (Ibañez García 2007).

4.3 Methods for Monitoring Environmental Pollutants

The different analytical techniques used focus on the subject and its characteristics; for this it is necessary to know and use the properties of the subject, and hence the qualitative or quantitative studies are possible. The above can be observed or provoked physically or chemically for further analysis.

Measurement instruments are therefore devices that can convert signals that cannot be detected directly by the human being, in a way that is. An analysis instrument consists of a signal generator, a signal detector, a signal processor, and a reading device, which are usually found in practically all chemical, organic, and inorganic analysis equipment (Faraldos 2011).

One of the most general classifications that are considered for chemical analyzes and that include the use of different methods and techniques of environmental analysis and monitoring are as follows:

Optical methods: Spectrophotometry—UV-visible, atomic absorption spectrophotometry with flame, turbidimetry.

Electrochemical methods: Detections potentiometric, voltammetry, coulometric, polarography, etc.

Chromatographic methods: High efficiency liquid chromatography (HPLC), nuclear magnetic resonance, etc.

According to the aforementioned analytical techniques, analytical methods can be classified according to the analytes in the following:

1. Selective electrodes (H^+, NH_4^+, Cd^{2+}, Cu^{2+}, Pb^{2+}, K^+, Ag^+, Na^+, total monovalent ions, total ions divalent, Br^-, Cl^-, CN^-, F^-, I^-, NO_3^-, ClO_4^-, S_2^-).
2. Visible spectrophotometry (anions, silica, Kjeldahl nitrogen, hydrogen sulfide, phosphorus, fluorine, iron, manganese, copper, zinc, aluminum, chromium, ammonium, residual chlorine, phenols, surfactants, COD.

3. Flame photometry: sodium, potassium, lithium, strontium.
4. Atomic absorption: metals.
5. Ionic chromatography: Br^-, Cl^-, NO_3^-, NO_2^-, SO_4^{2-}, PO_4^{3-}.
6. Gas chromatography: pesticides, aromatic hydrocarbons (Lucas 2002; Wang 2011)

4.4 Optical Methods in the Analysis of Environmental Pollutants

At present, the need to obtain data from natural or industrial systems continuously, in situ and in real time, remains latent. Hence, the methodology of continuous flow has arisen that has been widely used for the automation of analysis processes. Nowadays, this methodology is integrated into microanalytical systems through the development of microanalyzers where all the stages of the analytical procedure intervene, facilitating the manufacture of laboratories in microcircuits, known as *Lab-on-a-chip*, or microsystems of total analysis, *μ-TAS* (Ibañez García 2007; López 2017).

Glass and silicon are the materials commonly used in this type of systems. However, LTCC technology (*Low Temperature Co-fired Ceramics*), also called green ceramics, presents an alternative for the construction of microanalyzers that allow *on-site* testing and reliable results in real time; provides rapid analysis processes, simplicity in their manufacture, cost reduction, and portability; and likewise presents great versatility in the design of structures. One of the alternatives that has been used in this type of systems is the combination of flow techniques and flow detectors based on light-emitting diodes (LED) and photodetectors for the optical microsystems of analysis (Ibañez García 2007).

The optical methods of analysis are applied to a wide range of studies, and currently the use of these methods has expanded due to the speed, available instrumentation, and its possibilities of automation. In the analysis of polluted air, photometry is one of the methods that allow the detection and quantification of pollutants, and because it uses the absorption of infrared (IR), visible (VIS), or ultraviolet (UV) radiation, the lengths of wave with ranges of application are IR (1000–10,000 nm), VIS (400–800 nm), and UV (200–400 nm) depending on the nature of the analytes (Lucas 2002).

4.5 Optical Sensors

They represent a group of chemical sensors where the electromagnetic radiation (EM) is manipulated to generate an analytical signal in a transduction element. The interaction of radiation with the sample is evaluated from the change of parameters or opticals that is directly related to the concentration of the analyte (Hernandez 2016).

An optical chemical sensor consists of a receptor element that allows to identify a measurable parameter as the concentration of a compound, pH, etc., providing a signal proportional to the magnitude of that parameter. The receptor can in many cases fulfill an important function that allows the generation of the signal, such as interacting in a thin layer with the molecules of the analyte, catalyzing a reaction, or participating in a chemical equilibrium at the same time as the analyte. The transducer generates the signal produced in a measurable signal that is amplified, filtered, visualized, among others.

Sensors that have a receptor part based on a biochemical principle are called biosensors. Its use with biological compounds has great relevance in chemical analysis. The optical sensors can be based on absorbance, luminescence, reflectance, fluorescence, etc., covering different regions of the spectrum from UV–Visible–IR and allowing the measurement of various characteristics and properties of the material (Gründler 2007).

Optical chemical sensors have multiple applications, are safe to work with diverse samples, sensitive, economic, nondestructive and can continue to provide more advantages in their application, can be miniaturized, and allow multiple analyzes with a single control tool from a reference site (Wang 2011).

There is a variety of techniques that focus on optical phenomena, such is the case of colorimetry and the detections that are obtained using an indicator that allows the color to change when joining a certain analyte, has great application in heavy metal detection. These changes are also determined spectroscopically and are observed visually. For this case, there is a great diversity of organic chromophores, such as nitrophenols, azo dyes, sulfophthalein phthaleins, anilines, etc. The redox indicators are examples of chromophores, that is, all the organic colorants generate reversible redox reactions.

By means of metallic indicators it is possible to make ion detections, which form complexes colored with the metallic ones (Kaur et al. 2011), the ionophoresChromogenic can freely bind to ions, and are created to cause a color change when interacting with metal cations (Wen 2011).

There are new materials based on nanoparticles, and the colorimetric compounds are not as common unlike the luminescent nanoparticles. However, in recent years, examples of gold nanoparticles have been reported to detect Cd^{2+}, Fe^{3+}, Pb^{2+}, and nitrites, and nitrate ions were also used, and in 2007 carbon nanotubes were reported for nucleic acid detection (Lee et al. 2007; Ying 2011).

Next, a study of a microsystem using a device that allows detection of cobalt in aqueous solutions by a reaction using an organic chromophore is presented. The continuous flow system takes a flow rate of 0.3 mL/min and is added at a rate of 1.5 mL/min solution composed of ammonium acetate 0.5 M, 2 M ammonium citrate, and 60 μl nitrous–R–salt (NRS) at 1% by means of a syringe or pump. Detection is performed at a wavelength of 520 nm, once the general outline and arrangement for detecting where favorable results were obtained in the implementation and analysis of the analyte (as shown in Figs. 4.1 and 4.2) is presented (Bustos 2016).

The microsystem was designed according to the needs of the reaction itself, which implies that it is flexible and applicable to various analytes in aqueous solutions; in Fig. 4.3, the internal design of the devices is presented that allows them to be machined according to the system's need.

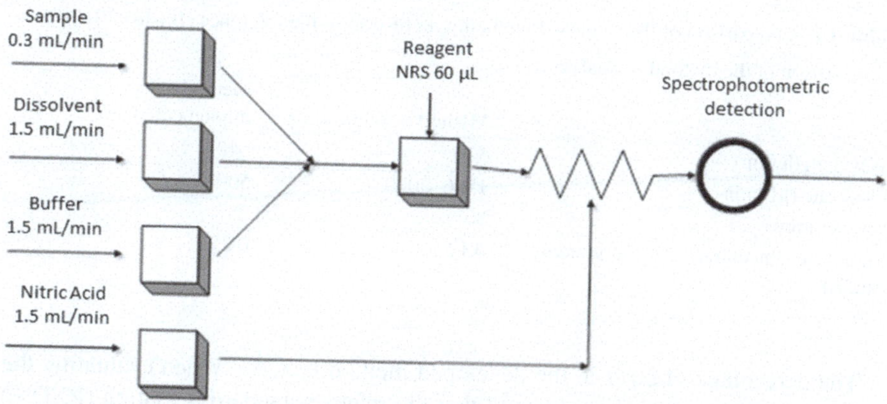

Fig. 4.1 General scheme of the continuous flow process for the detection of cobalt in a continuous flow system

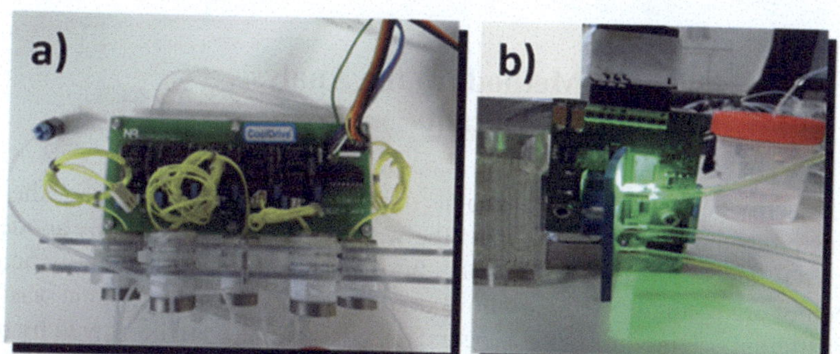

Fig. 4.2 (**a**) Peristaltic pump and injection valve. (**b**) Optoelectronic device adapted to the LTCC device

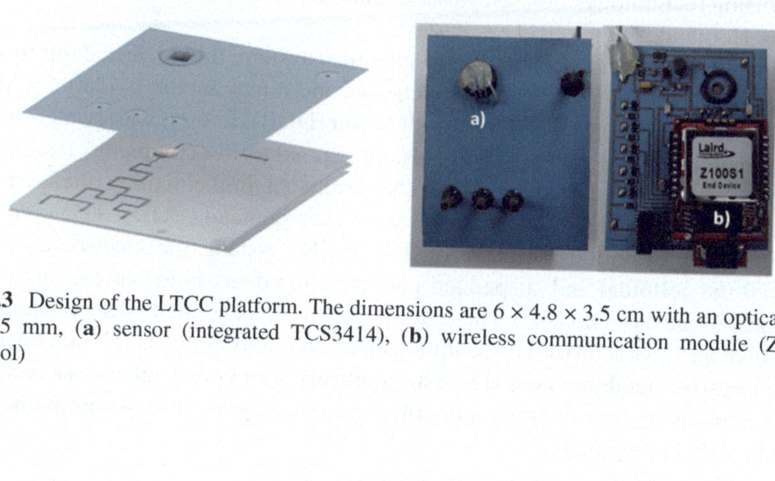

Fig. 4.3 Design of the LTCC platform. The dimensions are 6 × 4.8 × 3.5 cm with an optical path of 3.25 mm, (**a**) sensor (integrated TCS3414), (**b**) wireless communication module (Zigbee Protocol)

Table 4.1 Comparison of the proposed method with respect to the reference (Bustos 2016)

Comparison of the method of analysis		
	Method of reference	Method implemented
Wavelength (nm)	520	540
Flow rate (μL/min)	1500–2000	500
Temperature (°C)	80	25
Minimum concentration in the system (mg/L)	0.12	0.25

The percentage of error in the developed method is 5.2%. When evaluating the response of the system, it is obtained that the relative standard deviation (RSD%) for a series of data at a concentration of 3 mg/L is 4.28% ($n = 13$, 98% confidence) (Bustos 2014, 2016). Table 4.1 shows the advantages obtained from the proposed method.

4.6 Electrochemical Methods in the Analysis of Environmental Pollutants

In the electrochemical methods, anodic oxidation is found to generate the OH• radicals produced by the oxidation of the water at the anode, using as anodes Pt, PbO_2, TiO_2, and SnO_2, through the application of an electric current. Electrochemical procedures are methods used to remove organic pollutants and inorganic wastewater of different origin; during the process, the contaminants are destroyed by direct anodic oxidation (OAD) or indirect (OAI) that is carried out in the method used. However, electrocoagulation is one of the techniques that has been developed in wastewater treatment plants in almost all of Europe. Also some cities in the UK have adopted this technology with custom name of "Harness Targeted Electric WaterfusionTechnology" or electrocoagulation (Fóti and Comninellis 2007; Linares-Hernández et al. 2011).

following is an example of an electrochemical system; this is carried out by electrocoagulation methods, using iron electrodes and direct anodic oxidation (ADO), handling diamond electrodes doped with boron (DDB), to treat wastewater derived from a treatment plant whose downloads correspond to 144 companies from different areas and areas in Mexico City. The results of implementing the treatment showed a 99% elimination of the chemical oxygen demand (COD), 99% color, and 97% turbidity, in a time duration of 2 h. In the system, the electrocoagulation removed the colloidal and suspended particles, and degradation of organic matter was carried out though OAD. The amount of sludge generated was established in the system and characterized by scanning electron microscopy and elemental analysis (Linares-Hernández et al. 2011). This application and many others make electrochemical methods easy to apply and efficient in the degradation of compounds that are not easily biodegradable.

Another innovative application in recent years has been the development of electronic conductive materials with polymer matrix, where a heterogeneous process is carried out on the anode of an electrochemical cell containing a solvent and a salt, which initiates the flow of current and favors the oxidation of the polymer. The electrogeneration of conducting polymers is considered an effective and rapid method to obtain mixed materials through complex mechanisms. A mixed polymeric material is the result of an electrochemical polymerization. The physical and chemical properties of the polymer produced depend on the material composition, and in turn the composition is a function of the speeds of the electrochemical and chemical degradation processes (Fernández 2003).

Showing up next (Fig. 4.4), a study performed in recent years is presented, a microanalyzer for determination of ammonia in wastewater of different processes, with the intention of reducing the consumption of reagents and samples, the data acquisition and data processing miniaturized, as well as the fluid management system, enhancing the automation, autonomy, and reduction of energy consumption of the entire system. A solar-powered system, divided into three main parts, was implemented: microanalyzer, the fluid management system, and the system of acquisition and data processing. The flow system consists of three solenoid micropumps: a micropumpperistaltic, a revolutions controller, and four three-way solenoid microvalves. The controller used with its respective software, used for fluidic devices, was utilized to program the autocalibration process by means of multiconmutation

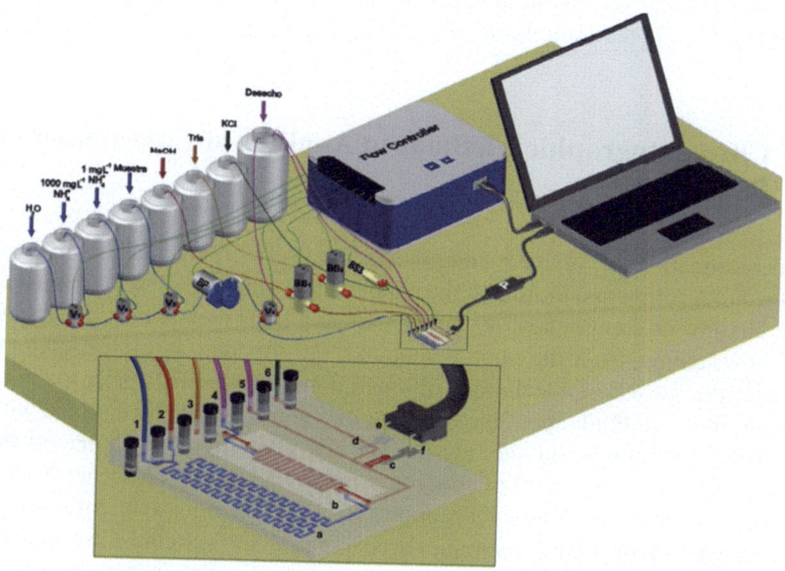

Fig. 4.4 Experimental microsystem assembly for the detection of ammonium: (*a*) micromixer, (*b*) diffusion membrane, (*c*) ammonium ion selective electrode, (*d*) Ag/AgCl integrated reference electrode, (*e*) reference electrode connector, (*f*) indicator electrode connector, (Vx) three-way solenoid valve, (BP) peristaltic micropump, (BSx) solenoid micro pump, (P) designed potentiometer. Liquid inputs to the system (1, 2, 3, 6). Outputs to waste (4 and 5) (Calvo 2017)

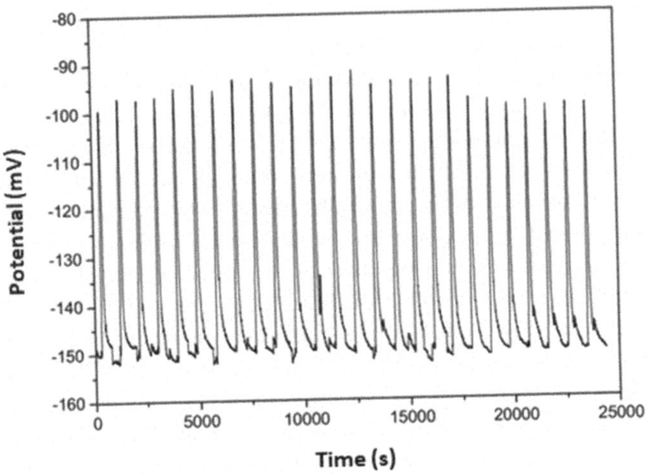

Fig. 4.5 Response of the microanalyzer signal corresponding to 26 injections every 15 min of residual water sample (Calvo 2017)

and to automate the process carried out without the requirement of human intervention. The acquisition of data and its processing were carried out by means of a customized potentiometer (TMI, Barcelona, Spain). All the components (controller, potentiometer) and the fluidic devices were controlled by a computer, obtaining favorable results in real samples (Calvo 2017) (Fig. 4.5).

4.7 Chromatographic Methods in Analysis of Environmental Pollutants

The determination of organic compounds in an environmental sample is not restricted to a compound or, to that the analyte, is in almost pure state, but is generally a mixture of compounds, or compounds and their metabolites generated, by different reactions, from the main compound. For this reason, almost always, we must resort to the use of a technique of separation of compounds prior to the determination; thus, we think about the handling of a chromatography coupled with the appropriate detection technique, considering the properties and characteristics of the compounds and their level of concentration (Alva Díaz and Angeles 2018).

Also, in general, environmental samples require a previous stage of purification and preconcentration, using techniques used as extraction systems, ion exchange, or new solid phase extraction techniques, both in macro- and microscale. The chromatographic techniques have the advantage of allowing the coupling of a very efficient separation technique in continuous flow, with sensitive detection techniques and universal or selective, according to the needs of the analysis to be performed (Spiegel and Maystre 2001).

Many of the chromatographic analyzes focus on ionic determinations, which can be differentiated into two types: (a) Anion-based chromatography with ion suppression consists in the separation of an anion mixture by ion exchange and its detection by electrical conductivity, where the elimination of the electrolyte that does not interest is carried out before measuring the conductivity. (b) The chromatography known as cationic with ion suppression is carried out in a manner analogous to the anion-based technique; however, here the suppressor replaces the Cl ion, which exits from a column with OH, through an anion exchange membrane (Girard 2010).

Now faster and more efficient ion chromatography is widely used from the analytical point of view since it has had improvements in its chromatographic components, more efficiency in columns and exchange resins, a sample volume is used smaller, and the detections are automatic. Using ion chromatography, studies have been conducted to quantify fluoride ions (F^-), bromide (Br^-), chloride (Cl^-), nitrate (NO^{3-}), nitrite (NO^{2-}), sulfate (SO_4^{2-}), and phosphate (PO_4^{3-}), present in drinking water (Alva Díaz 2018).

4.8 Evolution of Analytical Techniques in Environmental Analysis

The need to know both the composition and the possible contaminants of a given sample, the possibility of making a physical sample, or its impossibility, together with the need to know the precise variation over time of the contents in the most varied species and compounds, has led to that the conclusion that, at this time, analytical chemistry provides various methodologies appropriate to environmental needs (Rubio et al. 2016).

Within the analytical measures, it can be considered that most of the available analytical techniques can be used in continuous analysis, and automation is undoubtedly an example that is used in a water analysis in many laboratories for the control of drinking water. The need for continued analysis also applies for the regulation of compounds in the atmosphere and for the monitoring of the distribution of atmospheric pollutants such as CO_2 and CO, NO_2 and NO, SO_2, CH_4, and O_3 levels present in the troposphere or known as urban and O_3 level in the stratosphere, to mention the most important ones (Ionel and Popescu 2011).

In analytical techniques, one way to allow connectivity, partial or total, to unit operations is through the use of continuous flow systems. Of the various existing continuous flow techniques, the most commonly used is the so-called flow injection analysis (FIA), which is based on inserting a small volume of sample into a channel through which a solution that drives it to a detector circulates. In the trajectory, the sample can be subjected to pretreatment stages if required. This type of system allows the facility to integrate stages such as injection, mixing, gaseous diffusion, preconcentration, and electrophoretic separations and includes a wide range of detection techniques such as potentiometric, amperometric, optical, among others.

New concepts have arisen from the need to solve problems present in analytical systems, and in the 1990s, the concept of so-called μ-TAS (total analytical micro-systems) was presented for the first time, which have greatly evolved in the last years. The small size systems are designed to perform all stages of the analytical procedure in an integrated way in a microfluidic platform to obtain chemical information automatically, which involves stages of sampling, sample transport, sample pretreatment, separation, detection, and data analysis (Ibañez García 2007; Ionel et al. 2008).

Miniaturization in analytical systems offers the possibility of being portable, autonomous, reducing costs, and the possibility of in situ measurements. Some of the primary objectives of these devices are to have a more effective environmental analysis and monitoring, protecting natural water resources.

The nanosensors of ions have been developed recently. One of the most important analytes is lead (Pb^{2+}), because it is one of the most toxic and dangerous heavy metals, mainly for children, and the determination of low concentrations is of great importance. In 2007, a fluorescent nanosensor for lead was developed. High concentrations of copper can be toxic. The detection of traces of copper ions is very important due to its environmental implications. Contributions have been reported with nanoparticles that have allowed readings that trace copper ions (Wang 2011).

Nowadays, there are increasing harmful volatile organic compounds (VOC) available, which evaporate and cause great damage and air pollution; currently, there remains a great interest in the development of VOC sensors. With applications that involve environmental monitoring, until now gas chromatography systems have had applications for this purpose; however, it has a higher operating cost, and the response is not appropriate in real-time environmental monitoring applications (Alvarez et al. 2009).

4.9 Perspectives on the Methods of Analysis of Environmental Pollutants

onitoring has been defined as the systematic observation of parameters related to a specific problem, designed in such a way that they provide information on the characteristics of the problem to be treated and its changes over time. On the other hand, Roni (2005) defines it as "the systematic evaluation of something, with the purpose of collecting data to respond to specific objectives." Characterize the conditions (spatial variability) of the different pollutants (Gregorio and Irazustabarrena 2001).

It can be said that at present there are three ways to work with the problem of pollution prevention, control, and repair; you can find a hierarchy in the sense that the top priority is prevention, followed by action controls, leaving the separation considered in the last place. Pollution, monitoring, detection, and minimization encompass all the media that are related to it and usually, fall into a control. For a better control of the contamination, it is necessary to consider several strategies to avoid the generation of pollutants from the source that generates them, and this consists of prohibiting, eliminating, or gradually removing certain chemical products or groups of them. This strategy is reflected by laws or regulations adopted by different governments and nations. However, increasing awareness of the health and ecosystem risks of the different pollutants in the environment leads to a need to develop increasingly sophisticated control and analysis equipment that can reach detection limits lower, as well as obtaining fast and reliable results; it is also considered increasingly small, portable, automated, and multiparameter equipment, where the development of specific sensors is relevant. Moreover, the use of instrumentation for contaminant and selective species is of importance (OPTI 2001).

It has continued in the generation and construction of equipment that includes techniques of on-site analysis, which can characterize the waste in the places of origin and destination, through the miniaturization of equipment to assess environmental contaminants, knowing toxicity, content of organic matter and in solid state, among other characteristics (Cazes 2011).

On the one hand, the development of biosensors undoubtedly continues to boom for the coming years, allowing rapid and reliable analysis, through the use of physical, chemical, biological, and toxicological methods, which implies the development of technological tools according to the different types of pollution (Coopeer and Cass 2004).

On the other hand, the analysis techniques will tend toward automation, which is the use of robotics in multiparameter equipment that allows a simplification of the analytical process, minimizing the costs of execution and analysis; these techniques involve the development of the ideal instrumentation for the control and monitoring of existing systems, with the intention of having real information on environmental pollution systems (Harvey 2000).

Regarding the technologies of soil characterization, the problems associated with soil contamination are considered, and it involves the development of technologies for the characterization of these, associated with the normative development with regard to the typification of the contaminants present in soils and underground waters; these technologies are focused on determining the availability, speciation, mobility, and other characteristics of pollutants for the development of innovative and alternative methods for the characterization of contaminated soils (Smith 2008).

The monitoring and remote control equipments tend to have more information on environmental parameters that favor the development of equipment to obtain reliable information in real time, for decision making in different scenarios. These teams tend to be in continuous flow, and integrated monitoring and control networks are created, as well as reliable and safe wireless communication and data transmission protocols. Satellite detection and tracking equipment has also been generated using tools of simulation (Rubio et al. 2016).

Work has been done on the development of equipment for recycling and waste assessment; these teams consider obtaining fuels derived from waste (biofuels) and devices that can have equipment for the treatment of air pollution that generates valuable by-products, some with technology for gasification and energy use (Murray 2003).

New analysis technologies currently tend to the energy recovery which according to application can mention some innovations applied in incineration (fluid bed combined cycle, etc.), gasification, pyrolysis processes energy assessment through the efficient use biomass and energy use, advanced thermal processes, obtaining liquid fuels from industrial solid minimizing environmental pollution with the application of biotechnology, biological degradations aerobic, anaerobic co-digestion of waste fermentation, esterification, etc.

The recovery of materials is a tendency of the technologies associated to revalue the waste of contaminants and its minimization, through materials from waste with the intention of obtaining secondary raw materials, through a previous treatment that can serve as a basic element to another process, obtaining of materials, and the recovery of metals with added value through the application of advanced separation and extraction technologies, which can be intimately linked to hydrometallurgical, pyrometallurgical extraction processes, etc. (Ortiz 2001).

The new technologies will have a direct impact on industrial development, the use of catalysts with novel designs, the inclusion of bioprocesses, the use of membranes for the separation of products, microreactors, etc., which will change the way of elaborating many of the compounds that we know, allowing an optimization and minimization of contamination.

On the one hand, the design for the integration of processes will undoubtedly be one of the aspects that will be consolidated in the coming years, due to the wide range of applications, which allow the control of environmental emissions, process safety, etc. (Jorn and Laurindsen 2001).

On the other hand, the advances in the field of materials, analytics, and electronics available today allow to generate new sensors and measuring devices that permit to improve the information in the coming years and that is undoubtedly a large field of application. The analysis and decision making associated with the analysis processes are based on the use of models at different levels of simulation techniques. The use of simulation allows to reduce the development and start-up times, in addition to synthesizing the knowledge of a process and making decisions in a rational way; modeling is considered an important technology for future analysis techniques (Ionel and Francisc 2010).

The implementation of control systems and its optimization are based on the availability and clear and precise information of the processes, whether analytical, industrial, or monitoring. The instrumentation systems are based on spectroscopy, infrared, temperature, etc., and they will provide much better measurements than the actual systems. In order to interpret the information that they collect, and it be useful, a more sophisticated signal processing will be required and also the progress in topics such as estimation of variables not yet measured, using sensors and softwares developed for specific purposes (Rouessac 2004; Schwarzenbach 2007; Berna 2010).

4.10 Conclusion

Many developments and technological advances have been created without directly considering a real problem. However, it is necessary to see the real problems and the processes that generate high levels of contamination, minimizing waste, emissions, and therefore pollution, which implies a good control in the different anthropogenic processes of the current society. Additionally, another great challenge that should not be set aside is to build equipment and technologies that allow monitoring with solid, real, and practical foundations, capable of solving current and future environmental problems; however, going a step forward, with the right techniques, the correct instrumentation, and control in emissions, the society can augur a good future and monitor environmental pollution.

References

Alarcón-Herrera MT, Olmos-Márquez MA, Valles-Aragon C, Llorens E et al (2012) Assessments of plants for phytorremediation of arsenic-contaminated water and soil. Eur Chem Bull 2:121–125

Alloway BJ (2013) Heavy metals in soils, trace metals and metalloids in soils and their bioavailibility. Springer, Berlin, pp 587–592

Alva Díaz PS (2018) Estudio de iones presentes en el agua potable en sectores con suministro limitado de la provincia de Trujillo, por el método de cromatografía. Universidad Nacional de Trujillo, Trujillo, pp 6–9

Alva Díaz PS, Angeles RD (2018) Estudio de iones presentes en el agua potable, en sectores con suministro limitado de la provincia de Trujillo, por el método de cromatografía ionica. Universidad Nacional de Trujillo, Trujillo, pp 24–30

Alvarez S, Derfus A, Schwartz M (2009) Biomaterials 26–34

Berna A (2010) Metal oxide sensors for electronic noses and their application to food analysis. Sensors (Basel) 10(4):3882–3910

Bruguera AB (1996) Levels of DDT residues in human milk of Venezuelan women from various rural population. Elsevier Science, Amsterdam, pp 203–207

Bustos LO (2014) Analytical microsystem for the monitoring and analysis of cobalt. J Mex Chem 411–415

Bustos LO (2016) Detection and analysis of cobalt in continuous flow using an analytical microsystem. Sens Actuator B Chem, 11–16

Calvo LA (2017) Diseño, construcción y evaluación de analizadores miniaturizados para su aplicación aeroespacial, medioambiental, alimentaria, biomédica e industrial. Universidad Autónoma de Barcelona, Bellaterra, pp 158–162

Calvo LA et al (2017) Diseño, construcción y evaluación de analizadores miniaturizados para su aplicación aeroespacial, medioambiental, alimentaria, biomédica e industrial. Universitat Autonoma de Barcelona, Bellaterra, pp 157–179

Carvalho FZ (1998) Rastreo de paguicidas en los trópicos. Boletin OEIA:113–118

Cazes J (2011) Ewing's analytical instrumentation handbook. Marcel Dekker, New York, pp 50–62, 45–53

Coopeer J, Cass T (2004) Biosensors. Oxford University Press, Oxford

Da Rocha Z (2012) Compact and autonomous multiwavelength microanalyzer for in-line and in situ colorimetric determinations. Lab Chip 12:109–115

De Prada C (2004) El futuro del c ontrol de procesos. Rev Iberoam Autom Inform Ind 1:5–14
Dua VK, Pant C, Sharma VP (1996). Determination of level of HCH and DDT in soil, water, and whole blood from bioenvironmental and insecticide sprayed areas of malaria control. Indian Journal of Malariology. 33(1), 7–15 India
Faraldos MG (2011) Tecnicas de Análisis y caracterización de materiales. Consejo superior de investigaciones Científicas, pp 21–35, 735–737
Fernández OT (2003) Polímeros conductores: Síntesis, Propiedades y Aplicaciones Electroquímicas. Revista Iberoamericana de polímeros, pp 1–32
Fóti G, Comninellis C (2007) Investigatons of electrochemical oxygen transfer reaction on boron-doped diamond electrodes. Electrochim Acta:1954–1961
Girard JE (2010). Principles of environmental chemistry. Jones & Bartlett Burlington, 264–268
Gregorio O, Irazustabarrena A (2001) Tendencias de Futuro en el Medio Ambiente industrial. Economia Industrial, pp 87–94
Gründler P (2007) Chemical sensor. Springer, Berlin, pp 115–117
Harvey D (2000) Modern analytical chemistry. McGraw Hill, New York, pp 115–122
Hendy EJ, Peake BM (1996) Organochlorine pesticides in a dated sediment core from Mapua, Waimea Inlet, New Zealand. Marine Pollution Bulletin 32 (10):751–754
Hernandez DC (2016) Analytical techniques for environmental pollution control. Ciencia UNEMI, pp 118–131
Hernandez Lucas GC (2002) Introducción al análisis instrumental. Ariel Ciencia:15–19
Hernandez-Hernandez LG (2002) Introducción al Análisis Instrumental. Ariel Ciencia:88–96
Ibañez García N(2007) Miniaturización de analizadores químicos mediante la tecnología LTCC. Universidad Autonoma de Barcelona, Bellaterra, pp –
Ibañez García N (2017) Miniaturización de analizadores químicos mediente la tecnología LTCC. Universidad Autónoma de Barcelona, Bellaterra, pp 18–21
Ionel I, Francisc P (2010) Methods for online monitoring of air pollution concentration. Air Quality pp 85–90
Ionel I, Popescu F (2011) Methods for online monitoring of air pollution concentration. Air Quality, pp 82–114
Ionel I, Popescu F, Padure G (2008) Method for determination of an emission factor for a surface source. Optoelectr Adv Mater:851–854
Jerry S (2004) Control de la contaminación ambiental. El Medio Ambiente, 55–60
Jorn HN, Laurindsen PS (2001) Gestion y reciclado de residuos solidos. Control de la Contaminación Ambiental 55.44–55-46
Kaur P, Kaur S, Singh K, Sharma PR, Kaur T (2011) Indole-based chemosensor for Hg^{2+} and Cu^{2+} ions: applications in molecular switches and live cell imaging. Dalton Transactions, 40(41):10818–10821.
Kolb H (1999) Química para el nuevo milenio. Prentice Hall, Bergen, pp 93–95
Lee AC, Ye J-S, Tan N, Daniel P, Sheu F-S, Kiat Heng C, Meng Lim T (2007) Carbon nanotube-based labels for highly sensitive colorimetric and aggregation-based visual detection of nucleic acids. Nanotechnology 18(45):455102.1–455102.9
Linares-Hernández, Ivonne; Martínez-Miranda, Verónica; Barrera-Díaz, Carlos; Pavón-Romero, Sergio; Bernal-Martínez, Lina; Lugo-Lugo, Violeta (2011) Oxidation of persistent organic matter in industrial wastewater electrochemical treatments. Avances en Ciencias e Ingenieria 2:21–36
Lobnik AT (2011) Opticla chemical sensors: desing and applications. Adv Chem Sens:1–22
López FR (1993) Determination of chlorinated insecticides of agricultural workers. Elsevier Science, Amsterdam, pp 152–156
López AC (2017) Diseño, construcción y evaluación de analizadores miniaturizados. Universidad Autónoma de Barcelona, Bellaterra, pp 157–166
Lucas HH (2002) Introducción al análisis instrumental. Ariel Ciencia, pp 15–21, 33–39
Murray RM (2003) Future directions in control, dynamics and systems. EJC, pp 144–158
OPTI (2001) Medioambiente: Tendencias y Tecnologías a medio y largo plazo. MCYT-CDTI-OPTI, pp 50–62

Ortiz GI (2001) Tendencias de Futuro en el Medio Ambiente Industrial, Tecnologías y Escenarios. Economia Industrial. pp 87–95

Rouessac F (2004) Chemical analysis: Modern instrumental methods and techniques. Wiley, New York

Rubio JD, Hernández AJ (2016) Sensor system Based in Neutral Networks for the Environmental Monitoring. Ingeniería Investigación y Tecnología, pp 211–222

Rubio JD, Hernández Aguilar JA, Jacob ACF (2016) Sistema sensor para el monitoreo ambiental basado en redes neuronales. Ingeniería Investigación y Tecnlología, pp 211–222

Schwarzenbach PG (2007) Environmental organic chemistry. West Sussex, pp:135–149

Smith R (2008) Guide to environmental analytical methods. Genium, New York, pp 41–58

Spiegel J, Lucien Y. Maystre (2001) Control de la contaminación ambiental. El Medio Ambiente 55.1–55.2

Torres DC (2004) Uso del análisis químico como herramienta para el monitoreo ambiental. Asociación Española de Ecología Terrestre, pp 2–6

Torres D et al (2004) Agroquímicos un problema ambiental global: uso del analisis químico como herramienta para el monitoreo ambiental. Ecosistemas 13(3):2–6

Wang W (2011) Advances in chemical sensors. InTech, Rijeka, pp 3–6

Wen W (2011) Advances in chemical sensors. InTech, Rijeka, pp 11–15

Ying XH (2011) Colorimetric detection of Cd^{2+} using gold nanoparticles. Analyst 1:3725–3730

Chapter 5
Global Scenario of Remediation Techniques to Combat Environmental Pollution

Henna Hamadani, Shahzada Mudasir Rashid, Muneeb U. Rehman, Rameez Ali, Masrat Rashid, Manzoor ur Rahman, Ishraq Hussain, Gowhar Gul, and Zulfiqar ul Haq

5.1 Introduction

Advances in science and technology as well as rapid industrialization is leading to the development of the world at a very fast pace. Although development is a welcome step everywhere, it has also contributed significantly to environmental pollution. Pollution has serious impacts on ecosystem as well as human and animal health which need to be tackled as soon as possible. Polluted drinking water has been stated

H. Hamadani · R. Ali
Mountain Livestock Research Institute, Manasbal SKUAST-K, Jammu, India

S. M. Rashid (✉) · M. ur Rahman · I. Hussain
Molecular Biology Lab, Division of Veterinary Biochemistry, Faculty of Veterinary Sciences & Animal Husbandry, Sher-e-Kashmir University of Agricultural Sciences and Technology (SKUAST-K), Jammu, Jammu and Kashmir, India
e-mail: mudasir@skuastkashmir.ac.in

M. U. Rehman
Department of Biochemistry, Govt Medical College (GMC-Srinagar), Karan nagar, J & K, India

Department of Clinical Pharmacy, College of Pharmacy, King Saud University, Riyadh, Saudi Arabia

M. Rashid
Department of Drug and Food Control Organization, Govt of J&K, Jammu, Jammu and Kashmir, India

G. Gul
Division of Animal Nutrition, Faculty of Veterinary Sciences & Animal Husbandry, Sher-e-Kashmir University of Agricultural Sciences and Technology (SKUAST-K), Jammu, Jammu and Kashmir, India

Z. ul Haq
ICAR Poultry Seed Project, Division of LPM, FVSc, (SKUAST-K), Jammu, Jammu and Kashmir, India

K. R. Hakeem et al. (eds.), *Bioremediation and Biotechnology*,
https://doi.org/10.1007/978-3-030-35691-0_5

to be the cause of around 250 million water-based diseases (Kuppusamy et al. 2016). WHO (2013) has reported that in the USA alone, diesel emission-generated air toxicants contribute to around 70% cancer risk. It has been estimated that exposure of fine particles in polluted air causes death of around seven million people every year (WHO 2018). Exposure to toxic pollutants can have negative effects on health effects which include compromised immunity, physical, mental and neurological disorder, carcinoma, organ dysfunction, reduced life expectancy, and in some cases death (Yu et al. 2011; Huang et al. 2012; Kuppusamy et al. 2016). Various toxic organic and inorganic pollutants present in polluted environment include pesticides, polychlorinated naphthalenes, chlorinated solvents, perfluorochemicals, quaternary ammonium compounds, triclosan, polychlorinated alkanes, benzothiazoles, polydimethylsiloxanes, industrial byproducts, heavy metals, polycyclic aromatic hydrocarbons, engineered nanoparticles, and many others (Kuppusamy et al. 2016).

Remediation of pollution is a global concern, and many techniques each having their merits and demerits are available for combating pollution. Various techniques employed for remediation of environmental pollution have been described in this chapter.

5.2 Techniques for Remediation of Contaminated Soil and Groundwater

Various techniques are being used across the globe for remediation of environmental pollution. In general, these remediation techniques have been classified either as ex situ or in situ depending upon the site action. In ex situ technique, media is extracted physically from a contaminated site and then treated at a different location. For example, if the contaminant is present in the soil, then the soil is dug out, but if it has reached the groundwater, then both the water and the soil are excavated. As far as in situ remediation techniques are concerned, contaminants are treated on site only. Both the techniques have their advantages as well as disadvantages which are summarized in Table 5.1.

Table 5.1 Comparison between ex situ and in situ remediation techniques

S. No	Feature	Ex situ remediation	In situ remediation
1	Contaminant removal and transportation from the site	Required	Not required
2	Efficiency of contaminant removal	More	Less
3	Cost	High	Less
4	Exposure of excavators to health risks	More	Less
5	Time required for achieving effectiveness	Less	More
6	Monitoring	Easy	Difficult
7	Achievement of uniformity	More	Less

Kuppusamy et al. (2016)

5.2.1 *Ex Situ Remediation Techniques*

5.2.1.1 Dig and Dump Technique

Dig and dump, also known as excavation and disposal, is one of the most popular ex situ techniques for remediation of contaminated soil and water. The sites in which pollutants exceed the preset levels of risks are referred to as hot spots. These hot spots are targeted in this technique. The contaminated soils are dug and transported to specific locations (landfills) for dumping the wastes. A secure landfill generally has four elements consisting of natural hydrogeological setting, cap, leachate collection system, and bottom liner. Plastic material, clay, or a combination of both is used in the bottom liners, which are used for layering the pit dug in the ground and thereby prevents escape of waste into the environment. Leachates are prevented by covering the landfill with the cap. Another type of landfill is the bioreactor landfill, which is used to treat toxins with the help of microbial processes. It can be aerobic, anaerobic, or both. This system has many advantages (Kuppusamy et al. 2016) which include reduction in emission of greenhouse gases, non-requirement of land filling for the end products, reduced land filling and leachate treatment cost, and decreased concentrations of contaminant during the landfill operation. However, some drawbacks have also been listed by Campbell (2009) which includes human risk as well as the transportation cost from the site to treatment location.

5.2.1.2 Pump-and-Treat Technique

The contaminated groundwater is pumped out in this technique, and then granular activated charcoal is used for its treatment. This technique has two variants which include pulsed pumping and continuous pumping. Pulsed pumping has been reported to be more beneficial than the continuous one (Mackay et al. 2000). It has been suggested that cost-effectiveness and pollutant removal efficiency of this technique can be enhanced by the use of surfactant foam technology (Wang and Mulligan 2004). Various alternatives to this system have emerged which include the metallic iron technology, surfactant-enhanced remediation, reactive barriers, and nano-techniques.

5.2.1.3 Incineration Technique

Incineration remediation technique, used to treat contaminated soils, has grown importance over the last two decades. Incinerators are similar to closed burning rotary kilns having proper pollution control, quench, and afterburner units (Pavel and Gavrilescu 2008). Wastes are destroyed by subjecting them to a high temperature ranging from 750 °C to about 1200 °C. The technique treats large amount of waste and is effective against chlorinated hydrocarbons, dioxins, and explosives. Volume and weight of the waste materials are reduced to just 5 and 25% of their original volume and weight, respectively. Combustible carcinogens and toxic organic compounds are also detoxified by this method. In comparison to landfill technique, the greenhouse gas generation is less (Hutton 2009). The energy produced by the incineration process can be recycled and used as fuel for various other activities. However, Abbasi (2018) has debated over its negative impacts on the environment and practical infeasibility of energy recycling. Other disadvantages include high capital and operating costs.

5.2.1.4 Oxidation Technique

Oxidation technique destroys the target contaminant and reduces the pollutant toxicity to a significant level by chemical, biological, or other advanced process. Chlorine, hypochlorites, chlorine-di-oxide, ozone, permanganate, and peroxides are the common oxidizing agents used (Anonymous 2012). Advanced type of oxidation referred to as advanced oxidation processes (AOP) based on the formation of highly reactive radicals mainly hydroxyl radicals has the potential to many contaminants both inorganic and organic in nature. The free radical-based processes occur at higher rates than those based on other types of chemical oxidation (Rosenfeldt et al. 2007). Among the various types of materials capable of catalyzing these processes, perovskites have been reported to be very promising, because of their high stability and ability to stabilize unusual oxidation states of metals (Cervantes and Castillejos 2019). AOPs can be classified either as homogeneous or heterogeneous (Poyatos et al. 2010). Homogeneous AOPs can be subdivided into those using energy and those not involving energy. The heterogeneous AOPs processes are subgrouped into four types: (1) catalytic ozonation, (2) photocatalytic ozonation, (3) Fenton-like, and (4) photo catalytic oxidation (Cervantes and Castillejos 2019). Incapability to manage large quantities of materials and high costs includes some of the disadvantages of this technique.

5.2.1.5 Adsorption

Adsorption is the most commonly used technique for the remediation of pollution of air emissions, wastewater from industries, chemical spills, and groundwater. It is known to be the fast and inexpensive technology for removing various harmful chemicals such as xylene, dichloroethane, trichloroethene, tetrachloroethene, ethylbenzene, pesticides, herbicides, explosives, perchlorate, and heavy metals (Kuppusamy et al. 2016). Various types of adsorption are physical, chemical (chemisorption), and electrostatic. Activated carbon is the most commonly used adsorbent for water and air treatments. Other adsorbents include activated alumina, ion-exchange resins, forage sponge, and sorption clays.

5.2.1.6 Ion Exchange

Ion-exchange remediation technique is characterized by exchange of cations or anions between contaminants and media. Resins are mostly used as ion-exchangeable materials (Anonymous 2012). Ions present in the resins get exchanged with contaminated constituents of the polluted fluids after being passed over the resin bed. The exhausted resins can also be reused after regeneration but are often used only once. Both types of resins (i.e., cationic as well as anionic) are being used in this technique (Alexandratos 2008). These resins are insoluble, adaptable, and compatible with the environment and can be used for many years in most cases. This technique is effective against radionuclides, ammonia, dissolved metals, silicates, and nitrates from liquid media (Rengaraj et al. 2003).

5.2.1.7 Pyrolysis Remediation Technique

Pyrolysis remediation technique, also referred to as molten solid processing or plasma arc technology, is highly efficient and requires less time for action. It is effective in the remediation of organic as well as inorganic pollutants from soil and oily-sludges such as dioxins, mercury, cyanides, creosotes, and hydrocarbons. In addition to this, wastes from wood treating, coal tar, rubber processing, paint, and refinery are removed by pyrolysis (Arvanitoyannis et al. 2007). It acts by chemically decomposing hazardous substances by the use of thermal energy in absence of oxygen, under pressure at a temperature of more than 430 °C (Venderbosch et al. 2010). Targeted compounds are transformed into an insignificant quantity of liquid or solid residues or simply into gas. Nonproduction of carbon dioxide during the treatment is one of the important advantages of this technique (Inguanzo et al. 2002).

5.2.1.8 Physical Separation Technique

Soil washing is a cheap remediation technique requiring fewer investments on treatments, since it reduces the volume of contaminant materials requiring treatment to a significantly low level. The technique is also referred to as soil washing, soil scrubbing, or mechanical scrubbing or attrition scrubbing. It can be made applicable for heavy metals, semi-volatile organic compounds, polycyclic aromatic hydrocarbons, polychlorinated biphenyls, pesticides, petroleum, and fuel residues (Isoyama and Wada 2007). The polluted material is mechanically concentrated into smaller volumes by this technique for further treatments (Dermont et al. 2008). Although the technique is effective alone, it can also be combined with other techniques for increased pollutant cleaning efficiency.

5.2.1.9 Dehalogenation Technique

This remediation technique involves the removal of halogens from the halogenated compounds. It is classified into two types, i.e., chemical dehalogenation and biological dehalogenation. Chemical dehalogenation consists of halogen removal from the contaminant by heat, mixing the same with the chemical reagents such as sodium bicarbonate or polyethylene glycol. Biological dehalogenation comprises of dehalogenating the pollutants with the help of microorganisms such as anaerobic bacteria (*Dehalococcoides*, *Dehalobacter*, *Anaeromyxobacter*). Biological dehalogenation has been classified into following types depending upon the mode of action:

- Reductive dehalogenation.
- Oxidative dehalogenation.
- Dehalogenation by methyl transfer.
- Dehalogenation by hydration.
- Dehydro dehalogenation.
- Intramolecularsubstitution.
- Thiolytic dehalogenation.
- Hydrolytic dehalogenation.

5.2.1.10 Bioremediation Technique

Bioremediation is an environment-friendly and cost-effective remediation technique for treatment of environmental pollution by the use of microorganisms (Luka et al. 2018; Shafi et al. 2018). Bioremediation can be either solid phase bioremediation or slurry phase bioremediation (bioreactors).

Solid phase bioremediation: Solid-phase biological remediation techniques include biopiles, land farming, and composting. Biopiles, also referred to as biocells, static pile composts, bioheaps, and biomounds, are used to treat the polluted soils contaminated with petroleum products. It is a type of biodegradation, in which pile of contaminated is made first and then the same is aerated by artificially forcing air into it. Water is added to the pile after aeration and temperature, and pH is also controlled. This leads to activation of the microbiological processes which eventually degrades the petroleum compounds. Efficacy of the system is enhanced by using various structural materials such as woodchips, dry manure, straw, sawdust, and sand (Mohee and Mudhoo 2012). The efficacy of this technique is affected by the characteristics of soil, contaminant, and climate (Giasi and Morelli 2003). Land farming technique is used for the treatment of hydrocarbons, certain pesticides, diesel, wood preserving, and coke wastes. Contaminated material is spread into lined beds after excavation is done. Water and nutrients are added to trigger microbiological activity. Aeration, temperature, and moisture control are important aspects to be

Table 5.2 Level of various parameters during compositing (Semple et al. 2001)

S. No.	Parameters	Level
1	Oxygen (%)	10 to 15
2	Moisture (%)	50 to 55
3	C:N ratio	30:1
4	pH	6 to 9
5	Porosity (cm)	1 to 5

taken care of in this method. Hydrocarbon degrading bacteria are added to speed up the process of degradation. Composting is a technique which transformation of organic contaminants such as polycyclic aromatic hydrocarbons into stable innocuous products by microorganisms is done. The level of various parameters to be considered during compositing is given in Table 5.2.

In addition to this, a temperature of 50 °C to 65 °C needs to be maintained. Contaminated soil is extracted and then mixed with organic substances such as manure, vegetable wastes, and hay. This is one way to accelerate the thermophilic microbial activity as well as the porosity of the contaminated soil (Coker 2006).

Slurry phase bioremediation technique: Slurry phase bioremediation technique, also known as slurry bioreactor, is known to be one of the best biological remediation techniques, in which microbes are used to treat polluted wastewater. In this technique, in order to achieve bioremediation, polluted material is processed through a bioreactor. A bioreactor is a structure that is engineered to support a biologically active environment.

5.2.1.11 Solidification Remediation Technique

Solidification also referred to as stabilization, is a remediation technique that utilizes physical as well as chemical means to treat polluted materials contaminated with inorganic substances. The technique is effective against radionuclides as well, but it is less effective against pesticides and organic substances. Solidification of contaminants is carried out by binding them inside a stabilized mass. Stabilization of pollutants by reduction of their mobility is done through chemical reactions between them and the stabilizing agent.

5.2.1.12 Constructed Wetlands

Constructed wetland treatment is another remediation technique for wastewater which is known to a long-term technology and involves the use of microbial, natural, geo-chemical, and biological processes.

5.2.2 *In Situ Remediation Techniques*

In situ treatment techniques are those types of remediations in which contaminants within soil or groundwater is treated without removing them from the ground. They can be classified as chemical, physical, biological, thermal, or electrical techniques and are described below.

5.2.2.1 Biological Treatments

Biological treatment is a low-cost remediation technique in which microbes are used to degrade pollutants into innocuous substances such as fatty acids, water, biomass, and carbon dioxide. It is also employed for the remediation of co-contaminated soil with heavy metals and organic pollutants (Ye et al. 2017). The limitation to this technique is that it is a time-consuming and difficult process. Various types of biological treatments (Lodolo 2019) are enlisted in the Table 5.3.

Table 5.3 Types of biological treatments

S. No	Technique	Details	Target pollutant
1	Land farming	Reduction of contaminant concentrations through biodegradation	Petroleum hydrocarbons, polycyclic aromatic hydrocarbons, creosote, halogenated volatile, semi-volatile, non-halogenated-semi-volatile organic compounds, pesticides
2	Natural attenuation	Contaminants are degraded and mineralized by using natural subsurface processes such as dispersion, dilution, volatilization, biodegradation, and sorption	Nonhalogenated volatile organic compounds, semi-volatile organic compounds, fuel hydrocarbons, explosives
3	Bioventing	Soil microorganisms are stimulated to destroy contaminants in the polluted soil by making oxygen available	Wood preservatives, gasoline, non-chlorinated solvents, pesticides, fuel
4	Enhanced bioremediation	The process comprises of adding microorganisms to the subsurface environment so as to quicken the natural biodegradation	Petroleum hydrocarbons, semi-volatile organic compounds, volatile organic compounds, pesticides
5	Phytoremediation	Plants are used to treat contaminants in soil. Phytoremediation is classified into the following types depending the mechanism: (a) Rhizofiltration (b) Phytoextraction (c) Phytotransformation (d) Phyto-stimulation (e) Phytostabilization	Pesticides, insecticides, explosives, surfactants, various metals, radionuclides, chlorinated solvents, polycyclic aromatic hydrocarbons

5.2.2.2 Physical or Chemical Treatments

This technique involves the use of physical and chemical properties of contaminants to treat the contamination. The process is cost effective and less time consuming. The induction of phase transfer is carried out in the physical process, while as in chemical process, chemical reactions are carried out to change the chemical structure of the contaminant in order to make them less toxic. Various techniques (Lodolo 2019) involved in these types of treatment are enlisted in Table 5.4.

Table 5.4 Types of physical or chemical treatments

S. No	Technique	Details	Target pollutant
1	Solidification and stabilization	Binders and additives are used for treating the contaminated sites by reducing the mobility of harmful materials	Various inorganics, radionuclides, heavy metals, nonvolatile organics, semi-volatile organics
2	Soil flushing	Contaminants are extracted from the polluted soil by the use of water and water solutions which dissolves these contaminants	Metals, radioactive contaminants, fuels, pesticides, volatile organic compounds, semi-volatile organic compounds
3	LasagnaTM process	Electrokinetics is used to transfer contaminants in soil pore water into treatment zones where decomposition of contaminants occurs	Organic compounds, fuel trichloroethylene
4	Fracturing	Improves the efficiency of removal in situ treatment techniques by enlarging the existing fissures and introduction of new fractures, mainly in the horizontal direction. Commonly used soil fracturing techniques include the following: (a) Pneumatic fracturing (b) Blast-enhanced fracturing (c) Hydrofracturing	Variety of contaminants
5	Electroreclamation	Removal of pollutants from the soil by the use of electric or electrochemical processes	Heavy metals, polar organics, anions
6	Soil vapor extraction (enhanced volatilization)	Vacuum which is applied through extraction wells to create a concentration gradient and a zone of low vapor pressure induces gas phase volatiles, which are then removed through the extraction wells	Volatile metals, fuel contaminants, volatile organic compounds
7	Polymer adsorption	The use of water-soluble polymers, functionalized with those groups which have a strong affinity for a particular pollutant to treat contaminated soils	Heavy metals, radionuclides, inorganics, nonhalogenated volatile organic compounds

5.2.2.3 Thermal Treatments

Thermal treatment techniques are based on the principle of increasing the volatility, burning, decomposing, destroying, or melting the pollutants. It is the most costly treatment technique, and the time taken to accomplish the technique depends on various factors such as pollutant size and depth, type of soil, and the characteristics of the contaminant. Various types of thermal treatments (Lodolo 2019) used for pollution remediation is described in Table 5.5.

5.3 Techniques for Remediation of Air Pollution

Various ex situ remediation techniques used for the control of air polluted have been summarized (Kuppusamy et al. 2016) in Table 5.6.

Table 5.5 Types of thermal treatments

S. No	Technique	Details	Target pollutant
1	Soil vapor extraction	Volatilization rate of semi-volatile contaminants is increased to facilitate their extraction by the use of various processes such as electrical resistance, electromagnetic, fiber optic, radiofrequency heating, or steam injection	Volatile organic compounds, semi-volatile organic compounds
2	Vitrification	Based on the principle of subjecting polluted soil to temperatures high enough to allow melting of the same to form a glass after cooling	Organics, inorganics, radionuclides

Table 5.6 Remediation techniques for air pollution

S. No.	Technique	Procedure
1	Adsorption	Air pollutants are adsorbed onto activated carbon or zeolites
2	Filtration	Polluted air is passed through viscous substance-coated fibrous material
3	Ozonation	Pollutants are oxidized by the generation of ozone
4	Photolysis	Pollutants are oxidized by the application of UV alone or in combination with a photocatalyst
5	Biofiltration	Pollutants are degraded by passing the contaminated air through microbe-colonized packed bed of solid support
6	Membrane separation	Membranes are used in this technique to separate pollutants
7	Enzyme oxidation	Degradation of pollutants is done by catalysts or enzymes through the transfer of air emissions into aqueous phase
8	Botanical purification	Degradation of pollutants using enzymes or plants passing the air through the contaminated soil or on vegetation

5.4 Emerging Technologies

Many promising techniques for the remediation of environmental pollution are emerging, some of which are discussed below:

5.4.1 *Nanotechnology*

The use of nanotechnology in the remediation of environmental pollution is gaining importance nowadays (Kim et al. 2017; Gong et al. 2017; Guerra et al. 2018; Ratwani et al. 2018; Cai et al. 2018; Corsia et al. 2018). The property of having high surface-area-to-volume ratio of nanotechnology-based materials makes them suitable for such processes. Various types of nano-materials such as inorganic, carbon-based, and polymeric-based materials are used for the remediation of environmental pollution. The technique is effective against the contaminants such as heavy metals, dyes, halogenated herbicides, chlorinated organic compounds, volatile organic compounds, and organophosphorus compounds (Guerra et al. 2018).

5.4.2 *Microbial Fuel Cell Technology*

Microbial fuel cells, a sustainable and low-cost technology, have been promising for harvesting energy and treating wastewater (Chouler et al. 2016). They consist of two compartments, namely a cathodic and an anodic compartment, which are usually separated by a proton exchange membrane (PEM) so that the migration of electrolytes from one chamber to the other can be avoided (Ho et al. 2018). Organic matter is broken down to generate electrons and protons in the anode compartment by using bacteria as catalysts (Ezziat et al. 2019). Transfer of electrons to cathode compartment occurs via an external circuit, and the diffusion of protons takes place through the proton exchange membrane (Nimje et al. 2012; Mathuriya and Yakhmi 2014; Miskan et al. 2016). By serving as terminal electron acceptors in the cathode compartment, metals which serve as terminal electron acceptors in cathode compartment can be reduced electrochemically and recovered from the cathode surface (Ucar et al. 2017).

5.4.3 *Ultrasonic Technology*

Ultrasound remediation technology induces chemical reactions to cause degradation of the pollutants by using frequency of over 18 kHz which forms the source in developing cavitation bubbles leading to high localized pressures of more than

50 MPa and temperature of more than 4726 °C (Adewuyi 2001). This technique requires less space, less energy expenses, lower installation and maintenance cost, and is fast (Thangavadivel 2010).

5.5 Conclusion

Clean environment is vital for a healthy life, and no compromise in this regard can be afforded. Globally concerns have been shown toward the environmental pollution, and various remediation techniques to combat the same have been devised too. The techniques are diverse in nature depending upon the site of action, mechanism of action, cost-effectiveness, time consumption, effectiveness, target specificity, etc. However, all techniques have a common goal of combating pollution in their own specific manner.

References

Abbasi SA (2018) The myth and the reality of energy recovery from municipal solid waste. Energy Sustain Soc 8:36

Adewuyi YG (2001) Sonochemistry: environmental science and engineering applications. Ind Eng Chem Res 40:4681–4715

Alexandratos SD (2008) Ion-exchange resins: a retrospective from industrial and engineering chemistry research. Ind Eng Chem Res 48:388–398

Anonymous (2012) Remediation technologies screening matrix and reference guide version 4.0—remediation technology. Federal Remediation Technologies Roundtable, Washington, DC

Arvanitoyannis IS, Kassaveti A, Stefanatos S (2007) Current and potential uses of thermally treated olive oil waste. Int J Food Sci Technol 42:852–867

Cai Z, Dwivedi AD, Lee WN, Zhao WN, Zhao X, Liu W, Sillanpaa M, Zhao D, Huang CH, Fu J (2018) Application of nanotechnologies for removing pharmaceutically active compounds from water: development and future trends. Environ Sci Nano 5(1):27–47

Campbell KM (2009) Radionuclides in surface water and groundwater. In: Ahuja S (ed) Handbook of water purity and quality. Academic, New York, NY, pp 210–213

Cervantes MLR, Castillejos E (2019) Perovskites as catalysts in advanced oxidation processes for wastewater treatment. Catalyst 9(3):230

Chouler J, Padgett GA, Cameron PJ, Preuss K, Titirici MM, Ieropoulos I, Lorenzo MD (2016) Towards effective small scale microbial fuel cells for energy generation from urine. Electrochim Acta 192:89–98

Coker C (2006) Environmental remediation by composting. Bio Cycle 47:18–23

Corsia I, Nielsen MW, Sethi R, Puntad C, Della C, Torree D, Libralato G, Lofranog G, Sabatini L, Aielloi M, Fiordi L, Cinuzzi F, Caneschi A, Pellegrini D, Buttino I (2018) Echnologies and nanomaterials for environmental applications: key issue and consensus recommendations for sustainable and ecosafe nanoremediation. Ecotox Environ Safe 154:237–244

Dermont G, Bergeron M, Mercier G, Richer-Lafleche M (2008) Soil washing for metal removal: a review of physical/chemical technologies and field applications. J Hazard Mater 152:1–31

Ezziat L, Elabed A, Ibnsouda S, Abed SE (2019) Challenges of microbial fuel cell architecture on heavy metal recovery and removal from wastewater. Front Energy Res 7(1):1–13

Giasi CI, Morelli A (2003) A landfarming application technique used as environmental remediation for coal oil pollution. J Environ Sci Health A Tox Hazard Subst Environ Eng 38:1557–1568
Gong X, Huang D, Liu Y, Peng Z, Zeng G, Xu P, Cheng M, Wang R, Wan J (2017) Remediation of contaminated soils by biotechnology with nanomaterials: bio-behavior, applications, and perspectives. Crit Rev Biotechnol 38(3):455–468
Guerra FD, Attia MF, Whitehead DC, Alexis (2018) Nanotechnology for environmental remediation: materials and applications. Mol 23(1760):1–23
Ho NAD, Babel S, Sombatmankhong K (2018) Bio-electrochemical system for recovery of silver coupled with power generation and wastewater treatment from silver(I) diammine complex. J Water Process Eng 23:186–194
Huang Y, Wong C, Zheng J, Bouwman H, Barra R, Wahlstrom B, Neretin L, Wong M (2012) Bisphenol a (BPA) in China: a review of sources, environmental levels and potential human health impacts. Environ Int 42:91–99
Hutton B (2009) Waste management options to control greenhouse gas emissions—landfill, compost or incineration? Paper for the international solid waste association (ISWA) conference, Portugal, 12–15 Oct 2009, pp 1–10
Inguanzo M, Domınguez A, Menendez J, Blanco C, Pis J (2002) On the pyrolysis of sewage sludge: the influence of pyrolysis conditions on solid, liquid and gas fractions. J Anal Appl Pyrol 63:209–222
Isoyama M, Wada SI (2007) Remediation of Pb contaminated soils by washing with hydrochloric acid and subsequent immobilization with calcite and allophanic soil. J Hazard Mater 143:636–642
Kim DY, Kadam A, Shinde S, Saratale RG, Patra J, Ghodake G (2017) Recent developments in nanotechnology transforming the agricultural sector: a transition replete with opportunities. J Sci Food Agric 98:849–864
Kuppusamy S, Palanisami T, Megharaj M, Venkateshwarlu K, Naidu R (2016) Ex-situ remediation Technologies for Environmental Pollutants: a critical perspective. In: Voogt PD (ed) Reviews of environmental contamination and toxicology, vol 236. Springer International, Switzerland, pp 117–192
Lodolo A (2019) EUGRIS: portal for soil and water management. www.eugris.info/furtherDescription.asp?&ResourceTypes=True&eugrisid=26&Category=Content_Digests&Title=In%20situ%20treatment%20technologies&showform=&ContentID=3&CountryID=0&ResourceTypes=&DocID=&Tools=Further%20Description
Luka Y, Highina BK, Zubairu A (2018) Bioremediation: a solution to environmental pollution review. Am J Eng Res 7(2):101–109
Mackay D, Wilson R, Brown M, Ball W, Xia G, Durfee D (2000) A controlled field evaluation of continuous vs. pulsed pump-and-treat remediation of a VOC contaminated aquifer: site characterization, experimental setup and overview of results. J Contamin Hydrol 41:81–131
Mathuriya AS, Yakhmi JV (2014) Microbial fuel cells to recover heavy metals. Environ Chem Lett 12:483–494
Miskan M, Ismail M, Ghasemi M, Jahim MJ, Nordin D, Bakar AMH (2016) Characterization of membrane biofouling and its effect on the performance of microbial fuel cell. Int J Hydrog Energy 41:543–552
Mohee R, Mudhoo A (2012) Methods for the remediation of xenobiotic compounds. In: Mohee R, Mudhoo A (eds) Bioremediation and sustainability: research and applications. Wiley, Hoboken, pp 372–374
Nimje VR, Chen C, Chen H, Chen C, Tseng M, Cheng K (2012) A single-chamber microbial fuel cell without an air cathode. Int J Mol Sci 13:3933–3948
Pavel LV, Gavrilescu M (2008) Overview of ex-situ decontamination techniques for soil clean-up. Environ Eng Manag J 7:815–834
Poyatos JM, Munio MM, Almecija MC, Torres JC, Hontoria E, Osorio F (2010) Advanced oxidation processes for wastewater treatment: state of the art. Water Air Soil Pollut 205:187–204

Ratwani D, Khatri N, Tyagi S, Pandey G (2018) Nanotechnology-based recent approaches for sensing and remediation of pesticides. J Environ Manag 206:749–762
Rengaraj S, Joo CY, Kim Y, Yi J (2003) Kinetics of removal of chromium from water and electronic process wastewater by ion exchange resins: 1200H, 1500H and IRN97H. J Hazard Mater 102:257–275
Rosenfeldt EJ, Chen PJ, Kullman S, Linden KG (2007) Destruction of estrogenic activity in water using UV advanced oxidation. Sci Total Environ 377:105–113
Semple KT, Reid BJ, Fermor TR (2001) Impact of composting strategies on the treatment of soils contaminated with organic pollutants. Environ Pollut 112:269–283
Shafi S, Bhat RA, Bandh SA, Shameem N, Nisa H (2018) Microbes: key agents in the sustainable environment and cycling of nutrients. In: Environmental contamination and remediation. Cambridge Scholars, Cambridge. 152-179-188
Thangavadivel K (2010) Development and application of ultrasound technology for treatment of organic pollutants. PhD thesis. University of South Australia, Adelaide
Ucar D, Zhang Y, Angelidaki I (2017) An overview of electron acceptors in microbial fuel cells. Front Microbiol 8:643
Venderbosch R, Ardiyanti A, Wildschut J, Oasmaa A, Heeres H (2010) Stabilization of biomass derived pyrolysis oils. J Chem Technol Biotechnol 85:674–686
Wang S, Mulligan CN (2004) An evaluation of surfactant foam technology in remediation of contaminated soil. Chemosphere 57:1079–1089
WHO (2013) Cancer prevention. World Health Organization, Washington, DC
WHO (2018) World Health Organization releases new global air pollution data. www.ccacoalition.org/en/news/world-health-organization-releases-new-global-air-pollution-data
Ye S, Zeng G, Wu H, Zhang C, Dai J, Liang J, Yu J, Ren X, Yi H, Cheng M, Zhang C (2017) Biological technologies for the remediation of co-contaminated soil. Crit Rev Biotechnol 37(8):1062–1076
Yu MH, Tsunoda H, Tsunoda M (2011) Environmental toxicology: biological and health effects of pollutants. CRC, Boca Raton, pp 24–34

Chapter 6
Biopesticides: Clean and Viable Technology for Healthy Environment

Marian Butu, Ramona Stef, Ioana Grozea, Mihaela Corneanu, and Monica Butnariu

6.1 Background Biopesticides

In the last few decades, agriculture has been almost entirely dependent on synthetic organic insecticides or pesticides obtained through chemical synthesis. Agriculture has seen a major change by the use of crop protection chemicals that began in the late 1800s with the introduction of arsenic insecticides and Bordeaux mixtures as a grape fungicide (copper hydroxysulfate (II), basic copper sulfate, is a particulate suspension colloidal active substance-metallic copper containing up to 25% basic copper sulfate) and progressing to the very sophisticated compounds currently available. The effect of pesticides obtained through chemical synthesis on agriculture has been dramatic because today's agriculture means the use of chemicals. Despite the immense benefits, they are used in large quantities and are designed to kill organisms (Dawson and Buckley 2011).

However, the very useful properties of these pesticides obtained through chemical synthesis alongside the residual action, although giving high toxicity to a wide range of organisms, have given rise to serious environmental problems. In addition, the emergence and spread of resistance increase in many vector species refer to the concern of environmental pollution and the higher costs for new chemical insecticides. It is therefore obvious that pest control is not safe anymore, as long as it is based on the use of chemicals alone. Therefore, natural enemies such as predators, parasites, and pathogens attracted attention for extensive research actions. Unfortunately, none of the predators or parasites can be mass produced and stored for long periods of time, since all of them must be raised in vivo. It has become

M. Butu
National Institute of Research and Development for Biological Sciences, Splaiul Independentei, Bucharest, Romania

R. Stef · I. Grozea · M. Corneanu · M. Butnariu (✉)
Banat's University of Agricultural Sciences and Veterinary Medicine "King Michael I of Romania", Timisoara, Timis, Romania

K. R. Hakeem et al. (eds.), *Bioremediation and Biotechnology*,
https://doi.org/10.1007/978-3-030-35691-0_6

apparent that there is an urgent need for a biological agent that possesses the properties of a highly synthesized, highly toxic chemical pesticide for the target organism that can be mass produced on an industrial scale with a long shelf life and can be transported safely.

During the 1970s, WHO (World Health Organization) and other international organizations have initiated studies to identify new biological control agents and optimize the existing ones. Biological control is generally considered as a technique for controlling insects due to minimal impact on the environment and avoidance of resistance problems of vectors and agricultural pests (Gay 2012).

Biopesticides refer to the use of living biological organisms (fungi, bacteria, insect viruses, genetically modified organisms, natural enemies, etc.) or their biometabolites which are able to kill or inhibit pests. Examples of biometabolites can be considered as follows: 2,4-D (dimethylamine) and 2,4,5-trichlorophenoxyacetic acid (2,4,5-T), auxins (chemicals from the phytohormones category such as auxin A: $C_{18}H_{32}O_5$; auxin B: $C_{18}H_{30}O_4$; heteroauxin: $C_{10}H_9O_2N_{2,}$ 2-(1H-indol-3-yl) acetic acid, 2-(1H-indol-3-yl) ethanoic acid, indole-3-acetic acid, indolylacetic acid, 1H-indole-3-acetic acid, or indoleacetic acid), pheromones ((R)-hydroxydanaidal) and sodium 2-naphtholate ($C_{10}H_7NaO$) (Degenkolb and Vilcinskas 2016).

Also known as natural pesticides, they refer to nonchemical synthesis, natural chemicals, or living organisms and have the role of replacing pesticides and herbicides. Biopesticides include bacteria, insect-borne pathogenic nematodes, viruses and other microorganisms, plant-derived insects, and pheromones. Biopesticides play an important role in the integrated pest management system used in organic farming. The research results are supporting the development of biopesticides for the following reasons: chemical synthesis pesticides generate environmental pollution, pesticides chemically synthesized for groups, do not reach the target, and eventually exert a negative impact as the target population gradually develops resistance to chemicals (Pretali et al. 2016).

In the last decades, biochemists, chemists, experts in integrated pest management, and toxicologists have worked together to develop a new generation of natural herbal substances with a role in pest management and with relatively little impact on the environment.

The first generation of biopesticides contained alkaloids, nicotine ($C_{10}H_{16}N_2O_4S$) or 3-(1-methyl-2-pyrrolidinyl) pyridine, pyrethrins, rotenone ($C_{23}H_{228}O_6$), ryania, sabadilla (cevadine: $C_{32}H_{49}NO_{9,}$ veratridine: $C_{36}H_{51}NO_{11}$), and some vegetable oils (d-limonene, linalool) used as herbal repellents and insecticides (cedar, lavender, eucalyptus, pennyroyal, and citronella), etc. In 1690, water-soluble components of tobacco were used against grain pests. Pyrethrum and pyrethrins from *Chrysanthemum cinerariaefolium*, cinerin I ($C_{20}H_{28}O_3$), cinerin II ($C_{21}H_{28}O_5$), cinerin III ($C_{21}H_{30}O_3$), jamolin I ($C_{21}H_{30}O_3$), jamolin II ($C_{22}H_{30}O_5$), pyrethrin I ($C_{21}H_{28}O_3$), and pyrethrin II ($C_{22}H_{28}O_5$), are the main ingredients of bioproducts against mosquitoes (Chen et al. 2016).

The market share of biopesticides remains quite low. In 2015, biopesticides accounted for 1.7% of total pesticide sales worldwide. There are a number of factors limiting the growth of biological pesticides: biopesticides are not widely available and, compared to the effect of chemically synthesized pesticides that are relatively lethal, have a shorter duration and a higher cost of action. However, biopesticides have a higher developmental perspective in relation to the demand for pesticides obtained through chemical synthesis, which in many countries is stagnating or diminishing. In the following years, the estimated demand growth for pesticides obtained through chemical synthesis is about 2% and for biopesticides is 10–15%. By promoting organic farming, the demand for biopesticides is gradually increasing. For biopesticides, as with pesticides obtained through chemical synthesis, a prudent assessment of health, nutrition, ecosystem and environmental safety should be involved. Biopesticides are natural or genetically modified agents that differ from chemical synthesis pesticides (conventional pesticides) obtained by chemical synthesis through their unique mode of action, low dose and specificity of target species. In parallel with the development of science and technology, the field of biopesticides continues to expand, involving animals, plants, microbial species and a variety of biological substances associated with pesticides such as plant-derived substances, genetically modified pests, bionic synthesis or modification of synthetic compounds, artificial reproduction of the harmful organisms of the antagonist organisms, pheromones, etc. The use of biopesticides is growing worldwide and is increasingly present in integrated pest management and weed management, with the role of not only increasing the yields and quality of control programs but also of reducing the impact on the environment (Kraehmer et al. 2014).

6.2 Effective Use of Plant Protection Methods

Plant protection is the science that deals with the study of pests (phytopathogens, including arthropods, weeds, rodents, viruses, etc.) in order to establish effective measures to combat the damage/economic losses they produce. It is believed that about one third of the potential crop is destroyed by harmful organisms; hence, plant protection, as the applied biological discipline, contributes to the increase of the crop yields and the improvement of the harvest quality. For the above reasons, pest control is imperative for plant cultivation, plant protection procedures (including stored crops) against harmful organisms being a major component of all good agricultural practice guidelines.

Combating pests of agricultural crops is achieved through several methods: chemical (using pesticides), biological (through the use of antagonist and natural products), genetic (by improving the resistance of plants to harmful organisms), agro-technical (through soil works, including weed grass), and physico-mechanical (seed thermal disinfection, plant surgery, seed mowing, etc.) (Bianchi et al. 2013).

In agriculture, we are currently seeking to increase the quality by increasing the share of biological protection in order to eliminate chemical, toxic and polluting pesticides. In the attempt to reduce the use of chemicals in the control of major plant pathogens, safer alternatives have been developed that do not have negative consequences for the environment, humans and animals and are part of the concept of sustainable agriculture, with the assumption of deepening studies on integrated and rational pest control of plant diseases.

Due to the fact that the cereal seed is chemically treated and because there are no biological products to control the diseases on the seed, by designing and realizing a biological product, the research activities should respond to the urgent need for a commercial product to be used at large scale by farmers who want to set up or convert to the organic farming system.

Due to the fact that the bioproducts are applied only preventively, their efficacy increases by combining with a conventional fungicide. To improve the efficiency of a plant-based bioproduct, without the addition of classical fungicides, some researchers tried to introduce *Glomus* genus fungi in the technological process for obtaining the bioproduct.

These mycorrhizal fungi are recognized for their role in inducing increased resistance to pathogens, overcoming more than light stress conditions (drought, solar radiation) as well as obtaining high yield crops. The aim was to obtain unconventional biofungicides for the treatment of cereal grains (Togni et al. 2019).

The bioproduct would be used in the control of the major toxigenic fungi. It was also described the protocol for the production of this unconventional biofungicides as well as the appropriate method of applying this type of treatment, with impact on quality of wheat crops.

Due to the fact that most biofungicides tests were performed in vitro or in protected areas, the effect of seed treatment on crops throughout the vegetation period is not known. Therefore, it is recommended that the field efficacy of the treatments should be evaluated, and this can be quantified by the presence or absence of foliar and spice diseases and also by the quality of the obtained production (Liu et al. 2018).

6.2.1 Biological Control

Of the approximately one million known insect species, about 15,000 species are considered pests, and about 300 require control management. Most pest insects come together with associated pathogenic microorganisms. Insect pathology probably began in the nineteenth century with the stimuli of Basii and Pasteur. An important contribution to the microbiological control of insects was brought by Mechanikoff in 1879 and Krassilinikow in 1888, who were the first to document that an entomopathogen, *Metarhizium anisopliae*, can be produced and applied as a microbial insecticide to control some pests specific to cereals and sugar beet. d'Herelle investigated insect control with bacteria after Pasteur's description of silkworm diseases (Hopper et al. 2019).

All of these studies were not very consistent, and interest in the use of bacteria remained without the expected effects. Later in 1940, White and Dutky succeeded in demonstrating the control of the Japanese beetle by using *Bacillus popilliae* for a period of almost 30 years. This finding has led to increased research on antimicrobial properties of bacteria, and literature begun to provide more and more information about the efficacy of the strain of *Bacillus thuringiensis*. From 1960 to 1963, eight patents for the benefic action of *B. thuringiensis* were obtained. The use of viruses for the control of insect–pests was stimulated by the research of Balch and Bird in 1944 and Steinhaus and Thompson in 1949. This interest is currently of interest, information supported by recent registration of viral pesticide products in the United States by the Environmental Protection Agency (EPA). Thus, some bacteria, viruses and fungi were reproduced as commercial products due to their efficiency and lack of toxicity or pathogenicity to nontarget animals and plants. Biological control is generally the use of a particular living organism in the control of pests. The chosen organism could be a predator, a parasite or an infectious disease that attacks insect pests. Biological control methods can be used as part of an integrated pest management program to reduce the legal, environmental and public safety risks of chemicals. In addition, this may be an economic alternative to some pesticides obtained through chemical synthesis (Sui et al. 2019).

Unlike most insecticides, biological controls are often very specific for a particular parasite. The impact on the environment and water is small, and it offers a much more environmentally friendly alternative than pesticides obtained through chemical synthesis.

Biopesticides could be used if pests developed resistance to pesticides obtained through chemical synthesis. Research and development of biopesticides attract little financial support compared to financial support for pesticides obtained through chemical synthesis. Thus, increased attention should be paid to broad-spectrum biopesticides and improvements in the production, training and application of technologies. Efforts should be made to optimize the impact of these agents by integrating them with other new crop production strategies.

The use of biological control requires an understanding of the biology of the pests and the damage they produce. Thus, biological control can be more costly than using pesticides obtained through chemical synthesis. In general, the results of the use of biological control are more efficient and faster than the use of pesticides obtained through chemical synthesis.

Most biopesticides attack only certain types of insects, as opposed to pesticides obtained by chemical synthesis, with a broad spectrum, which can kill a wide range of insects (Yamada et al. 2019).

6.2.2 Temperature, Timely Spraying, Improved Control Effect

The active ingredients of biopesticides are mainly crystals of proteins, and for best results, the temperature should be above 20 °C. If the temperature drops, the harmful organism have low reproductive speed, and the biopesticide crystals are difficult to apply. Research indicates that at 20 °C–30 °C the biological control effect of biopesticides is greater 1–2 times than at 10–15 °C (Bellefeuille et al. 2019).

6.2.3 *Advantages of Biopesticides*

Biopesticides are biological or biological agents typically derived in a similar way to chemical pesticides, with the exception that they achieve pest management while being environmentally friendly (Bhat et al. 2018). For microbial biopesticides, effective control requires appropriate formulation and application. Biopesticides used against crop diseases have already been established for a variety of crops. The major research interest area for biopesticides is for seed treatment and soil sterilization. Advantages of biopesticides as follows:

- Are usually less toxic than pesticides obtained by chemical synthesis.
- Generally, affect only target pests and closely related organisms, as opposed to the broad spectrum of chemical synthesis pesticides that may affect different organisms (birds, insects and mammals, etc.).
- Are effective in small quantities and decompose rapidly, resulting in lower exposures and avoiding pollution problems caused by pesticides obtained by chemical synthesis.
- When used as a component of integrated pest management programs, biopesticides can reduce the use of conventional pesticides, while yields of crops remain high (Reil et al. 2018).

6.2.4 *Types of Biopesticides*

According to composition and sources of production, biopesticides can be:

6.2.4.1 Botanical Biopesticides

They are obtained from plants and are used to combat losses due to pests of agricultural importance, with particular attention being paid to the use of higher plants. The studies reported, so far, a large number of useful plant species such as bael (*Aegle marmelos* L.), begonia family Begoniaceae, karanj (*Millettia pinnata* L.), mahua (*Madhuca longifolia* L.), neem (*Azadirachta indica* L.), pyrethrum (*Chrysanthemum cinerariifolium* L. and other 424 plant names of Pyrethrum species) and tobacco (*Nicotiana tabacum* L.). Botanical biopesticides can minimize undesirable side effects of pesticides obtained through chemical synthesis, helping to conserve the environment. Plant extracts have long been used to control insects. L-(−)-nicotine, (S)-3-(1-methyl-2-pyrrolidinyl) pyridine ($C_{10}H_{14}$) is an alkaloid (naturally synthesized by several plant species) used as an insecticide since 1763. Nicotine is a toxic compound, symptoms of acute nicotine poisoning occur rapidly, and death can occur within minutes (death occurs due to respiratory arrest and due to respiratory paralysis) (Medina-Romero et al. 2017).

Phytopathogenic biopesticides are one of the first options for organic biopesticides, due to their advantages in the natural environment. These are mainly plant insecticides, plant fungicides, plant herbicides and photoactivated plant molds. The plants contain a series of pesticide-active compounds, nicotine ($C_{10}H_{14}$), pyrethrin II ($C_{22}H_{28}O_5$), rotenone ($C_{23}H_{22}O_6$), etc. Pyrethrin ($C_{22}H_{28}O_5$) is an extract of several types of chrysanthemums and is one of the oldest insecticides used (there are six esters and acids associated with this bioinsecticide). Pyrethrin is applied at low doses and is considered impervious.

The toxicity of pyrethrins for mammals is quite small, apparently because they are rapidly degraded by enzymes in the liver. The acute LD50 dose for rats is approx. 1500 mg/kg. The most common reactions to pyrethrin are contact dermatitis and respiratory allergic reactions, possibly as a result of other constituents of the composition. They have low toxicity for mammals (Zhang et al. 2019).

6.2.4.2 Microbial Biopesticides

Microbial biopesticides are based on the discovery of new microbial genes with insecticidal properties that will reduce the use of chemical pesticides. The selected genes have been validated in experiments that prove their effectiveness against pests.

The first set of tested genes with insecticidal properties were validated against insects in the family *Coleoptera* and *Lepidoptera*. These insect families include some of the most devastating pests for corn crops. When introduced into the genome of the plant of interest, the genes discovered by their bioinsecticide properties provide protection against a number of pests. Most of these bioinsecticides are based on microbial genes derived from *B. thuringiensis*, to which insects have become resistant over the years. The next generation of bioinsecticide products will most likely be based on the discovery of microbial genes that are not derived from *B. thuringiensis*. With such a large microbial genetic background, there are hundreds of millions of potential genes that can form the basis for future insect control products, everything will depend on quantifying the immense amount of relevant data and analyzing them. *Suyun* sp. and *Bacillus* sp. are the most used bacterial species for their practical application as successful biopesticides (Calvo-Garrido et al. 2019).

6.2.4.3 Mycotoxins

The alkaloids produced by the ergot affect the nervous system and are vasoconstrictors. There are known outbreaks caused by the ergot (but which are no longer due to the knowledge of causes and a more varied diet), *Fusarium* species and their analogous mycotoxins (deoxynivalenol, fusarenon-X and nivalenol).

Aflatoxins are produced by species of the genus *Aspergillus*, common fungi that appear as contaminants of cereals, corn, peanuts, etc. These are carcinogenic for animals and humans. Aflatoxin B1 ($C_{17}H_{12}O_6$), the most toxic in this class, must be enzymatically activated to exert its carcinogenic effect (Matumba et al. 2015).

6.2.4.4 Toxins Generated by Algae

Toxins produced by freshwater or saltwater algae (cylindrospermopsin-alkaloid cyanotoxin, microcystin, lipopolysaccharides, saxitoxin, group toxins: azaspiracid, okadaic acid, pectenotoxin, yessotoxin) often accumulate in fish and shellfish present in water, causing poisoning to humans and animals as well as deadly poisoning of fish. Unlike microbial toxins, algae toxins are generally thermally stable and are not removed by cooking, which increases the likelihood of toxic exposure in humans. Thus, this type of toxin could be researched to get biopesticides (Gerssen et al. 2011).

6.2.4.5 Toxins Produced by Plants

Phytotoxins are believed to have evolved as defense mechanisms against herbivores, especially insects and mammals. These compounds may be repellent (can reject) without being very toxic or can be acutely toxic for a wide range of organisms. Phytotoxins include sulphur compounds, lipids, phenols, alkaloids, glycosides and many other types of chemical compounds.

Many alkaloids, such as caffeine ($C_8H_{10}N$), cannabinoids ($C_{42}H_{60}O_4$), cocaine ($C_{17}H_{21}NO_4$), morphine ($C_{17}H_{19}NO_3$) and nicotine ($C_{10}H_{14}N_2$), are phytotoxins. Many chemical compounds that have proven to be toxic are constituents of plants that are part of the human diet as the carcinogenic compound safrole ($C_{10}H_{10}O_2$) and related substances found in black pepper. Solanine ($C_{45}H_{73}NO_{15}$) and chaconine ($C_{45}H_{73}NO_{14}$), which are cholinesterase inhibitors and possibly teratogens, are found in potatoes, and quinines and phenols have a wide spread in food. Plant poisoning of livestock in livestock remains an important veterinary issue in some parts of the world (Youssef and Saenger 1998).

6.2.4.6 Animal Biopesticides

Some animal species produce toxins for offensive or defensive purposes. Some toxins are poisonous (they are toxic when swallowed accidentally), and others are venomous (the species actively injects the toxin through parts specifically adapted to the body: pins, tongues, etc.). Animal toxin chemistry extends from enzymes, peptides and cardiotoxic and neurotoxic proteins to small molecules such as amines, alkaloids, glycosides, terpenes, among others (bee venom contains a biogenic amine, histamine, three peptides and two enzymes). The effects of snake venoms are generally due to 60–70 amino acid peptide toxins.

These toxins are cardiotoxic or neurotoxic, and their effects are exacerbated by the enzymes present in the venom, which can affect the blood clotting mechanisms and can damage the blood vessels. Fish species, over 700 in the world, are either directly toxic or become poisonous to humans by ingestion. A classic example is the toxin produced by *Sphaeroides*, called tetrodotoxin ($C_{11}H_{17}N_3O_8$), who lives in the Asia–Japan area.

Death occurs within 5–30 min of ingestion, at a rate of approx. 60% (Yotsu-Yamashita et al. 2017).

Natural enemies can provide adequate biological control of aphids if broad-spectrum insecticides are not used, especially after mid-July. Common natural pests are *lady beetle adults, lady beetle larvae, lacewing larvae, hover fly larvae, damsel bugs, minute pirate bugs*, and *parasitic wasps*. Several natural enemies of the Colorado beetle can diminish the population. Unfortunately, it was not yet discovered a natural enemy to properly control the population of this pest. A species of *lady beetle* feeds beetle eggs, and some flying parasites and a fungus can attack the population. These natural enemies can be preserved using *New Leaf* variety of potatoes or by microbial insecticides such as *B. thuringiensis* that are toxic to Colorado beetle but not for beneficial insects. Research continues on using exotic parasites that can be raised in the laboratory and then released into experimental potato fields, but none has yet been found to be effective enough to be produced for commercial use. Pesticides of animal sources include mainly animal toxins, such as spider toxins, toxins from the hornet (*Vespa velutina*), Naproxen (aka Aleve or Midol), among others. Beneficial insects are widely used in the United States, Britain, France, Russia, Japan and India, and there are over 40 types of products for insect control (Tan et al. 2012).

6.3 Pheromones

The pheromones are substances that are used in plant protection. The pheromones (Greek *φέρειν—to wear, to inform*) are also referred to as "external hormones," are chemical substances, molecules that, even in very large dilutions, are meant to transmit signals and messages of different species to their living environment. These substances play a role in the process of communication between species, in marking of territories, in finding partners in the breeding process or finding food.

The term "pheromone" was defined by Peter Karlson, Martin Lüscher, and Adolf Butenandt. It was not identified any pheromone of *Homo sapiens*. *Olfactory footprint* is not synonymous with "pheromone." Pheromones produce "traps" to destroy harmful insects or disrupt the communication system, preventing males from finding females for reproduction. In the integrated insect control, various pheromone devices can be used. Pheromones of more than 1000 species of insects are known. Pheromones serve to locate and attract the sexual partner, stimulate and regulate the copulation process, mark the territory, regulate social behavior, trigger defense, and alarm behavior as a stimulator of egg deposition. With these properties, researchers have discovered that they can handle and combat insect pests by using them in the field of plant protection (Brezolin et al. 2018).

Mode of action of pheromones is as follows:

- Messages are transmitted with great efficacy, using a small amount of substance.
- The response of the receiving organism is completely preprogrammed. By receiving sexually attractive pheromone, the males are unable to distinguish whether it is transmitted from a female or an artificial source; this feature is the basis for using pheromone baits to capture males of the harmful species, in their own habitat.
- Each pheromone transmits the message strictly intraspecific, the information contained therein being received and decoded only by individuals of the same species.

However, there are also cases when the substances in the pheromone composition are also received by individuals belonging to other species, but the information transmitted is only partially deciphered or has other meanings. Natural pheromones are chemicals that facilitate intraspecific relationships. They are produced by exocrine glands (mostly located on the surface of the body) and detected with the help of olfactory organs located on the antennas. Pheromones represent a group of insect-borne (or animal species) substances through which a message exchange takes place between individuals of the same species. Pheromones encode a large amount of information related to various biological activities and determine adequate behavior in the receptor individuals. In the social insects, the pheromone range is wider than the solitary insects. As the range of messages transmitted by pheromones is expanded, a complete and especially definitive classification remains difficult to achieve. The most accessible and operative is the classification based on the nature of the pheromone-triggered response in the receiving organism (Niogret and Epsky 2018).

There are two large categories of pheromones:

- Development pheromones, which cause changes in development or metabolism in recipient organisms. This category includes pheromones that regulate the relationships within the colonies of social insects. Emissions by the queen insect determine the atrophy of the genital organs of working bees or the differentiation of the termites in casts.
- Action pheromones, triggering some actions or behavioral changes in receiving bodies. Among them, several important categories are:
  - Marking pheromones by which social insects trace the route of the colony to a new source of food.
  - Aggregation pheromones, characteristics for social insects, but also for locusts and other insect species. For example, grasshopper emit aggregate pheromones when they find suitable places for depositing eggs.

- Alarm pheromones, emissions at the onset of an enemy. For example, it may determine the immediate dispersion of the aphid colony or may trigger mobilization to fight enemies attacking colonies of bees or wasps.
- Pheromones of oviposition. *Aedes atropalpus* mosquito larvae secrete a substance that stimulates the approach and deposition of eggs by females belonging to the same species, the emitted substance actually indicating the presence of a suitable egg place, where future larvae have all the conditions to reach their full development.
- Necrophorous pheromones. During the process of decomposing, the ant's body produces mixtures of fatty acids with the function of mortal pheromones, indicating to the living members of the colony the presence of the dead individuals and thus determining their removal from the area.
- Pheromones for recognition of the nest and of the members of the colony: they were identified in ants (*Solenopsis invicta*), and it appears that such substances are generally present in all species of social insects.
- Pheromones for thermoregulation of the atmosphere in the nests were identified in wasp species.
- Sexual pheromones, mediating sexual relationships between physiologically capable individuals for reproduction; they are the most numerous and well-studied group of pheromones. They are especially known for insects, but they were also reported in crustaceans, fish, batrachians, reptiles, mammals and even plants. Of the sex pheromones, three subgroups are delimited:

  Sex attractants: they are pheromones emitted by individuals of a gender and have the role of attracting the opposite sex for mating.

  Aphrodisiacs: they are pheromones that have the role of excitement of the sexual partner and of inducing him to accept copulation.

  Sexual repellents: it was found that in some mosquitoes, males do not attempt to mate with females who have made earlier copulations because the sperm fluid transferred on the first copulation contains substances that are warnings or repellents for other males looking for mating partners (Akotsen-Mensah et al. 2018).

Pheromones are provided in the form of pheromone traps, which then associate with other types of traps. This feature gives ecological value to pheromone "treatments" that will only affect the species against which they are applied, the ecosystem as a whole being unaffected. (R)-hydroxydanaidal component of males *Utetheisa ornatrix* (Lepidoptera: Arctiidae) produces from dietary pyrrolizidine alkaloids obtained by larvae from the host plant *Crotalaria spectabilis* (e.g., monocrotaline) (Köblös et al. 2018).

Most of the pheromones are long-chain, C-C, long-chain carbon atoms, acids, alcohols or ketones, making it easier to isolate them. Molecular conformation of pheromones plays an important role in their biological activity. In any insect, the conformational structure of the main pheromone has a high specificity. Sometimes, minor changes in the molecular structure of pheromones annihilate their activity. Sometimes the *cis* and *trans* isomers of the same compounds behave differently in

Table 6.1 Sexual pheromones of insects

Name	Structure	Species producing
Valeric acid	$C_5H_{10}O_2$	*Limonius californicus*
Trans-9-keto-2-decenoic acid	$C_{10}H_{16}O_3$	*Apis mellifera*
Cis-7-dodecenyl acetate	$C_{14}H_{26}O_2$	*Cabbage looper*
Cis-8-dodecenyl acetate	$C_{14}H_{26}O_2$	*Grapholita molesta*
Cis-11-tetradecenyl acetate	$C_{16}H_{30}O_2$	*Archips semiferanus*
Hexadecanyl acetate	$C_{18}H_{36}O_2$	*Lycorea ceres*
Cis-11-octadecenyl acetate	$C_{20}H_{38}O_2$	

terms of pheromone activity. One of the best-known pheromones is, undoubtedly, *9-keto-2-decenoic acid* or *9-keto-2-decenoic acid* ($C_{10}H_{16}O_3$), the bee queen's substance, which attracts the males to mating with the queen. This acid represents only one compound of the 32 compounds of similar structure existing at the head of the bee queen. Sexual pheromones produced by females have a double role: both to attract distant males and to excite them when mated. The main insect sex pheromones are shown in the Table 6.1.

The pheromones produced by males are intended to excite females. These compounds are sometimes referred to as aphrodisiacs. In chemical terms, natural pheromones of insects are usually formed from a single chemical compound, very rarely from mixtures of several compounds. In the latter case, the specificity of the pheromones depends on the proportion of the various compounds. The simplest sex pheromone in insects is *valeric* acid produced by females of the beet wormhole, *Limonius californicus* (Binyameen et al. 2018).

Some of the pheromones, such as *frontaline* and *exobrevicomine*, have a cyclic structure. They facilitate copulation to cockroaches of genres *Brevicomis* and *Ips*. Some cyclohexane derivatives are Boll weevil pheromones, *Anthonomus grandis*. Until now, few pheromones with aromatic structure (cyclic organic structure with system of conjugated π-bonds), with functional group (acetate, alcohol or aldehyde), and with double-bond position and configuration (E or Z), stabilized by, through, or from the conjugation, were identified. It appears that the phenol acts as a pheromone for larvae of grass beetles. Phenol can form in the body of larvae from tyrosine present in their food. For *Leucania impuris*, the role of pheromone is met by benzoic aldehyde and to butterflies of the genre *Danaus*, by some alkaloids.

Sex pheromones are found low quantity in insects, which requires the sacrifice of a large number of individuals when determining their molecular structure. Each larva of the female *Orygia pseudotsugata* (the hairy caterpillar of the fir tree) contains approximately 40 ng of pheromones in the abdominal cavity. A total of 6000 larvae were needed to isolate and identify their sex hormone. Some insect sex hormones are found in various plant species. Thus, D-acetatul de borneol is found in *Gymnospermae*. This pheromone is active at concentrations below 0.07 mg/cm. Borneol L-acetate has 100% hormonal activity in the activity of the D-isomer. Numerous species of *Angiospermae* produce volatile compounds with pheromone effects on American Swift (beetle) *Periplaneta americana*. Of the synthetic pheromones commercially used, "alarm substances" are the most used. They attract male flies that populate fruit and melons in the Mediterranean region (Li et al. 2016).

An interesting case of specificity is the polyphenic moths, such as *Antheraea polyphemus*, which can be mated exclusively in the presence of *Quercus rubra* leaves (red oak). The analyzes performed showed that the oak leaves emanate a volatile compound (trans-2-hexenal and trans-2-hexenol), which stimulates the sensory receptors in the female antennae, thus triggering the release of their sexual pheromone, which then excites and causes the male to mating. Another interesting case is the sugar cane laurel, *Rhabdoscelum obscurus*, who eliminate his own sexual pheromone only after feeding on sugar cane. In this case, the pheromone elimination trigger must be an unidentified chemical compound present in the sugar cane. Pheromones of the oak moth, *Archips semiferanus*, are found as such in the oak leaves, which hurries the destruction of natural oaks by this moth. The larvae of the females *Archips semiferanus* take their pheromones, already synthesized from the oak leaves, which directly contribute to attracting males. After the arrival of the males on the leaves, the females remove a larger amount of pheromone to cause the male to move toward the copulation organ. Insect sexual pheromones are increasingly used in agriculture as biologically active substances effective for biological control of insect pests, especially as chemical insect pest control has been shown to have undesirable side effects: environmental pollution and biocenosis imbalance due to lack of specificity of some phytopharmaceutical compounds, including numerous insecticides (Hassemer et al. 2019).

In Table 6.2 are presented some examples of herbal compounds and sex pheromones with synergic activity.

Pheromones are preferable to insecticides in pest control because they have high specificity and are produced in small quantities. Insect sexual pheromones are used in agriculture for both the assessment and surveillance of the pest population in order to improve the prognosis in the application of chemical treatments and to combat them directly by preventing gender mating, which will result in a gradual reduction of the total population and, finally, the disappearance of pests (Oliveira-Hofman et al. 2019).

In Table 6.3, there are some examples of plant volatile compounds and sex pheromones with synergistic activity.

Preventing insect mating can be accomplished by either mass capture of males by means of traps containing natural or artificial pheromones from the natural population, before mating, or by atmosphere impregnation with a high concentration of pheromone in order to disorient males, who thus fail to spot the females for mating. In theory, some insect pests may be removed from any area by catching male with natural or artificial pheromones. In practice, there are factors that reduce the effectiveness of sexual baits. Pheromones transmit information related to certain biological insect activities, according to which they are classified as:

– *Sexual pheromones*, emitted by individuals of a sex (typically females), having the role of attracting individuals of the opposite sex, for mating.

Table 6.2 Examples of intensifying the action of two associated substances (volatile plant compounds and sex pheromones)

The host plant	Volatile compounds from host plants	Insects	Sexual pheromone
Beta vulgaris	Linalool ($C_{10}H_{18}O$), myrcene ($C_{10}H_{16}$), benzaldehyde (C_7H_6O)	*Spodoptera exigua*	(Z)-9-tetradecenol ($C_{14}H_{28}O$), (Z, E)-9,12-tetradecadienyl acetate ($C_{16}H_{28}O_2$)
Brassica oleracea subsp. *capitata*	(Z)-3-hexenyl acetate ($C_8H_{14}O_2$)	*Plutella xylostella*	(E)-2-hexenal ($C_6H_{10}O$), (Z)-11-hexadecenol ($C_{16}H_{32}O$), (Z)-11-hexadecenyl acetate ($C_{18}H_{34}O_2$), (Z)-3-hexenol ($C_{24}H_{42}O$)
Cotton, tobacco, tomato	(Z)-3-Hexenyl acetate ($C_8H_{14}O_2$)	*Heliothis virescens*	(Z)-11-Hexadecenal ($C_{16}H_{30}O$), (Z)-9-tetradecenal ($C_{14}H_{26}O$), hexadecanal ($C_{16}H_{32}O$), tetradecanal ($C_{14}H_{28}O$)
Japanese cedar (*Cryptomeria japonica*), Japanese cypress (*Chamaecyparis obtuse*)	Methyl phenyl acetate ($C_9H_{10}O_2$)	*Anaglyptus subfasciatus*	(R)-3-hydroxy-2-octanone ($C_8H_{16}O_2$), (R)-3-hydroxy-2-hexanone ($C_6H_{12}O_2$)
Papaya	Host fruit odors: Methyl butanoate ($C_5H_{10}O_2$), ethyl butanoate ($C_6H_{12}O_2$), 3-methyl-1-butanol ($C_{11}H_{16}O_2$), 1-butanol ($C_4H_{10}O$)	*Toxotrypana curvicauda*	2-Methyl-6-vinylpyrazine ($C_7H_8N_2$)
Pine	Pine bolt odors: α-terpineol ($C_{10}H_{18}O$), cyclic terpene alcohols terpene hydrocarbons, ethers, and esters	*Pissodes nemorensis*	Grandisal ($C_6H_6O_2$), Grandisol ($C_{10}H_{18}O$)
Prunus padus	Benzaldehyde (C_7H_6O)	*Rhopalosiphum padi*	Nepetalactol ($C_{10}H_{16}O_2$)
Zea mays	(Z)-3-hexenyl acetate ($C_8H_{14}O_2$)	*Helicoverpa zea*	(Z)-11-hexadecenol ($C_{16}H_{32}O$), (Z)-7-hexadecenal ($C_{16}H_{30}O$), (Z)-9-hexadecenal ($C_{16}H_{30}O$), hexadecanal ($C_{16}H_{32}O$)
Zea mays	(Z)-3-hexenyl acetate ($C_8H_{14}O_2$)	*Cydia pomonella*	(E, E)-8,10-dodecadienol (codlemone) ($C_{12}H_{22}O$)

Table 6.3 Examples of intensifying the action of two associated substances (plant volatile compounds and aggregation pheromones)

The host plant	Volatile compounds from host plants	Insects	Aggregation pheromone
Apple, orange, stored grains, spices, seeds	2-Propanol (C_3H_8O), butanoic acid ($C_4H_8O_2$), methanol (CH_4O), methyl butanoate ($C_5H_{10}O_2$), n-heptanol ($C_7H_{16}O$), propanoic acid ($C_3H_6O_2$), propanol (C_3H_8O)	*Carpophilus hemipterus*	(2E,4E,6E,8E)-3,5,7-trimethyl-2,4,6,8-decatetraene ($C_{13}H_{20}$)
Cereal grain, flour	Maltol ($C_6H_6O_3$), valeraldehyde ($C_5H_{10}O$), vanillin ($C_8H_8O_3$)	*Sitophilus oryzae*	Sitophinone: 4,8-dimethyldecanal ($C_{12}H_{24}O$)
Coconut, palm, banana	Ethyl acetate ($C_4H_8O_2$)	*Rhynchophorus palmarum*	(2E)-2-Methyl-5-hepten-4-ol ($C_{15}H_{22}O_2$)
Coconut, palm, banana	Ethanol (C_2H_6O), ethyl acetate ($C_4H_8O_2$)	*Rhynchophorus palmarum, Dynamis borassi*	Rhynchophorol ($C_8H_{16}O$)
Cotton	Trans-2-hexenol ($C_6H_{12}O$), cis-3-hexenol ($C_6H_{12}O$), n-hexanol ($C_6H_{14}O$)	*Anthonomus grandis*	Grandisol ($C_{10}H_{18}O$), grandisal ($C_{10}H_{18}$), papayanol ($C_{10}H_{18}O_2$)
Fermenting aspen (*Populus tremula*) bark	Host odors: Acid resin of a hop-like *odor, populin* ($C_{20}H_{22}O_8$), *salicin* ($C_{13}H_{18}O_7$), *chrysin* ($C_{15}H_{10}O_4$) *acetyl-benzoyl-phloroglucin* ($C_{15}H_{20}O_8$), *tectochrysin* ($C_{16}H_{12}O_4$), *saligenin* ($C_6H_4OH.CH_2OH$), salicylic aldehyde ($C_6H_4.OH.CHO$)	*Drosophila borealis, Drosophila littoralis*	Ethyl tiglate ($C_7H_{12}O_2$)
Oil palm (*Elaeis oleifera*)	Fruit bunches: 9-Octadecenoic acid ($C_{18}H_{34}O_2$), L-(+)-ascorbic acid-2,6-dihexadecanoate ($C_{38}H_{68}O_8$), 14-methyl-8-hexadecenal ($C_{17}H_{32}O$), 4-hydroxyl-4-methyl 2-pentanone ($C_6H_{12}O_2$), 3-ethyl-2, 4-dimethylpentane ($C_9H_{20}O_3$)	*Oryctes rhinoceros oil palm*	Ethyl-4-methyloctanoate ($C_{11}H_{22}O_2$)
Palm (*Elaeis guineensis*)	Alcohols and esters: Methyl ester-hexadecanoic acid ($C_{17}H_{34}O_2$)	*Rhynchophorus phoenicis, Rhynchophorus vulneratus*	3-Methyl-4-octanol ($C_9H_{20}O$), 4-methyl-5-nonanol (rynchophorol) ($C_{10}H_{22}O$)
Palm (*Acrocomia aculeata*)	Ethyl acetate ($C_4H_8O_2$)	*Rhynchophorus cruentatus*	5-Methyl-4-octanol (cruentol) ($C_9H_{20}O$)
Palm (*Astrocaryum vulgare*)	Host plant volatiles: 9-octadecenamide ($C_{18}H_{35}NO$)	*Rhynchophorus ferrugineus*	Ferrugineol (4-methyl-5-nonanol) ($C_{10}H_{22}O$)

(continued)

Table 6.3 (continued)

The host plant	Volatile compounds from host plants	Insects	Aggregation pheromone
Palm (*Elaeis oleifera*)	Palm tissue volatiles: Oleic acid ($C_{18}H_{34}O_2$), L-(+)-ascorbic acid ($C_9H_{15}NO_8S$), 2,6-dihexadecanoate ($C_{18}H_{36}O_4$), 9-octadecenoic acid ($C_{18}H_{34}O_2$), methyl esterhexadecanoic acid ($C_{18}H_{36}O_2$), 9-octadecenamide ($C_{18}H_{35}NO$)	*Rhynchophorus phoenicis*	Rhynchophorol ($C_8H_{16}O$)
Palm, sugarcane (*Saccharum officinarum*), banana, pineapple	Ethyl esters ($C_{46}H_{70}O_4$)	*Metamasius hemipterus sericeus*	2-Methyl-4-heptanol ($C_8H_{18}O$), 5-methyl-4-nonanol ($C_{10}H_{22}O$)
Palm, sugarcane	Ethyl acetate ($C_4H_8O_2$)	*Rhabdoscelus obscurus*	(2E)-6-methyl-2-hepten-4-ol ($C_8H_{16}O$), 2-methyl-4-heptanol ($C_8H_{18}O$), 2-methyl-4-octanol ($C_9H_{20}O$)
Wheat (*Triticum vulgare*)	Fermenting whole wheat bread dough: Gluten ($C_{32}H_{52}O_2$), fibrine ($C_5H_{11}N_3O_2$)	*Carpophilus mutilatus*	(3E,5E,7E)-5-Ethyl-7-methyl-3,5,7-undecatriene ($C_{14}H_{24}$),
Wheat (*Triticum aestivum*)	Propyl acetate ($C_5H_{10}O_2$)	*Carpophilus obsoletus*	(2E,4E,6E,8E)-3,5,7-trimethyl-2,4,6,8-undecatetraene ($C_{14}H_{22}$)

- *Aggregation pheromones*, which give the gathering signal, find feed sources or places favorable for the deposition of eggs (grasshoppers).
- *Alarm pheromones,* emitted at the appearance of an enemy (lice and wasps).
- *Path marker pheromones* (to ants).
- *Necrophorus pheromones* for recognizing dead individuals and removing them from a colony.

Irrespective of the information they encode and the behavior they induce, pheromones have the following common characters: they are produced and are active in very small quantities, very volatile and can cause a behavioral response with a single compound or mixture of compounds. Some compounds have synergistic, inhibitory or multifunctional effect.

Multifunctional role compounds have synergistic effect in very small amounts and inhibitor in larger amounts (this is the case of some compounds of aggregation pheromone in bark beetles). The transfer rate and the amount of information transmitted via pheromones can be maximized in different ways: through the existence of complex pheromones, chemical complexity or substance mix, concentration changes, behavioral changes caused by photoperiodism and annual periodicity. Pheromone communication is influenced to a large extent by endogenous factors and external factors. Based on the results obtained so far, it is estimated that the importance of pheromones will increase and are already considered "pesticides" of the future. Pheromones are real compounds that cannot be sensitively detected just like fragrances and essences, but when they come into contact with the brain and with the subconscious, a chemical reaction occurs that results in action (Bell et al. 2018; Yang et al. 2014).

6.4 Plant Protection Products

Plant protection products are classified as chemicals (pesticides) and biological products (bioproducts). Current research aims at the realization and use of biological products to ensure effective control of both toxigenic microorganisms and microorganisms that affect the qualitative parameters of crops. Bioproducts are biological means based on microorganisms useful for crop plants or on natural compounds (originating from plant extracts, suggestively called "botanicals"). Due to their biological nature, bioproducts have a complex action on crop plants. The most appropriate term would be biopreparations for agricultural use. An already classic example, illustrative of this complex action, is that of antagonist fungal bioproducts of the genus *Trichoderma*. Approved as biofungicides, a number of bioproducts were shown to be stimulants of plant growth, and this stimulation of plant growth has been shown to be due to this biofungicides intervention in plant nutrition (Bernat et al. 2018).

"Formulation" is the form under which a pesticide is marketed and represents a combination of various compounds (solvents, surfactants (surface-active agent), cosurfactants, adhesive, suspending agent, penetration enhancers of the plant cuticles, etc.) whose ultimate goal is to make the product usable in an effective way. The compounds used for pesticide conditioning are also important chemical pollutants (organic solvents, surfactants that are similar to water pollutants detergents, etc.); hence, it is of great importance to develop a code of practice for the use of pesticides. Several examples of bioproducts are as follows:

- *Trichodermin-BL* is based on the *Trichoderma lignorum* and used for combating white, gray and root rot of vegetable, ornamental, leguminous crops, as well as tobacco seedlings and vegetable crops, reducing the pathogens attack by 2–3 times and stimulating plant growth by 25–30%.
- *Trihodermin-F7* is based on *Trichoderma harzianum* in granular and liquid form. It is used to combat root crop rotations, reducing root rotations by 1.5–2 times.
- *Nematofagin-BL* is based on *Arthrobotrys oligosporum* and used to combat gallium nematodes in vegetable and technical crops.
- *Verticilin* is based on *Verticillium lecanii* in the form of a wet table powder. It is used to control the greenhouse mussel, with the effectiveness of 95%.
- *Rizoplan* is based on *Pseudomonas fluorescens* AP-33 and is used to combat root rot in agricultural crops.

- *Pentafag-M* is intended to combat bacteriosis in stubble and chick peas. The preparation is based on five bacteriophage strains that are effective in controlling plant diseases caused by *Pseudomonas* sp. A number of environmentally friendly viral preparations have been developed to combat pests that cannot be combated by other biological means.
- *Virin-ABB-3* is for combating hairy caterpillar of mulberry tree in orchards, parks, and forestry. The preparation is based on cumulative and synergistic nuclear polyhedrosis viruses, showing epizootic and post-action effects.
- *Virin-MB* is intended for the control of the green bollworm and is based on the nuclear polyhedrosis virus of *Mamestra brassicae*.
- *Virin-OS* is for combating sowing bollworm and moths of the genus *Agrotis*. It is based on synergetic nuclear granulosis and nuclear polyhedrosis viruses.
- *Virin-HS-2* combating cotton caterpillar and bollworm from genus *Heliothis* is based on the nuclear polyhedrosis virus of a non-specific host.
- *Virin-CP* is intended for combating apricot worm and based on granulosis virus *Carpocapsa pomonella* (Taghdisi et al. 2019).

Ensuring the effectiveness of nonchemical plant protection systems becomes a reality in the implementation of integrated plant protection systems with the predominant application of biological protection methods. Biological plant protection as an effective method of avoiding conflicts between plant protection and environmental quality is based on the continued use of information on the monitoring of populations of harmful and useful organisms and on the use of offsetting and control measures for the application of entomophagous, bioproducts and biologically active substances (Marczewska et al. 2019).

6.5 Agricultural and Secondary Products Can Be Used for Production and Processing

At present, costs are relatively low for the production and processing of biopesticides, as the main raw materials are renewable natural resources (such as agricultural products and by-products of maize, soybean flour, fish flour, wheat bran, etc.). Biopesticide production generally does not involve nonrenewable raw materials as it is the case when producing chemical synthesis products, that involve the use of nonrenewable resources (such as oil, coal, natural gas, etc.). Biopesticides offer additional benefits, such as complex and new action modes for managing resistance (Mushtaq et al. 2018), therefore helping to extend the life span of conventional pesticide products. They also add flexibility to a classic combat program with reduced intervals from the last treatment to harvest, a very good pesticide residue management for exported products and an excellent ecotoxicological profile for humans, animals and useful entomofauna (Heimbach et al. 2016).

6.6 Transgenic Products

Wuhan virus experts have extracted from the African scorpion body toxic compounds to produce biopesticides—the recombinant cotton bollworm virus—that can reduce the death of cotton worms in 2 days. In 2004, Sun Xiuling isolated unique insect-selective scorpion toxins from African scorpion by extraction of scorpion venom genes. He developed a biological pesticide to control the cotton bollworm, *Helicoverpa armigera*. Cotton bollworm is one of the more common cotton pests, but too much use of chemical pesticides can cause environmental pollution. Field test performed during a period of 4 years confirmed that "recombinant cotton bollworm virus" is safe for natural enemies of *Helicoverpa armigera*, being an environmentally friendly biopesticide. The abovementioned study has won the environmental protection award in the field of environmental protection and the first technology award of the Chinese state (Jin et al. 2018).

Although the last decades have been marked by success in improving various plant species of economic interest, there remain heavy losses to cultivated plants due to biotic and abiotic stress. Despite the popularity of potatoes as food, they are very difficult to cultivate. Farmers make significant losses each year due to the Colorado beetle (*Leptinotarsa decemlineata*), as repetition of insect spraying operations is absolutely necessary for pest control. In a similar situation is corn, whose major pest in North America and Europe is corn borer (*Ostrinia nubilalis*). Modern corn hybrids show some resistance to *Ostrinia nubilalis* attack, but in the case of medium or strong infestation, losses are estimated at approx. Five percent of the total production or even more, depending on year and location. Reduction of these losses would be possible by increasing the cultivated area, but, in the presence of a strong impact on the environment and on natural resources, this is a limited option. Due to the fact that we relied too much on only a few plant species, monocultures have been created, where most of the cultivated plants are often severely affected by harmful insects. In order to counteract the negative action of the insect pests and to reduce dependence on chemical insecticides, a range of insecticidal natural proteins have been identified in bacteria and plants, and the genes encoding them have been isolated and transferred to a group of crop plants. As already mentioned, it is well known that certain strains of *B. thuringiensis* produce proteins that cause dysfunctions of the digestive system of insects with alkaline digestive tract, resulting in slowing growth and ultimately death (Gnepe et al. 2014).

The different *B. thuringiensis* strains are highly selective, being effective only against certain insect species such as *Ostrinia nubilalis* and *Helicoverpa zea* (Lepidoptera), *Leptinotarsa decemlineata* (Coleoptera) and certain flies and mosquitoes (Diptera). *B. thuringiensis* was identified in 1911 when it was found to kill the larvae of *Anagasta kuehniela* and was registered as a biopesticide in 1961 in the United States. At present, the different strains of *B. thuringiensis* are used to control lepidoptera pests such as those in the families *Tortricidae*, *Pieridae*, *Noctuidae*, *Plutellidae*, among others. At least 1% of all pesticides used in the United States contain *B. thuringiensis*. However, as an ingredient for insecticides, *B. thuringiensis* is relatively expensive and has some drawbacks (Wang et al. 2018).

Although some insecticides have a lethal effect through simple contact with pests, *B. thuringiensis* must be ingested to be effective and should therefore be applied exactly when insects feed. In addition, the rainwater cleans the plant, and the sun's rays can destroy it.

Thanks to genetic engineering, it is possible to identify genes encoding *B. thuringiensis* protein synthesis, and transferring them to culture plants that will synthesize the same *B. thuringiensis* protein with a lethal effect on harmful insects. To this end, the *B. thuringiensis* genes encoding the cry1Ac and cry1Ab proteins in cotton, the cry3Bb1 gene for the control of the populations of *Diabrotica* spp. and the cry1Fa2 gene for the control of a wide spectrum of maize lepidopteran pests (the latter two were introduced in 2003). Attention is focused specifically on the possible transfer by pollination of genes for insect resistance from transgenic plants to related wildlife species. However, the emergence of wild plants with a superior genotype could alter the composition and abundance of plant and animal species in natural or agricultural ecosystems. Also worrying is the possibility of installing genetically modified, annual or perennial plants, such as weeds, in natural or controlled habitats. In this context, the introduction of transgenic plants into environments where there are wild correspondents must be properly assessed before cultivating them on an industrial scale. Genetically modified plants that produce *B. thuringiensis* toxin as its own constituent throughout the growing season exert a high pressure on the selection of insect pests due to the spread of varieties of transgenic plants in areas with numerous pests and due to the fact that at least a generation of pests will be dependent on plants for survival from 1 year to another (Baum et al. 2004; Anil and Podile 2012).

The first transgenic cotton plants to which the gene coding for *B. thuringiensis* toxin synthesis was transferred were obtained in 1996. Since cotton has a wide range of pests, insecticides cannot be discarded, especially due to *Lepidoptera* which are not susceptible to endotoxin expressed in plants, either in the field or in the laboratory, probably due to the continuous expression of the toxin in the plants, which led to a strong selection pressure.

Scientists have also noticed a "good behavior" of insect resistance, because the expression of the toxin in the tissues of plants is unequal, and they will attack those tissues or portions of tissue where the toxin concentration is low. Moreover, because the concentration of the toxin often decreases in leaf and stem as the plant reaches maturity, the low doses can kill or weaken the susceptible larvae (homozygotes), and therefore adaptation to the *B. thuringiensis* toxin occurs more rapidly when the concentration always remains high (Hyakumachi et al. 2013).

Over time, over 500 cases have been reported when insects have developed resistance to the conventional insecticide spectrum or insecticide products containing *B. thuringiensis*. In fact, recent research in England has shown that *Plutella xylostella* larvae show resistance to the cry1Ac toxin, while at the same time there is a faster development and a higher weight of the pupae in the presence of the toxin. This may be a genetic effect, indirectly linked to the presence of an allele gene that confers resistance to the cry1Ac toxin, or can be determined by their increased ability to survive and digest this toxin. Therefore, the presence of the toxin can have favorable, but undesirable, nutritional effects. However, it cannot be concluded that all insect pests of crop plants will exhibit the resistance to insecticides, including the current transgenic plants. More questions, but also many answers, are also available for the effects of transgenic plants on insect pests and more. By keeping pest populations at extremely low levels with insecticides, *B. thuringiensis* plants can completely destroy their enemies, while they only need small amounts of food to survive in agroecosystems. The possibility that *B. thuringiensis* toxins move into the Arthropod's food chain has serious implications for the agroecosystem equilibrium (Cao et al. 2014).

Studies in Scotland suggest that aphids are capable of retaining the *B. thuringiensis* toxin and transferring it to the coccinelides that consume them, further affecting the reproduction and longevity of the quails (*Melolontha vulgaris*).

Recent research has shown that the pollen of the *B. thuringiensis*-bearing maize corn can be worn a few meters to the leaves of *Euphorbia, Asclepiadaceae*, with a potentially damaging effect on Monarch butterfly populations, this being a new dimension to the unexpected impact of transgenic plants on organisms on target. Moreover, no scientific analysis of the possible adverse effects following the release of transgenic plants into the environment could guarantee the absence of "ecological surprises" in the future. Recently, it has been discovered that the *B. thuringiensis* toxin, which normally degrades rapidly, under certain circumstances can bind to the clay soil particles, remaining biologically active for at least 230 days. It accumulates over time and reaches much higher concentrations than expected, thus endangering the life of organisms in the soil. For centuries, people have tried to obtain plants that survive and develop despite the insect pest attack. Consciously or unconsciously, old farmers selected genes for pest resistance by simply collecting seeds, only from high-yielding crops in their crops.

Nowadays, due to genetic engineering that complements the mastery of breeders, genes for pest insect resistance can be transferred from one organism to another more rapidly and deliberately, and certainly the resistance of transgenic plants to specific pests would be a privilege for agriculture. However, obtaining and cultivating such genetically modified organisms is only part of an equation with many unknown parts (Mahmoud et al. 2017; Armada et al. 2014).

6.7 Natural Biopesticide

Pesticides used for pest control are very toxic chemicals that ultimately affect plants and pollute the environment. That's why, in organic farming, specialists are trying to find biological substitutes for chemicals.

6.7.1 Garlic: A Biopesticide

German biologists believe that garlic can become an important means of plant protection in organic farming. They found that allicin contained in garlic is an extremely effective antibiotic against a wide range of phytopathogens—bacteria and fungi. A diluted garlic extract was effective against a typical potato and tomato disease caused by *Phytophthora infestans*. Some researchers believe that the new biopesticide can be used to treat seeds in order to be protected from pests.

Representatives of the seed industry have already shown interest in this treatment.

However, in order to use the natural product in field crops, it is to be determined how resistant garlic-based biopesticide is to rain and temperature oscillations. It is also not known whether such a biological product in high doses will or will not influence the taste of the vegetables. It can be assumed that a slight taste of garlic will be maintained. The issue of authorization of the active substance allicin as a means of plant protection has not yet been finalized (Bharadwaj et al. 2015).

6.7.2 Lemon Eucalyptus Oil

It is a powerful repellent, effective against mosquitoes, deer ticks and other pests. There is also a synthetic form of lemon eucalyptus oil, known as PMD, which is also effective. Both compounds are found in several brands of repellents that are marketed as natural. Although chemically sounding, IR3535 is an herbal compound that has been used in Europe for decades as an insecticide.

There are dozens of recipes available for insect repellents, which contains an alcohol base or 'carrier oil' and one or more of the following ingredients: cedar oil, tea tree oil, geranium oil, rosemary oil, lemonade oil, lemon oil, eucalyptus, cinnamon oil. They may not have the same effects as commercial preparations (Ghosh et al. 2012).

6.8 Technology for the Production of Entomopathogenic Bacterial Biopesticides

As mentioned before, of all entomopathogenic bacteria, the most widely is *B. thuringiensis*, which, in addition to spore formation, is causing insect septicemia. It produces a number of toxic compounds that increase the effectiveness of the various preparations. These toxic products are is divided into four components:

- Phospholipase C, an exotoxin. It causes the breakdown of essential phospholipids into the tissues of insects leading to their death.

- β-exotoxin is a thermostable toxin. It accumulates during the vegetative growth of the cells. Its action is due to the inhibition of DNA-dependent RNA polymerase, resulting in termination of RNA synthesis in insect cells and their death. The action of this toxin is quite slow.
- δ-endotoxin is a crystal protein (in the form of ordinary crystals). Crystal proteins are formed as parasporal crystalline inclusions during the stationary phase of growth. It exerts its toxic effects in the gastrointestinal tract of the insect, destroying the enzyme system (Scheckhuber 2019).

6.8.1 General Mode of Action

A pest control product acts through different molecular interactions with a certain type of protein. The body of insects, according to heredity, forms many types of proteins, which make up tissues, organs, insect exoskeleton, etc. Other proteins act as catalysts in the storage or transport of nutritional products, in the movement of molecules through cell membranes. Most insecticides, acaricides or some metabolites (biopesticides) act on target proteins contained in the signaling nervous system (by disrupting neurotransmitter receptors), participating in cellular respiration (as breathing disturbers) or on those having a role in the development of insects (by disrupting the activity of growth regulators). Insecticides can bind to proteins at one or more target sites, with activating or inhibiting functions, with disruptive effects whose symptoms are called action mechanisms.

Depending on the mode of binding to the target protein, the following characteristics are determined: the mode of action of the control product, the selectivity in use, the rate of action and the resistance of the host. Since animal organisms contain many similar proteins with identical function, there is a risk that the products will act on the target protein, on the harmful insects but also on beneficial insects or human beings. The mode of action of biological preparations includes the modes of action of bacteria, baculoviruses, fungi and their metabolites (Dos Santos et al. 2019).

6.8.2 The Bacteria

Generally, diseases produced by bacteria are characterized by penetration into hemocel and multiplication of the pathogen, resulting in septicemia of insects. *B. thuringiensis* strains, which are most used in biological control, produce insecticidal endotoxins. They bind to the proteins in the insect's middle intestine.

Arrived in the membrane, they form channels that allow leakage of ions and destruction of cells. The production of these toxins is also accomplished by genetic engineering, by introducing genes producing endotoxins into the genome of some plants, resulting in genetically modified organisms (Guo et al. 2019).

6.8.3 *Baculoviruses*

There are vibrions assembled in protein formations called supraovaryocapsides that enter the body of the insects by ingestion, reaching the middle intestine and under the action of enzymes it protects and releases the vibrions that develop into the epithelial cells destroying them (Cuartas et al. 2019).

6.8.4 *Fungi*

Fungi produce losses in vegetal crops. Toxins and other compounds released by fungi are deemed to cause adverse health effects. Compounds produced to combat these negative effects are called fungicides. For the production of a mycosis, the fungal pathogen must penetrate the host through mechanical and enzymatic action. After piercing the tegument, the fungal strain penetrates into the hemocel, where it multiplies.

Insect colonization is reached by blastoporal proliferation and mycelial growth, which blocks blood circulation and completely disintegrates tissues. A good example regarding the mode of action on the nervous system, spinosad, a natural metabolite obtained from the actinomycetes *Saccharopolyspora spinosa*, opens up a new class of natural products derived from microorganisms. Spinosad binds to one or more nervous system proteins.

Compared to other products, it acts on a different nicotinic receptor site and, by activating nicotinic–acetylcholine receptors, produces an influx of sodium ions that depolarize the hyperactive neurons excitement of the body muscles (manifested by leg extension, tremors, beatings from the wings, swallowing air). Neuromuscular fatigue results in insect paralysis. By discovering such biopesticides with a different mode of action, some more toxic insecticides can be replaced for ecosystems and cross-resistance (Xiao et al. 2016).

6.9 What Alternatives Exist if the EU Says NOT to Pesticides

What are the alternatives? Will it be possible to practice intensive farming and plant protection, given that it is possible to have a smaller number of active substances to combat pests? Over time were reported numerous researches for biological control. In the following lines will be presented some study directions on the use of entomopathogenic fungi, as well as the advantages and disadvantages of such method (Latré et al. 2015).

6.10 How Viable Are Biological Control Methods?

- Inoculation of phytopathogenic fungi. In recent years, a new method has been developed, namely, 'self-inoculation' of insect pests with a phytopathogenic fungus. The method implies that in the open field are placed devices that attract harmful insects without capturing them. The insects come in contact with the entomopathogenic fungus, and soon they do not feed and die in a few days. The results are satisfactory compared to the classical method (chemical treatments), but in the case of strong attack (or high pest density), it is ineffective.
- Immunization of plants. Another new method of biological control is the inoculation of plants with endophytes, microorganisms that activate in the plant certain defense reactions against the attack of bacteria or phytopathogenic fungi. Practical results on plant immunization are promising (a case study on vegetables in protected areas), but the method does not provide protection against insect pest attack.
- A new innovative method to combat soil pests consists of incorporating into the soil microcapsules containing an entomopathogenic fungus that kills the insect pests. In order for the larvae of these insects to be attracted, the microcapsules, once introduced into the soil, eliminate carbon dioxide. In this way, the activity of the root system of the plant is simulated, to attract the larvae and to bring them into contact with the entomopathogenic fungus.
- Another method of control involves an association of biological and chemical control by using a combination of low-dose insecticide with a biological insect control agent. The low-dose stresses the insect, weakening the immune system and making it more sensitive to the biological agent (entomopathogenic fungus, virus, etc.). The preliminary results on this method are promising.
- Biotreatments in vegetation phase. Research on the formulation of biopreparations based on *Beauveria bassiana* and *Metarhizium brunneum* applied as a treatment in vegetation to combat the insect pests of the main crops and horticultural crops is being conducted in Germany. In order to develop this new formulation, it is necessary to isolate entomopathogenic fungi, cultivate them in an artificial environment to ensure good sporulation and research into the use of different adjuvants to protect the biopreparation from the action of environmental factors (ultraviolet radiation). Laboratory experiments have been conducted on the degree of penetration of the formulated bioproducts into plant leaves. Experimental results have highlighted that the use of titanium dioxide as an adjuvant was beneficial, protecting *Beauveria bassiana* spores from ultraviolet radiation, their viability increasing by 77% ± 11%. In this way, the biopreparations are more effective in combating pests and can solve one of the problems encountered by bio-insecticides, namely, the very short time that these products have efficacy (3–7 days). Further, research is geared towards finding innovative materials to be used in new formulations of bioproducts to improve the penetration of entomopathogenic fungi into plants, to better protect them (Thomas 2003).

6.11 Biopesticides Are Far from the Efficiency of Chemical Treatments

All these research methods in Germany and other European countries are looking to be a viable alternative to chemical control, given the limitation of the range of insecticides as a result of increasingly restrictive EU regulations. Briefly reviewing the current state of the art of pest control with entomopathogenic fungi and the challenges faced by the sector, such as the sensitivity of bioproducts to climatic conditions, the short life span of bioproducts, increasing problems on the effectiveness of bioproducts (currently at most 7 days) and problems raised by the formulation of biopreparations, it can be concluded simply that if the characteristics of bioproducts used in the biological control of harmful insects are not improved in the future, they will not be adopted by farmers (Kathage et al. 2018).

6.11.1 Biological Control of Pests with Entomopathogenic Fungi Metarhizium anisopliae *and* Metarhizium brunneum

Research on the biological control of pests has been extended with entomopathogenic fungi *Metarhizium anisopliae* and *Metarhizium brunneum*. There is a research project that approached this topic with the objective of finding innovative alternatives to chemical control of soil pests. In the medium and long term, it is desirable to formulate a long-acting bioproduct in the soil and at the same time reduce the dose (low amount of *M. brunneum* spores), as the costs are small, and the biopreparations have high efficacy. These simultaneous goals, listed above, are a great challenge for researchers in the field. Bioproducts formulation took the form of capsules that are incorporated into the soil. These capsules have two roles, one to attract insects, by releasing CO_2 (similar to the root system of plants attacked by these insects), and another to combat (the strategy "draws and fight"). After incorporation of the capsules into the soil, the spores of the fungus germinate and the fungus develops inside the capsules, thus resulting the carbon dioxide. Various drying processes of the spores of the *M. brunneum* have been tested, so that they have a long-lasting viability. Emphasis has also been put on the materials from which the capsules are made which will include the spores of *M. brunneum* so that they have a substrate for the germination and development of entomopathogenic fungi after the capsules are introduced into the soil and at the same time the capsules release a high amount of CO_2 to attract pests of harmful insects (Clifton et al. 2019).

Disadvantages of this bioproduct: Spores of *M. brunneum* do not withstand temperatures above 25 °C. In the future, it is desirable to produce a very stable, long-lasting product for the biological control of soil pests in order to control their populations and, at the same time, reduce the amount of chemicals used to combat insects. The experimental studies resulted in biological control agents for potato crops (Colorado beetle—*Leptinotarsa decemlineata*) and maize (worms—*Agrotis* ssp.) The results are satisfactory, but its costs are far higher than classical control (seed treatment) (Huang et al. 2017).

6.11.2 Use of Entomopathogenic Microorganisms

Entomopathogenic microorganisms are used in the form of biological insecticides that have as active principles viruses, bacteria, fungi, protozoa. Advantages: they are harmless, nontoxic to the environment and leave no residue; some have a high specificity of host insects, ensuring the protection of useful fauna; and are compatible with chemical insecticides because infections sensitize insects to their action. Disadvantages: the need to apply the product as prevention method, respecting the incubation period of the diseases, and the specificity of a microorganism for a particular species (Gross et al. 2010). In caterpillars, symptoms of viral infections occur 48 h after infection and are manifested by reduced feeding. The insect's body becomes lighter by the accumulation of protein in the adipose tissue, and the larvae from the early ages climb to the top of the plants, where they die. The body is liquefied, and, by touching the skin, the liquid spreads and disperses the infection. If viral infections occur in larvae ages 5 and 6, before dying, they move on tree trunks to a height of 2.5–3 m and head down. The hostages of infected *Hyphantriacon* leave the nests and move in the crown and then die dangling on silk threads. Hosts infected with cytoplasmic granulosis virus reduce their growth rate by loss of appetite. On the ventral side of the abdomen appears a whitish or whitish-yellow color. In the latter stages, it eliminates oral fluid with polyhedrons developed in the intestinal area. From a pathological point of view, in the case of polyhedroses, the most obvious changes occur in the nuclei of the initially attacked cells, in those of the hemolymphs or fat body cells. The infected nuclei increase, being filled with polyhedrons, occupying the entire cell, which it destroys and pass into the hemolymph. In cytoplasmic polyhedrosis, infections develop in the middle intestine and then spread throughout the digestive tract. *Lymantria* presents nuclear polyhedrosis induced by the *Borrelinavirus reprimans* virus, which produces epizootics during the collapse of the pest. After 10–12 days of viral infection, death occurred. The infection can also be transmitted by laying eggs, with the food ingested and the polyhedra dissolved by the alkaline gastric juice, observed on the sixth day in the hemolymph (Shi and Bode 2018). The virus is specific to caterpillar and can also be transmitted through entomophagous insects. Characteristic is the concentration of diseased caterpillars, ages 5 and 6 (80–90%) at the base of the stems, fixed to the bark, 2–3 m from the ground. The diseased hosts cease feeding and die within 1–2 days. At the *Lymantria monacha* caterpillars, the nuclear polyhedrosis produced by the *Borrelinavirus efficiens* virus appears. The disease is manifested by the presence in the nucleus of cells of tetrahedral and cubic polyhedrons, measuring 2.5–10 microns, visible on a microscope. Within the polyhedrons, virus particles are grouped 2–4 in bundles, in a common development membrane. The characteristic feature of these epizooties is the migration of viral pests to the top of the trees ('spine disease').

The identification of epizootic diseases is done during the eruption phase by observing the dead caterpillars at the top of the crown. Especially *Operophtera brumata* and less *Erannis defoliaria* are susceptible to nuclear polyhedrosis, which, after a 6–7-day incubation, produces a 97% mortality. The virus is located in the nuclei of fat cells, hypoderm and tracheal tissue. Diseased caterpillars are concentrated in the trees crown (90–95%), where they die fixed to the lower part of the leaves. Viral infection is visible from the third age of the caterpillars when they stop feeding and their body gets a yellowish color. In the crisis phase, the mortality rate of the caterpillars may reach 40–50%. *Malacosoma neustria,* present in the cervinaceae, and, in particular, those of the cerebro-spiny type, by the fourth and fifth larva infection with the nuclear polyhedrosis virus, exhibit high mortality epizooties.

The pathogen produces hypertrophy of the hypodermis, adipose and tracheal cell nuclei (Lester et al. 2015). Disease is manifested by reduced feeding, the caterpillars get a dark brown color, a phenotype of negative phototropism at the caterpillars in the last ages, migrating from the crown to the base of the trees and die soon. The caterpillars *Choristoneura murina* are attacked by *Borrelinavirus* which produces a nuclear polyhedrosis. Granulosis of the caterpillars is produced by *Bergoldiavirus calypta*. The infected caterpillars have a whitish color, and after death they become gray and wrinkle. Death occurs 15–20 days after infection. For *Neodiprion sertifer* appears nuclear polyhedrosis produced by *Borrelina diprionis* virus, and larvae die after 7–9 days of infection. The virus is located only in the intestine and is eliminated with excretions that become infectious. Symptoms of the disease include reducing larval feeding and browning of dead larvae, which remain attached to the substrate with the last or first pair of legs. The disease is spread by insects and birds before the mortality. In the strong epizooties, in short time, almost 80–85% of the larvae are destroyed. This viral disease extinguishes mass multiplications and is passed on to subsequent generations. It is spread across Europe and occurs in the downgrade period (Kollberg et al. 2014).

6.11.3 Spreading Viral Infections

Spread of viruses can be done primarily by physical factors, such as winds, rains, and entomophagous species and birds. It applies against many defoliators: *Lymantria dispar, Malacosoma neustria, Euproctis chrysorrhoea, Leucoma salicis, Thaumetopoea processionea, Drymonia ruficornis, Yponomeuta* sp., *Tortrix viridana, Operophtera brumata, Erannis* sp., etc. Morphologically, it is a gram-positive bacterium, bacillary form. After 20–24 h, they form an oval spore and a parasporal crystal of a protein nature. After lysis of cell walls, growth and crystal pass into the environment.

There are 40 varieties/serotypes of *B. thuringiensis* strains known. Since 1950, it has been used as a biological insecticide to combat defoliation caterpillars. In addition to the successes of using the bacteria, there are inconveniences that limit its use: the product's inability to control most pests in a crop; lack of action on insects feeding in the galleries because it is a stomach poison; and short residual activity, as protein crystals are unprotected against climatic factors, bacterial enzymes (Mäkinen and De 2019), and secondary plant compounds.

Applying treatments to bacterial preparations requires a device that ensures optimal protection of the protected plant and a concentration that ensures maximum efficacy (Maruthi et al. 2019).

6.12 Bacterial Preparations Used in Biological Control

Dipel-8 L is based on *B. thuringiensis* var. kurstaki, with pathogenicity against defoliant caterpillars, with no toxic action on entomofage insects. It is an emulsifiable, tan liquid that forms a homogeneous solution in water that can be applied at doses of 2–3 l/ha. It works by ingestion, endotoxin being activated by alkaline stomach secretions. It is not pathogenic to humans, vertebrate animals, bees, etc. The retention time is 12–14 days, and the maximum mortality of the caterpillars is recorded after 7–8 days after application; at low temperatures and rain, mortality occurs after 18 to 20 days.

6.12.1 Application of Bacterial Bioproducts

Bioproducts act on caterpillars by ingestion, mortality occurs after 1–2 days of feeding with treated leaves. At the age of four, the caterpillars become resistant to the action of bacteria, except those of *Tortrix viridana, Archips xylosteana,* and *Geometridae*. Because it acts by ingestion, it applies when most of the caterpillars are elderly and the buds are open, with individualized leaves (Ghirardo et al. 2012).

6.12.2 Fungi

They were the first organisms identified as producing diseases to insects. The first signals of entomopathogenic a fungus was made in 1835 by *A. Bassi*, mentioning a disease of silkworm caused by fungus *Beauveria bassiana*. More than 500 entomopathogenic species are known to produce natural epizooties and maintain balance in forest biocenoses. Fungi primarily infect the tegument or various body orifices (stigmas) by feeding, being favored by heat and humidity. The death of the hosts occurs 5–14 days after infection. After the death of the hosts, the mycelium develops strongly, fills the general cavity, then goes out through the cuticle, and covers the whole body and fructifies, with different colorings depending on the species; at first, the insect's body is soft, and then becomes rough and mummified. Fungi-induced diseases are called "mycosis" and were first observed because they are generally macroscopic (Serna-Domínguez et al. 2019).

6.12.3 Symptomatology and Mode of Action

After infection, insects become agitated, feed less, and tend to migrate to high ground or to the surface of the soil. Disintegration of infected tissue occurs before or after insect death. Lepidopteran larvae infected with *Entomophthora* species become flaky, have an aqueous, and brittle skin content.

After the insect's death, the fungus continues to develop saprophytes, forming micelle-shaped masses, which turn into a dense sclerot. Micelium comes to the surface of the insects through intersegmental regions, develops, and covers the body and spores (*Entomophthora beauveria*) (Wraight et al. 2018).

6.12.4 The Role of Climatic Factors

Humidity is the parameter that influences the most the evolution of mycosis epizootics, because entomopathogenic fungi are hydrophilic. For spore germination and mycelial growth at the insect's surface, a moisture content of 60–100% is required; thus, *Beauveria bassiana* conidiospores germinate at 92–94% humidity, *Paecilomyces farinosus* at 81%, etc. Generally, optimal development temperatures range from 20 °C to 30 °C. An important role in the occurrence of insect pests in the soil is the composition and structure of the soil. Soils rich in organic and well-structured substances have a high capacity of water retention and favor the appearance of mycoses (González-Mas et al. 2019).

6.12.5 The Protozoa

They are unicellular animal organisms that parasite insects; an important role is played by microsporidies, which are endocellular parasites. The infection is caused by parasitoid entomophagous insects or by ingestion of spores by host larvae; to some lepidopteran species, it is transmitted from one generation to the next through the eggs deposited by the infected females. Protozoa have a broad spectrum of action. In some cases, it can cause infections in vertebrates. They have a slow action on the host (from a few days to a few months) and have great prospects for being used in the regulation of harmful populations. So far, no biopreparations could be produced from them because they have a short biological cycle, ranging from 3 to 9 days. There are over 200 species of protozoa that affect lepidopteran larvae. There are few external signs of protozoa infections, with a decrease in feeding and mobility, sometimes seizures and diarrhea. Microsporids are varied and active, causing the death of insects at different stages, within a few days (Suaste-Dzul et al. 2019).

6.12.6 *Symptomatology and Mode of Action*

In most insects, infection occurs orally, causing infections of the digestive tract, Malpighi tubes, etc. Those that grow in the fat body are transmitted through predatory insects. Spores eliminated by excrement are sources of infection. Microsporides are transmitted by females by infecting deposited eggs, which is manifested in defoliators *Hyphantria cunea, Euproctis chrysorrhoea,* and *Choristoneura murinana.* They may have a narrow spectrum of action, attacking species of one genus (*Nosema lymantriae*), by development only in the digestive system, in adipose tissue, hemolymph, or muscles. Parasitic protozoan infections cause hypertrophy of the cells and nuclei. Microsporids develop intracellularly, without affecting the nucleus, spores, and breaks cell walls, spores being scattered in the hemolymph. Attacked organs are completely destroyed because the digestive tract breaks gently, white spots appear in the muscles, and Malpighi's tubes become brittle (Wang-Peng et al. 2018).

6.13 The Entomophages

They are animal organisms that live on insect species. Entomophages are part of different systematic groups, such as arthropods (mites, insects), reptiles, birds, mammals. For the first time, pheromones were used to study *Lymantria monacha* species by Dyk in 1932.

The young, non-fetus females were enclosed in metal sieve cages and fitted with glue-plated panels, and the panels captured males attracted by pheromones emitted by females, the method being known as the Dyk method. Females had to be raised in the lab and replaced at intervals of several days as they ceased to be attractive to males in 3 to 8 days. This shortcoming was later removed when, with the discovery of modern methods of chemical analysis (chromatography), it was possible to identify the composition of pheromones of hundreds of insect species and synthesize them artificially. The use of pheromone substances in the field requires the existence of devices to intercept and retain the insects, called pheromone traps. There are different types of pheromone traps, varying by insect intercept and capture (Suaste-Dzul et al. 2019). Traps commonly used to capture butterfly species:

6.13.1 Adhesive Panels

Adhesive panels retain individuals attracted to the habitat by fixing on a nonstick glue layer so positioned as to capture as many insects as possible. In practice, a 30/40 cm (30/40 cm) panel is used with one-sided adhesive, the rubber stopper with the bait fastened to the middle of the panel with a pin on the stem of the shaft.

6.13.2 The Tetratap Trap

The Tetratap trap with an adhesive bottom panel and a roof plate has the advantage of providing access holes with different shapes and sizes, and periodically, the adhesive panel, if cluttered with insects caught, can be changed.

6.13.3 Traps Commonly Used to Capture Coleopterans (Especially Bark Beetles) They Are Barrier (for Flying Insects) and Tubular (for Insects that Rest)

6.13.4 The Glass Trap

It is made up of a glass panel or transparent foil under which a wooden or sheet metal gutter with a length of 60 cm, a width of 20–30 cm, and a depth of 15–20 cm is mounted. Permanent gutter will contain half water.

6.13.5 Trap with Wings

Two rectangular panels, assembled in cross, above protected by a pyramid-shaped lid, are made to protect against the wind and sun. In the middle, at the joining of the two panels, there is a niche in which the pheromone substance is attached. At the bottom, the panels are continued with a pyramidal funnel for collecting insects and guiding them into a plastic jar for insect retention, provided with a wire mesh cap to drain the water.

6.13.6 *The Tubular Traps*

The tubular traps attract insects that land on the outer grooves of the tube and penetrate through holes in the grooves on the tube. Insects slip on the inner walls of the tube, which are smooth and reach a collecting jar located at the bottom of the tube. The pheromone envelope is inserted with a wire up to the half of the tube (Pappas et al. 2017).

6.14 Territorial Planning Methods to Increase the Role of Biological Predators in Combating Crop Pests

Many of the agroecosystems are an unfavorable environment for the natural enemies of pests (and especially for predators/parasites of harmful insects) because of the high degree of imbalance resulting from disruptions and anthropogenic interventions. Territorial management is a way of fostering the biological protection of crops, being an ecological approach to stimulate the activity of insect natural enemies. The best way to meet ecological and economic requirements is organic farming systems. The development of organic farming ("organic," "biodynamic") is one of the fundamental orientations in developed countries and especially in the European Union:

- Increasing concerns for enhancing the quality and safety of the food chain.
- Developing an ethic of sustainable use of resources in agriculture.

As an illustration of this interest, a programmatic document was created entitled "European Action Plan for Organic Food and Farming" (Annex COM415/2004). The number of organic farmers is on the rise, according to a recent study by the National Federation of Organic Farming (FNAE). According to FNAE, organic farming attracts more and more investors as a result of a profit of up to 400% that can be obtained from organic crops. Organic farming develops in complementary directions to agriculture in the European Union. One of these directions aims to increase the role of predators/parasitoids in combating crop pests. The main purpose is to create an environmentally friendly infrastructure that provides additional resources for entomophagic adults, namely, food and shelters, against unfavorable conditions.

These resources must be integrated into a territory so that they are favorable in space and time for natural and practical enemies, at the same time to be implemented by agricultural producers. Increasing heterogeneity of vegetation around cultivated areas favors an overall increase in the abundance and diversity of predatory and parasitic organisms. Techniques available to increase the role of parasitic and predatory arthropods through this increase in biodiversity/vegetation heterogeneity are presented below (Murrell 2017).

6.14.1 Intercropping/Strip Cropping

Two or more plant species are grown together on the same site in parallel strips or adjacent plots. Data from the literature is relevant. Of 209 studies on intercalated crop systems reviewed, it was found that 65% of the 130 species of natural enemies studied increased in density in mixed crops. In another study, it was found that parasitoids were more abundant in 72% of the intercalated crop cases studied. In 64% of other studies, the parasitism rate was found to be higher in intercalated crops. Intercalated crops are a way of reducing pests in that the mixture of species physiologically interferes with the ability of the pests to find or react on the host plant, and in that the plant, mixture is a refuge for natural enemies that of the pests. The intercalated culture system practiced on cabbage with white clover bands has been shown to be an effective means of managing the root fly (*Delia radicum*), due to the increased activity of the predatory carabiders. Another study on the impact of territorial settlement on Carabid activity was carried out in maize crops (Pioneer 3573), intercalated with clover strips, and a mixture of perennial plants with flowers to supplement the predators and parasites. The use of these habitats for refuge has led to an increase in the number of predatory carabiders in maize crops during the summer. Other predators such as staphilinides and arachnids have benefited from these cropping systems (Ferry et al. 2007). Also, grass strips have reduced the negative effects of insecticides on carabiders, by providing refuge during application of insecticide treatments.

6.14.2 Undersowing

A second crop is sown in the first crop, at the same time or later, resulting in two crops at the same time. Usually the bands in which the second crop is grown are transformed into a plant mulch (by mowing, herbicide with total herbicides, mulching with suffocating plastics, etc.). The natural fertility of the soil is also improved.

6.14.3 Conservation Headlands

A strip of 6 m outside the plots receives only selective spraying with short-acting pesticides, which reduces drift and deposition in the boundaries of parcels.

6.14.4 Weed Strips Within the Crops

Sowing a few strips close to weeds with flowers or herbs at certain intervals across the cultivated area increases the abundance of aphids predatory insects.

6.14.5 Field Margins and Beetle Banks

This system gains importance on large crop areas. Such a system increases the number of habitats available for predators and parasitoids for wintering, reproduction during spring, and foraging in summer, thus enhancing the potential of biological crop protection. The invasion of weeds from such systems is very low, and sometimes pest breeding situations are created. The edges of the raygras are important nesting places for birds, solitary wasps, bees, and bumble bees. Those containing wildflowers provide pollen and nectar for a number of invertebrates, including bumblebee species. The botanical interest of this system is that they act as important buffer strips between cultural practices and sensitive habitats such as hedges and watercourses. The margins of wild plants attract small mammals that consume owls. Carabide areas are created in the middle of the culture and are areas similar to those on the edges. They are grassy areas (usually transversally in the center of the crop) where predators can win, thus acting as nests of predatory insects that spring easily migrate into culture. Carabide areas are effective in fields of over 20 hectares with a good network of grass edges. Moreover, they are about 0.4 m high and 2 m wide, created in one opposite direction. They do not communicate, creating the effect of the island, which means they will be preferred by predators. Carabide areas are sown with perennial grass species mixed sometimes with perennial legumes. It is mentioned that such perennial grass cultures mixed with perennial legumes are encouraged by the common agricultural policy (Kim et al. 2008).

6.14.6 Insectarium Plants

Insectarium plants can be added to the crop as interstitial strips or as individual plants in the nursery. Insectarium plants may also involve introducing a cover crop between plant rows. A wider spectrum of vegetal resources (nectar, pollen) for natural enemies can be provided by cultivating the territory in insectarium bands of the species of fam. *Apiaceae* (parsley), *Cruciferae* (mustard), *Lamiaceae* (mint), and *Compositae*. Attracting and preserving natural enemies involves understanding their basic needs for food, behavior, and hosting. To support and increase populations, many of the biological protection agents need extra nectar, pollen, and food.

By providing land with diverse vegetative resources, farmers can increase the number and diversity of predators and parasites, while improving their fertility and reducing the cost of pesticide treatments (Razavi et al. 2011).

6.14.7 *Use of Pheromones in Forest Protection Actions*

In the ecosystem is circulating a vast flow of chemically encoded information, which mediates most inter- and intraspecific relationships. These substances are called ecomones or telergones. Ecomones are divided into allomones, which interfere with interspecific communication and pheromones which are responsible for intraspecific communication. Pheromones are produced by exocrine glands, localized on the surface of the body, among the epidermal cells. At bark beetles, the pheromones are found in the intestine epithelium and Malpighi tubules (pheromones are eliminated with excrement). The reception, decoding of these signals is done by the olfactory organs located on the antennas. Pheromone communication has three fundamental features:

- Messages are transmitted by small amounts of chemicals but with a large amount of information (0.33 mg of the marker pheromone of *Atta texana* is likely to leave a pheromone trail surrounding the earth).
- Insects receive unconditionally the information transmitted by the stimuli; males do not detect the source that conveys the information, which may be a female or pheromone bait.
- The information has an intraspecific character and can only be decoded by individuals of the same species (except for some predators on which the pheromone precursors have kairomonal effect).

The sex pheromones mediate relationships between sexual partners before, during, and after mating. Three groups are known: sex attractants, produced by a sex to attract the opposite sex (usually issued by females); aphrodisiacs, male products in order to excite females and accept their copulation; and sexual repellents produced by the male genital apparatus and transferred with the sperm, thus marking fecundated females. Tracing pheromones is used by which social insects (ants, bees) pave the way for a new source of food; alarm pheromones cause different behaviors when enemies appear (for aphids the colonies disperse immediately; for bees and wasps, it triggers mobilization to fight); pheromones of aggregation are characteristics of social insects, but also locusts, used to indicate suitable places for the laying of eggs; in bark beetles, males emit pheromones indicating finding favorable places for feeding and reproduction; necrophorous pheromones to remove dead individuals from ants colonies. In forest protection activities, the most used are pheromones in the categories of sex and aggregation (Serrano et al. 2018).

6.15 Plant Protection Products Management

Effective use of plant protection methods involves selecting the optimal method and minimizing the use of plant protection products and pesticides.

Figure 6.1 illustrates the main basic requirements for selecting optimal biopesticides to be applied in accordance with the principles of good agricultural practice and maintaining the land in good condition for agriculture and the environment. Prognosis of the occurrence of the attack of a particular harmful organism or groups of harmful organisms and the subsequent issuance of warnings for a plant protection intervention is of great importance for the proper conduct of the combat activities in time and space. It is good to know how harmful organisms exist in an area for agricultural crops in vegetation and potential levels of attack to know which spectrum of control means to use, which machine system should be used to apply them, and optimum moments intervention to make the damage as small as possible. This way the economic effects will be maximum and the ecological effects will be minimal (Shrestha and Reddy 2019).

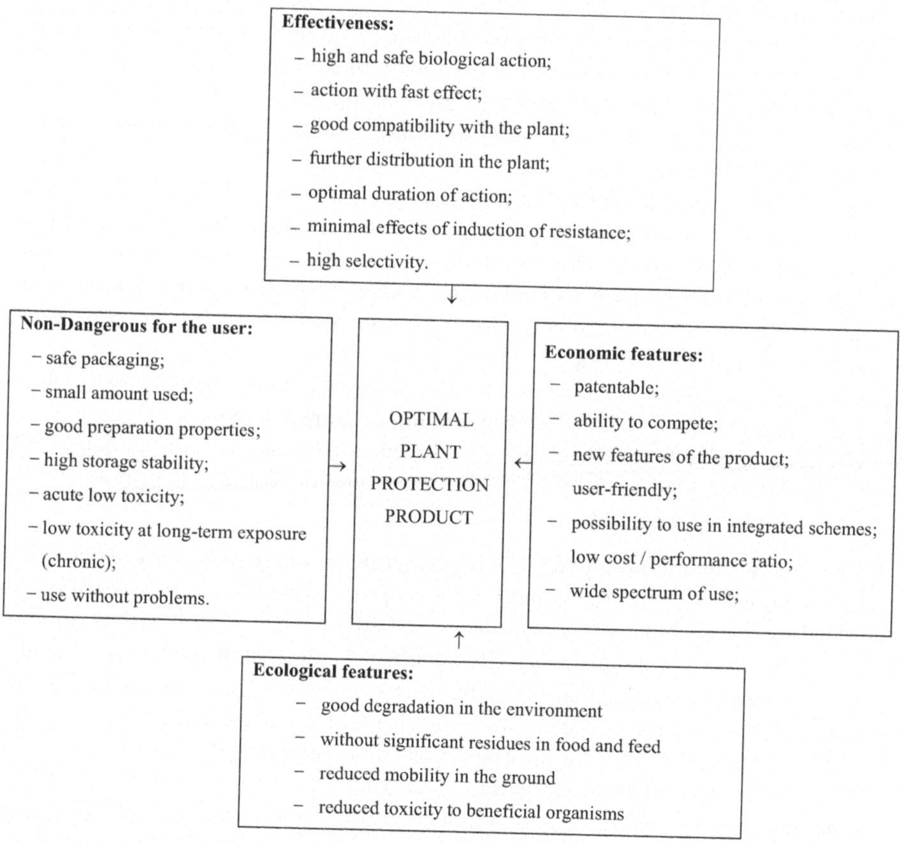

Fig. 6.1 Main requirements for selecting optimal biopesticides

6.16 Modern Methods in the QSAR/QSPR Study of Biopesticides

In the last period of time, structural indexes using QSPR/QSAR (quantitative structure-property/activity relationship) studies are increasingly being calculated for steric (geometric) and/or electrostatic (partial load) considerations as compared to past topological considerations. Semiempirical and quantum structural calculations are performed by programs such as Hondo95, Gaussian94, Gamess, Icon08, Tx90, Polyrates, Unichem/Dgauss, Allinger's MM3, Mopac93, Mozyme, HyperChem. Regular class regression, linear regression, multiple linear regression, nonlinear regression, or, for large databases, expert systems or neural networks are used in the property/structural index regression analysis. As a preliminary method of analysis, some authors align the set of molecules (Wang et al. 2017).

Furthermore, CoMFA38 introduces a six-step algorithm for QSAR analysis:

- Builds the set of molecules with known activity and generates the 3D structure of the molecules (possibly with one of the programs: Mopac, Sybyl, HyperChem, Alchemy2000, MolConn).
- Chooses a method of overlapping (overlapping of fragments chosen from molecules or overlapping pharmaco-scopes) and overlay virtual spatial coordinates.
- Builds a network of points surrounding the overlapping molecules at (B) in standard (grid38) or altered (curvilinear) form and chooses a sample atom for interaction with network points.
- Employs an empirical method (Hint51), a specific model (pharmacophore overlaps), classical potential energy (Lennard–Jones, Coulomb), the potential of hydrogen bonds, molecules generated by molecular orbitals or any other field defined by model 49 user, and calculates values interaction of the field induced in the network (C) by the interaction field chosen with a sample atom (C) placed at the points of the network.
- Uses the calculated interaction values (D) between network points and the sample atom and performs the QSAR prediction of known activity.
- Uses the obtained QSAR parameters (E) and performs activity prediction for molecules that lend themselves to the same type of overlay with those of the school set (A) (Nendza et al. 1991).

The CoMFA method is a good tool in predicting a variety of biological activities such as cytotoxicity, inhibition, and formation properties. The method is also used in modeling compounds with a pharmaceutical effect and analyzing HIV60 inhibitors. An important problem in QSAR modeling is the search in the biologically active molecules of the active substructures that give most of the measured biological response. The search for molecular invariants is particularly useful in the case study. The WHIM (weighted holistic invariant molecular) method calculates a set of statistical indices derived from steric and electrostatic properties of molecules. The original method was modified and assigned the MS-WIHM name (molecular surface—weighted holistic invariant molecular) and was successfully applied in molecular surface analysis65. MS-WHIM is a collection of 36 statistical indices derived from steric and electrostatic properties and oriented toward molecular surface parameterization. A new model structure proposed property based on the molecular topology obtained from the structural formula and the molecular topography obtained from the quantum calculations. For molecular modeling, a new class of indices is used: fragmental property index family (FPIF) containing 61,440 index members calculated on the basis of:

- Eight topological fragmentation methods called MI, MA, SzDi, SzDe, CfDi, CfDe, CjDi, and CjDe.
- Four physical interaction models called RG, DG, RT, and DT.
- Eight property descriptors p depending on the distance d: p, d, 1/p, 1/d, p; d, p/d, p/d2, and p2/d2.
- Five patterns of overlapping fragmentary interactions: S, P, A, G, and H.
- Four types of summation indices on matrices with resulting fragmentary properties: P_, P2, E_, and E2.
- Index scaling operators: id, 1/ln.
- Four default properties C, M, E, and Q and generate the entire set of 61,440 indices for a given molecule based on the topological (atoms and links) and topographic structure (spatial coordinates and partial loads) (Liu et al. 2011).

Not all of the indices obtained are generally distinguished. Degenerations arise from the degeneration of the values of the atomic properties and the descriptors chosen. Following the elimination of identities across the set, there are approximately 15,000 distinct indices. Outstanding results are obtained when recognizing ownership patterns. The construction of the indices allows, following the selection made in correlation, to identify the structural cause of the measured or calculated macroscopic property. An example is QSAR and QSPR analysis of a set of 17 3-(phthalimidoalkyl)-pyrazolin-5-one substitution compounds with inhibitory activity on *Lepidium sativum* L. (Creson) (Xie et al. 2018).

6.17 Conclusions and Recommendations

Biopesticide soil damage is negligible compared to chemical ones because biopesticides do not contain heavy metals and are entirely composed of biodegradable substances. The use of biological pesticides allows to preserve the soil's good state and to reduce its pollution indices, with good consequences not only for human health but also for land utilization and the economic efficiency.

It can be seen that preparations, whether chemical or biological, act by binding to certain target proteins, which disrupt the transmission of signals through the nervous system, mitochondrial respiration, digestion in the middle intestine, and insect growth and development. To prevent the occurrence of resistance phenomena, it is of great importance to rationally use products acting on the same target protein. For example, it is appreciated that most biopesticides act on the acetylcholinesterase target. Thus, with all their diversity, using preparations with the same target protein will increase the risk of cross-resistance. To avoid the phenomenon of resistance or cross-resistance, the use of products of the same chemical class over long periods of time should be avoided because they act on the same target protein.

The rotation of preparations of different chemical groups, with different modes of action, ensures effective control. By discovering new classes of natural products with insecticidal action, with different mechanisms of binding to the target protein and their modes of action (see spinosad), the cross-resistance phenomenon is eliminated, and their use is maintained over long periods of time. Regarding insect resistance to different insecticides, it develops through four mechanisms:

- Changing the behavior of insects by avoiding the consumption of treated feed or contact with insecticides by the phenomenon of repellency.
- Changing metabolism, by quantitative increase and by the effectiveness of enzymes that break down insecticides. This mechanism of metabolic resistance can give rise to cross-resistance.
- Lowering the penetration rate of insecticides, during which they break down metabolic enzymes.
- Through a major resistance mechanism, where the target protein changes its shape and reduces the action of the insecticide. This type of reaction produces cross-resistance to all products in the same chemical class and to products acting on the same target sites.

From the modes of action of insecticides or biological preparations, the following conclusions can be drawn:

- Integrated pest management must be managed by plant protection specialists with appropriate training.
- In control schemes, repeated use of products of the same chemical class or even other chemical classes, but acting on the same target protein, will be avoided in order to avoid cross-resistance.
- Observance of the prescribed use doses removes the pressure of product resistance phenomena.
- Research provides new natural products, which eliminates cross-resistance and low toxicity to the environment.
- To achieve the desirability of managing sustainable ecosystems, research efforts must be directed toward the use and discovery of new metabolites, insecticidal, and with a special way of action by binding to the target protein.

The biological alternative is relatively difficult and consists of biological conversion to the production of organic chemicals and enzymes. These microbial transformations can be accomplished via growth cells, spores, or dried cells. For medical applications, a range of synthetic drugs can be replaced with herbal and algae pharmaceuticals. As we highlighted above, the 1940s were the beginning of the antibiotic pharmaceutical industry that immediately reached the maximum due to the high demand caused by war and poor living conditions. Antibiotic applications have not been limited to treating infectious diseases. Thus, half of the quantity produced was used in agriculture and zootechnics.

References

Akotsen-Mensah C, Kaser JM, Leskey TC, Nielsen AL (2018) Halyomorpha halys (Hemiptera: Pentatomidae) responses to traps baited with pheromones in peach and apple orchards. J Econ Entomol 111(5):2153–2162

Anil K, Podile AR (2012) HarpinPss-mediated enhancement in growth and biological control of late leaf spot in groundnut by a chlorothalonil-tolerant *Bacillus thuringiensis* SFC24. Microbiol Res 167(4):194–198

Armada E, Roldán A, Azcon R (2014) Differential activity of autochthonous bacteria in controlling drought stress in native Lavandula and *Salvia plants* species under drought conditions in natural arid soil. Microb Ecol 67(2):410–420

Baum JA, Chu CR, Rupar M, Brown GR, Donovan WP, Huesing JE, Ilagan O, Malvar TM, Pleau M, Walters M, Vaughn T (2004) Binary toxins from *Bacillus thuringiensis* active against the western corn rootworm, *Diabrotica virgifera* virgifera LeConte. Appl Environ Microbiol 70(8):4889–4898

Bell MJ, Sedda L, Gonzalez MA, de Souza CF, Dilger E, Brazil RP, Courtenay O, Hamilton JGC (2018) Attraction of *Lutzomyia longipalpis* to synthetic sex-aggregation pheromone: effect of release rate and proximity of adjacent pheromone sources. PLoS Negl Trop Dis 12(12):e0007007

Bellefeuille Y, Fournier M, Lucas E (2019) Evaluation of two potential biological control agents against the foxglove aphid at low temperatures. J Insect Sci 19(1). https://doi.org/10.1093/jisesa/iey130

Bernat P, Nykiel-Szymańska J, Gajewska E, Różalska S, Stolarek P, Dackowa J, Słaba M (2018) *Trichoderma harzianum* diminished oxidative stress caused by 2,4-dichlorophenoxyacetic acid (2,4-D) in wheat, with insights from lipidomics. J Plant Physiol 229:158–163

Bharadwaj A, Hayes LE, Stafford KC 3rd (2015) Effectiveness of garlic for the control of *Ixodes scapularis* (Acari: Ixodidae) on residential properties in Western Connecticut. J Med Entomol 52(4):722–725

Bhat RA, Beigh BA, Mir SA, Dar SA, Dervash MA, Rashid A, Lone R (2018) Biopesticide techniques to remediate pesticides in polluted ecosystems. In: Wani KA, Mamta (eds) Handbook of research on the adverse effects of pesticide pollution in aquatic ecosystems. IGI Global, Hershey, pp 387–407

Bianchi FJ, Ives AR, Schellhorn NA (2013) Interactions between conventional and organic farming for biocontrol services across the landscape. Ecol Appl 23(7):1531–1543

Binyameen M, Ejaz M, Shad SA, Razaq M, Shah RM, Schlyter F (2018) Eugenol, a plant volatile, synergizes the effect of the thrips attractant, ethyl iso-nicotinate. Environ Entomol 47(6):1560–1564

Brezolin AN, Martinazzo J, Muenchen DK, de Cezaro AM, Rigo AA, Steffens C, Steffens J, Blassioli-Moraes MC, Borges M (2018) Tools for detecting insect semiochemicals: a review. Anal Bioanal Chem 410(17):4091–4108

Calvo-Garrido C, Roudet J, Aveline N, Davidou L, Dupin S, Fermaud M (2019) Microbial antagonism toward botrytis bunch rot of grapes in multiple field tests using one *Bacillus ginsengihumi* strain and formulated biological control products. Front Plant Sci 10:105

Cao D, Stewart CN Jr, Zheng M, Guan Z, Tang ZX, Wei W, Ma KP (2014) Stable *Bacillus thuringiensis* transgene introgression from *Brassica napus* to wild mustard *B. juncea*. Plant Sci 227:45–50

Chen X, Zhang X, Jia A, Xu G, Hu H, Hu X, Hu L (2016) Jasmonate mediates salt-induced nicotine biosynthesis in tobacco (*Nicotiana tabacum* L.). Plant Divers 38(2):118–123

Clifton EH, Gardescu S, Behle RW, Hajek AE (2019) Asian longhorned beetle bioassays to evaluate formulation and dose-response effects of *Metarhizium microsclerotia*. J Invertebr Pathol 163:64–66

Cuartas PE, Villamizar LF, Barrera GP, Ruiz JC, Campos JC, León-Martínez G, Gómez-Valderrama J (2019) Novel biopesticide based on Erinnyis ello betabaculovirus: characterization and preliminary field evaluation to control *Erinnyis ello* in rubber plantations. Pest Manag Sci 75(5):1391–1399

Dawson AH, Buckley NA (2011) Toxicologists in public health-following the path of Louis Roche (based on the Louis Roche lecture "an accidental toxicologist in public health", Bordeaux, 2010). Clin Toxicol (Phila) 49(2):94–101

Degenkolb T, Vilcinskas A (2016) Metabolites from nematophagous fungi and nematicidal natural products from fungi as alternatives for biological control. Part II: metabolites from nematophagous basidiomycetes and non-nematophagous fungi. Appl Microbiol Biotechnol 100(9):3813–3824

Dos Santos DS, Rosa ME, Zanatta AP, Oliveira RS, de Almeida CGM, Leal AP, Sanz M, Fernandes KA, de Souza VQ, de Assis DR, Pinto E, Belo CAD (2019) Neurotoxic effects of sublethal concentrations of cyanobacterial extract containing anatoxin-a(s) on *Nauphoeta cinerea* cockroaches. Ecotoxicol Environ Saf 171:138–145

Ferry A, Dugravot S, Delattre T, Christides JP, Auger J, Bagnères AG, Poinsot D, Cortesero AM (2007) Identification of a widespread monomolecular odor differentially attractive to several *Delia radicum* ground-dwelling predators in the field. J Chem Ecol 33(11):2064–2077

Gay H (2012) Before and after silent spring: from chemical pesticides to biological control and integrated pest management—Britain, 1945-1980. Ambix 59(2):88–108

Gerssen A, Mulder PP, de Boer J (2011) Screening of lipophilic marine toxins in shellfish and algae: development of a library using liquid chromatography coupled to orbitrap mass spectrometry. Anal Chim Acta 685(2):176–185

Ghirardo A, Heller W, Fladung M, Schnitzler JP, Schroeder H (2012) Function of defensive volatiles in pedunculate oak (*Quercus robur*) is tricked by the moth *Tortrix viridana*. Plant Cell Environ 35(12):2192–2207

Ghosh A, Chowdhury N, Chandra G (2012) Plant extracts as potential mosquito larvicides. Indian J Med Res 135(5):581–598

Gnepe JR, Tyagi RD, Brar SK, Valéro JR, Surampalli RY (2014) Corrosion and stability study of *Bacillus thuringiensis* var. kurstaki starch industry wastewater-derived biopesticide formulation. J Environ Sci Health B 49(11):889–896

González-Mas N, Ortega-García L, Garrido-Jurado I, Dembilio O, Jaques JA, Quesada-Moraga E (2019) Which came first: the disease or the pest? Is there a host mediated spread of *Beauveria bassiana* (Ascomycota: Hypocreales) by invasive palm pests? J Invertebr Pathol 162:26–42

Gross J, Eben A, Müller I, Wensing A (2010) A well protected intruder: the effective antimicrobial defense of the invasive ladybird *Harmonia axyridis*. J Chem Ecol 36(11):1180–1188

Guo Z, Sun D, Kang S, Zhou J, Gong L, Qin J, Guo L, Zhu L, Bai Y, Luo L, Zhang Y (2019) CRISPR/Cas9-mediated knockout of both the PxABCC2 and PxABCC3 genes confers high-level resistance to *Bacillus thuringiensis* Cry1Ac toxin in the diamondback moth, *Plutella xylostella* (L.). Insect Biochem Mol Biol 107:31–38

Hassemer MJ, Borges M, Withall DM, Pickett JA, Laumann RA, Birkett MA, Blassioli-Moraes MC (2019) Development of pull and push-pull systems for management of lesser mealworm, *Alphitobius diaperinus*, in poultry houses using alarm and aggregation pheromones. Pest Manag Sci 75(4):1107–1114

Heimbach F, Russ A, Schimmer M, Born K (2016) Large-scale monitoring of effects of clothianidin dressed oilseed rape seeds on pollinating insects in northern Germany: implementation of the monitoring project and its representativeness. Ecotoxicology 25(9):1630–1647

Hopper JV, McCue KF, Pratt PD, Duchesne P, Grosholz ED, Hufbauer RA (2019) Into the weeds: matching importation history to genetic consequences and pathways in two widely used biological control agents. Evol Appl 12(4):773–790

Huang YS, Higgs S, Vanlandingham DL (2017) Biological control strategies for mosquito vectors of Arboviruses. Insects 8(1):E21. https://doi.org/10.3390/insects8010021. Review

Hyakumachi M, Nishimura M, Arakawa T, Asano S, Yoshida S, Tsushima S, Takahashi H (2013) *Bacillus thuringiensis* suppresses bacterial wilt disease caused by *Ralstonia solanacearum* with systemic induction of defense-related gene expression in tomato. Microbes Environ 28(1):128–134

Jin L, Wang J, Guan F, Zhang J, Yu S, Liu S, Xue Y, Li L, Wu S, Wang X, Yang Y, Abdelgaffar H, Jurat-Fuentes JL, Tabashnik BE, Wu Y (2018) Dominant point mutation in a tetraspanin gene associated with field-evolved resistance of cotton bollworm to transgenic Bt cotton. Proc Natl Acad Sci U S A 115(46):11760–11765

Kathage J, Castañera P, Alonso-Prados JL, Gómez-Barbero M, Rodríguez-Cerezo E (2018) The impact of restrictions on neonicotinoid and fipronil insecticides on pest management in maize, oilseed rape and sunflower in eight European Union regions. Pest Manag Sci 74(1):88–99

Kim HK, Lee T, Yun SH (2008) A putative pheromone signaling pathway is dispensable for self-fertility in the homothallic ascomycete *Gibberella zeae*. Fungal Genet Biol 45(8):1188–1196

Köblös G, François MC, Monsempes C, Montagné N, Fónagy A, Jacquin-Joly E (2018) Molecular characterization of MbraOR16, a candidate sex pheromone receptor in *Mamestra brassicae* (Lepidoptera: Noctuidae). J Insect Sci 18(5)

Kollberg I, Bylund H, Huitu O, Björkman C (2014) Regulation of forest defoliating insects through small mammal predation: reconsidering the mechanisms. Oecologia 176(4):975–983

Kraehmer H, Laber B, Rosinger C, Schulz A (2014) Herbicides as weed control agents: state of the art: I. weed control research and safener technology: the path to modern agriculture. Plant Physiol 166(3):1119–1131

Latré J, Dewitte K, Derycke V, De Roo B, Haesaert G (2015) Integrated weed control in maize. Commun Agric Appl Biol Sci 80(2):241–249

Lester PJ, Bosch PJ, Gruber MA, Kapp EA, Peng L, Brenton-Rule EC, Buchanan J, Stanislawek WL, Archer M, Corley JC, Masciocchi M, Van Oystaeyen A, Wenseleers T (2015) No evidence of enemy release in pathogen and microbial communities of common wasps (*Vespula vulgaris*) in their native and introduced range. PLoS One 10(3):e0121358

Li W, Evans JD, Huang Q, Rodríguez-García C, Liu J, Hamilton M, Grozinger CM, Webster TC, Su S, Chen YP (2016) Silencing the honey bee (*Apis mellifera*) naked cuticle gene (nkd) improves host immune function and reduces nosema ceranae infections. Appl Environ Microbiol 82(22):6779–6787

Liu XH, Pan L, Ma Y, Weng JQ, Tan CX, Li YH, Shi YX, Li BJ, Li ZM, Zhang YG (2011) Design, synthesis, biological activities, and 3D-QSAR of new N,N′-diacylhydrazines containing 2-(2,4-dichlorophenoxy)propane moiety. Chem Biol Drug Des 78(4):689–694

Liu K, McInroy JA, Hu CH, Kloepper JW (2018) Mixtures of plant-growth-promoting rhizobacteria enhance biological control of multiple plant diseases and plant-growth promotion in the presence of pathogens. Plant Dis 102(1):67–72

Mahmoud SB, Ramos JE, Shatters RG Jr, Hall DG, Lapointe SL, Niedz RP, Rougé P, Cave RD, Borovsky D (2017) Expression of *Bacillus thuringiensis* cytolytic toxin (Cyt2Ca1) in citrus roots to control *Diaprepes abbreviatus* larvae. Pestic Biochem Physiol 136:1–11

Mäkinen K, De S (2019) The significance of methionine cycle enzymes in plant virus infections. Curr Opin Plant Biol 50:67–75

Marczewska P, Miszczyk M, Płonka M, Kronenbach-Dylong D, Szeremeta D, Sajewicz M (2019) Application of different chromatographic techniques and chemometric analysis in authenticity testing of plant protection products containing azoxystrobin as an active substance. J Environ Sci Health B 3:1–8

Maruthi MN, Whitfield EC, Otti G, Tumwegamire S, Kanju E, Legg JP, Mkamilo G, Kawuki R, Benesi I, Zacarias A, Munga T, Mwatuni F, Mbugua E (2019) A method for generating virus-free cassava plants to combat viral disease epidemics in Africa. Physiol Mol Plant Pathol 105:77–87

Matumba L, Sulyok M, Njoroge SM, Njumbe Ediage E, Van Poucke C, De Saeger S, Krska R (2015) Uncommon occurrence ratios of aflatoxin B1, B 2, G 1, and G 2 in maize and groundnuts from Malawi. Mycotoxin Res 31(1):57–62

Medina-Romero YM, Roque-Flores G, Macías-Rubalcava ML (2017) Volatile organic compounds from endophytic fungi as innovative postharvest control of *Fusarium oxysporum* in cherry tomato fruits. Appl Microbiol Biotechnol 101(22):8209–8222

Murrell EG (2017) Can agricultural practices that mitigate or improve crop resilience to climate change also manage crop pests? Curr Opin Insect Sci 23:81–88

Mushtaq N, Bhat RA, Dervash MA, Qadri H, Dar GH (2018) Biopesticides: the key component to remediate pesticide contamination in an ecosystem. In: Environmental contamination and remediation. Cambridge Scholars, Cambridge, pp 152–178
Nendza M, Dittrich B, Wenzel A, Klein W (1991) Predictive QSAR models for estimating ecotoxic hazard of plant-protecting agents: target and non-target toxicity. Sci Total Environ 109-110:527–535
Niogret J, Epsky ND (2018) Attraction of *Ceratitis capitata* (Diptera: Tephritidae) sterile males to essential oils: the importance of linalool. Environ Entomol 47(5):1287–1292
Oliveira-Hofman C, Kaplan F, Stevens G, Lewis E, Wu S, Alborn HT, Perret-Gentil A, Shapiro-Ilan DI (2019) Pheromone extracts act as boosters for entomopathogenic nematodes efficacy. J Invertebr Pathol 164:38–42
Pappas ML, Broekgaarden C, Broufas GD, Kant MR, Messelink GJ, Steppuhn A, Wäckers F, van Dam NM (2017) Induced plant defences in biological control of arthropod pests: a double-edged sword. Pest Manag Sci 73(9):1780–1788
Pretali L, Bernardo L, Butterfield TS, Trevisan M, Lucini L (2016) Botanical and biological pesticides elicit a similar induced systemic response in tomato (*Solanum lycopersicum*) secondary metabolism. Phytochemistry 130:56–63
Razavi SM, Zarrini G, Rad FG (2011) Isoarnottinin 4′-glucoside, a glycosylated coumarin from *Prangos uloptera*, with biological activity. Bioorg Khim 7(2):269–272
Reil JB, Doorenweerd C, San Jose M, Sim SB, Geib SM, Rubinoff D (2018) Transpacific coalescent pathways of coconut rhinoceros beetle biotypes: resistance to biological control catalyses resurgence of an old pest. Mol Ecol 27(22):4459–4474
Scheckhuber CQ (2019) Studying the mechanisms and targets of glycation and advanced glycation end-products in simple eukaryotic model systems. Int J Biol Macromol 127:85–94
Serna-Domínguez MG, Andrade-Michel GY, Rosas-Valdez R, Castro-Félix P, Arredondo-Bernal HC, Gallou A (2019) High genetic diversity of the entomopathogenic fungus *Beauveria bassiana* in Colima, Mexico. J Invertebr Pathol 163:67–74
Serrano JM, Collignon RM, Zou Y, Millar JG (2018) Identification of sex pheromones and sex pheromone mimics for two north American click beetle species (Coleoptera: Elateridae) in the genus Cardiophorus esch. J Chem Ecol 44(4):327–338
Shi YM, Bode HB (2018) Chemical language and warfare of bacterial natural products in bacteria-nematode-insect interactions. Nat Prod Rep 35(4):309–335
Shrestha G, Reddy GVP (2019) Field efficacy of insect pathogen, botanical, and jasmonic acid for the management of wheat midge *Sitodiplosis mosellana* and the impact on adult parasitoid *Macroglenes penetrans* populations in spring wheat. Insect Sci 26(3):523–535
Suaste-Dzul AP, Rodríguez-Vélez JM, Rodríguez-Vélez B, Arredondo-Bernal HC, Gallou A (2019) Non-destructive DNA extraction methods for entomophagous insects with emphasis on biological control. Genome 62(4):287–293
Sui G, Song X, Zhang B, Wang Y, Liu R, Guo H, Wang J, Chen Q, Yang X, Hao H, Zhou W (2019) Design, synthesis and biological evaluation of novel neuchromenin analogues as potential antifungal agents. Eur J Med Chem 173:228–239
Taghdisi MH, Amiri Besheli B, Dehdari T, Khalili F (2019) Knowledge and practices of safe use of pesticides among a group of farmers in northern Iran. Int J Occup Environ Med 10(2):66–72
Tan K, Yang M, Wang Z, Radloff SE, Pirk CW (2012) The pheromones of laying workers in two honeybee sister species: *Apis cerana* and *Apis mellifera*. J Comp Physiol A Neuroethol Sens Neural Behav Physiol 198(4):319–323
Thomas MR (2003) Pesticide usage in some vegetable crops in Great Britain: real on-farm applications. Pest Manag Sci 59(5):591–596
Togni PHB, Venzon M, Lagôa ACG, Sujii ER (2019) Brazilian legislation leaning towards fast registration of biological control agents to benefit organic agriculture. Neotrop Entomol 48(2):175–185
Wang BL, Zhang LY, Liu XH, Ma Y, Zhang Y, Li ZM, Zhang X (2017) Synthesis, biological activities and SAR studies of new 3-substitutedphenyl-4-substitutedbenzylideneamino-1,2,4-triazole

Mannich bases and bis-Mannich bases as ketol-acid reductoisomerase inhibitors. Bioorg Med Chem Lett 27(24):5457–5462

Wang X, Cai M, Zhou Y (2018) Biological influence of cry1Ab gene insertion on the endophytic bacteria community in transgenic rice. Turk J Biol 42(3):231–239

Wang-Peng S, Zheng X, Jia WT, Li AM, Camara I, Chen HX, Tan SQ, Liu YQ, Ji R (2018) Horizontal transmission of *Paranosema locustae* (Microsporidia) in grasshopper populations via predatory natural enemies. Pest Manag Sci 74(11):2589–2593

Wraight SP, Galaini-Wraight S, Castrillo LA, Griggs MH, Keith LM, Matsumoto TK (2018) Collection, isolation, in vitro culture, and laboratory transmission of *Hirsutella eleutheratorum* (Hypocreales: Ophiocordycipitaceae) from coffee berry borer on Hawai'i Island. J Invertebr Pathol 157:53–66

Xiao Y, Liu K, Zhang D, Gong L, He F, Soberón M, Bravo A, Tabashnik BE, Wu K (2016) Resistance to *Bacillus thuringiensis* mediated by an ABC transporter mutation increases susceptibility to toxins from other bacteria in an invasive insect. PLoS Pathog 12(2):e1005450

Xie Y, Peng W, Ding F, Liu SJ, Ma HJ, Liu CL (2018) Quantitative structure-activity relationship (QSAR) directed the discovery of 3-(pyridin-2-yl)benzenesulfonamide derivatives as novel herbicidal agents. Pest Manag Sci 74(1):189–199

Yamada T, Hamada M, Floreancig P, Nakabachi A (2019) Diaphorin, a polyketide synthesized by an intracellular symbiont of the Asian citrus psyllid, is potentially harmful for biological control agents. PLoS One 14(5):e0216319

Yang CY, Kim J, Ahn SJ, Kim DH, Cho MR (2014) Identification of the female-produced sex pheromone of the plant bug *Apolygus spinolae*. J Chem Ecol 40(3):244–249

Yotsu-Yamashita M, Toennes SW, Mebs D (2017) Tetrodotoxin in Asian newts (Salamandridae). Toxicon 134:14–17

Youssef T, Saenger P (1998) Photosynthetic gas exchange and accumulation of phytotoxins in mangrove seedlings in response to soil physico-chemical characteristics associated with waterlogging. Tree Physiol 18(5):317–324

Zhang Y, Li T, Liu Y, Li X, Zhang C, Feng Z, Peng X, Li Z, Qin S, Xing K (2019) Volatile organic compounds produced by *Pseudomonas chlororaphis* subsp. aureofaciens SPS-41 as biological fumigants to control *Ceratocystis fimbriata* in postharvest sweet potatoes. J Agric Food Chem 67(13):3702–3710

Chapter 7
Inoculum Addition in the Presence of Plant Rhizosphere for Petroleum-Polluted Soil Remediation

Farhana Maqbool, Zulfiqar A. Bhatti, Muhammad Faisal Siddiqui, Ibrar Khan, Yang Guo Zhao, Muhammad Sajid, Umm-e-Kalsoom, Qaiser Mehmood, and Faiza Nawaz

7.1 Introduction

Composting of contaminated soil is one of the feasible methods to clean up petroleum hydrocarbon in the biopiles by the addition of nutrients and amendments, which enhances the hydrocarbon degradation and improve soil quality (Van Gestel et al. 2003). However, the contents of the petroleum hydrocarbons and their decomposed matter present in the soil after composting are often above environmental standards (Zhang et al. 2008). Bioremediation is the conversion of complex organic contaminant into simpler inorganic compounds including carbon dioxide, water, and cell biomass by biological agents like microorganisms (Das and Chandran 2011). Due to the presence of pollutant in the soil, microorganism including bacteria, fungi, and yeast that prefer the chemicals as a source of food and energy can be flourished. In the soil, mineralization rate contaminant depends on the microbial activity and abundance in the soils. Pollutant degradation is extremely dependent on the presence of electron provider or acceptors, presence of co-metabolites, absence

F. Maqbool (✉) · M. F. Siddiqui · F. Nawaz
Department of Microbiology, Hazara University, Mansehra, KPK, Pakistan
e-mail: drfarhana@hu.edu.pk

Z. A. Bhatti · Q. Mehmood
Department of Environmental Science, COMSATS University Islamabad, Abbottabad Campus, Abbottabad, KPK, Pakistan

I. Khan
Department of Microbiology, Abbottabad University of Science and Technology, Abbottabad, KP, Pakistan

Y. G. Zhao
College of Environmental Science and Engineering, Ocean University of China, Qingdao, China

M. Sajid · Umm-e-Kalsoom
Department of Biochemistry, Hazara University, Mansehra, KPK, Pakistan

K. R. Hakeem et al. (eds.), *Bioremediation and Biotechnology*,
https://doi.org/10.1007/978-3-030-35691-0_7

of inorganic nutrients, plant vitamins and hormones, as well as microbial competition (ITRC 2009).

Bioremediation methods, which enhance the biodegradation rates, include natural attenuation without any interference, use of seed culture of hydrocarbon degraders, biostimulation with nutrient addition for improved microbial growth, and rhizospheric biodegradation (Odokuma and Dickson 2003; Bhatti et al. 2017). Microbes spread in the rhizosphere, and their number increases due to the availability of nutrients which are released by plants including C, N, S, and P. Few studies used bioaugmentation in the presence of plant to remediate the petroleum-polluted soil (Kirk et al. 2005; Baek et al. 2004).

Through bioaugmentation desired microbes and their enzymes increase, but this process does not work always due to the inability of culture to adjust in new environment and to compete with native microbes (Gerardi 2016; Ramos et al. 2010). Microbial entrapment in solid media is used for the protection and survival of inoculated culture, and the most effective microbial immobilization was cell entrapment into a porous matter (Partovinia and Rasekh 2018). The immobilized cells have proved stability under different environmental conditions for petroleum-contaminated soil (Dongmei et al. 2011). The present chapter reveals different strategies of complete biodegradation of highly petroleum-polluted soil, limitation, and outcomes of these techniques.

7.2 Different Strategies of Bioremediation

7.2.1 Composting of Highly Petroleum-Polluted Soil

Composting is the addition of biomaterial and biogenic bulking agents used for aeration and to adjust water, nutrient, and pH of the contaminated soil (Kästner and Miltner 2016). Composting of highly petroleum-contaminated soil is a very encouraging technique, and it can be performed at the first step before phytoremediation to lower down the concentration to acceptable level for plant growth. Maize straw, pine wood chips, and soybean cake were used as bulking agent in composting to lower down the total petroleum hydrocarbon (TPH) concentration from 17,900 to 3700 mg kg^{-1} (Wang et al. 2011). This composted soil was further used to lower down the concentration up to 673 mg kg^{-1} using *Sesbania cannabina*, a coastal halophyte (Table 7.1 shows all results of the project) (Farhana et al. 2012). Jørgensen et al. (2000) used bark chips as the bulking agent with inoculum to compost the lubricating oil-contaminated soil; 70% removal rate was observed with organic matter addition and no effect of inoculum addition. Another study used cow bed and potato peelings for high asphaltenic fuel oil composting, and the result showed that with autochthonous bacteria, fragmentation into easily degradable smaller structure took place (Martin-Gil et al. 2008).

Table 7.1 Different bioremediation approaches for the Shengli Oil Field petroleum-polluted soil and their maximum removal percentages

S. no.	Technique	Maximum bioremediation duration (days)	Removal percentage of petroleum hydrocarbon (%)
1	Bioaugmentation with free culture inoculation	70	39
2	Bioaugmentation with immobilized culture inoculation	70	47
3	Addition of compost in contaminated soil	90	45
4	Phytoremediation with seepweed	90	42
5	Phytoremediation with *S. cannabina*	90	73
6	*S. cannabina* in the presence of free culture inoculation	120	73
7	*S. cannabina* in the presence of immobilized culture inoculation	120	68

[1, 2]Wang et al. (2012), [3, 4]Wang et al. (2011), [5,6,7]Farhana et al. (2012)

7.2.2 *Enhanced Degradation of Pollutants with Bioaugmentation*

Biodegradation can be less effective due to insufficient indigenous oil degrader population or their incapability of degrading the broad range of toxic fractions present in the environment (Hussein 2006). In this situation, bioaugmentation was found to be a promising low-cost technique in which effective bacterial culture or microbial community capable of degrading hydrocarbons is added to the contaminated soil (Zawierucha and Malina 2011). Bioaugmentation can be achieved in three ways, first, autochthonous bioaugmentation in which seed cultures are obtained from the original contaminated site, re-cultured in the lab, and reused in the same soil which needs treatment (Ueno et al. 2007). Secondly seed culture is taken from one site and used in different contaminated site, and the third method is the use of genetically modified organism with degradative potential (Vogel and Walter 2001). Among these methods; autochthonous bioaugmentation is of great interest because it is found to be best adopted with the contaminated environment.

Many researchers used bioaugmentation alone for crude oil degradation (Bento et al. 2005; Yu et al. 2005), but few researches are found in the field of rhizosphere bioaugmentation (Peng et al. 2009; Ma et al. 2010; Cai et al. 2010; Tang et al. 2010; Thangarajan et al. 2011) (Table 7.2). The use of higher plants to enhance removal of contaminants from soil although low cost than other remediation techniques often does not result in complete removal of contaminants (Banks et al. 2003); therefore some researchers tried to use inoculum addition in the presence of plant. Bioaugmentation required successful and safe introduction of microbial cells into uncontained environments (Trevors et al. 1993).

Table 7.2 Researches on bioaugmentation in bioremediation of petroleum-contaminated soil

S. no.	Strategies	Pollutants	Outcome	References
1	Natural attenuation	TPH	Effective than biostimulation	(Mishra et al. (2001)
2	Phytoremediation (grasses)	Diesel and heavy oil	Noneffective in old contaminated soil	Banks et al. (2003)
3	Bioaugmentation	TPH	Effective than NA*/B[a]	Bento et al. (2005)
4	Bioaugmentation	PAH	Noneffective than NA*	Yu et al. (2005)
5	Bioaugmentation	PAH	Noneffective	Herwijnen et al. (2006)
6	Phytoremediation	TPH	Effective than NA*	Peng et al. (2009)
7	Phytoremediation	PAH	Effective in freshly contaminated soil than old	Ma et al. (2010)
8	Phytoremediation (*Impatiens balsamina L*)	TPH	Effective than natural degradation	Cai et al. (2010)
9	Phytoremediation and bioaugmentation	Petroleum	Combine treatment effective than single	Tang et al. (2010)
10	Bioaugmentation	TPH	Noneffective than NA*/B[a]	Thangarajan et al. (2011)
11	Bioaugmentation	Aromatic and asphaltic	Effective than indigenous	Gonzalez et al. (2005)
12	Bioaugmentation	Crude oil	Mix culture effective than single spp.	Rahman et al. (2002)

**NA* natural attenuation, *B[a]* bioaugmentation

Ramos et al. (2010) explained that inoculated cultures sometimes fail to resist and survive in contaminated soil or in their native living place soil. Seed culture should be highly stable and resistant to different environmental conditions (Burken 2004). These obstacles can be overcome using some suitable inoculation techniques.

7.2.2.1 Liquid Microbial Culture Inoculation

In previous studies for bioaugmentation in soil, broth culture inoculation such as free culture spraying procedures or culture in semisolid media, i.e., immobilization techniques, was used (Trevors et al. 1994). The use of inoculated culture is not always acceptable, since inoculation experiments have shown ambiguous results in comparison to degradation by native microbial population (Leahy and Colwell 1990). Kastner et al. (1998) used several strains of PAH-degrading bacteria, but after introduction into artificially PAH-contaminated soil, no degradation was observed; further experiment suggested that inoculation with mineral salt medium inhibited bacterial survival and growth as well as their degradation activity; while using water without increasing salinity, no inhibition of autochthonous bacterial strain took place (Hosokawa et al. 2009).

7.2.2.2 Inoculation with Immobilized Microbial Culture

Captivity of enzyme or whole bacterial cell inside or outside the carrier matrix is known as immobilization (Partovinia and Rasekh 2018). For petroleum hydrocarbon biodegradation, microbial attachment on solid surface, entrapment in a gel or membrane, or encapsulation into carrier was used (Hussein 2006). Microbial culture for bioaugmentation in an immobilized form may offer more complete and more rapid degradation, which is easy to handle, can be reutilized, and has high tolerance to pH and temperature changes (Partovinia and Rasekh 2018). Immobilized matrices provide sustained microbial population by continuous release of bacteria in the soil, water, and sediments. Although in many studies culture used in immobilized beads form, still most of the studies have not addressed the growth dynamics of different bacterial strains inside the beads (Bazot and Lebeau 2009).

There are two types of hydrogel material for cell immobilization, natural and synthetic. Natural material includes agar, agarose, polyacrylamides, carrageenan, and alginate (Leenen et al. 1996) (Table 7.3). Alginate is used widespread due to its nontoxic and nutritional behavior for bacterial cell immobilization. The incorporation of some adsorbents into alginate beads is used to transport pollutants inside and outside of the beads (Zhang et al. 2008). Sodium alginate matrix has hindrance to mass and air transfer inside the dense gel layer (Mikkelsen and Elgsaeter 1995). Diatomite was incorporated to overcome this problem (Farhana et al. 2012). Diatomite ($SiO_2\ nH_2O$) is off-white to white color, soft, light in weight sedimentary rock composed mainly of silica microfossils of aquatic algae. Diatomite is highly porous with good sorption ability, is nonreactive with other chemicals, and has low density and large surface area; due to these unique properties, this material is used in cell immobilization for hydrocarbon degradation (Nenadovic et al. 2009). Wang et al. (2012) in their study found that after 20 days of oil degrader inoculation in the soil, the maximum degradation rate in the sodium alginate diatomite (SAD)-immobilized systems reached up to 29.8%, significantly higher ($P < 0.05$) than free cells (21.1%). Moreover, both microbial number and total microbial activity reached to highly significant level ($P < 0.05$) in the immobilized culture than free culture inoculation systems at a same initial inoculation amount. In the presence of plant rhizosphere, this immobilized inoculum was ineffective (Table 7.1) (Farhana et al. 2012) because sodium alginate diatomite carrier has short life and degraded within 20 days and plant growth takes more time (Mikkelsen and Elgsaeter 1995).

Table 7.3 Summary of the researches on different immobilization strategies for bioaugmentation

Material for immobilization	Target pollutant	Microorganisms type	Effect	Reference
Biofix and Drizit	Petrol	*Pseudomonas fluorescens*	Best results than free system	Wilson and Bradley (1996)
Sodium alginate	Phenol, ortho- and Para-cresol	Methanogenic consortium	Twofold higher than free culture	Guiot et al. (2000)
Polyvinyl alcohol and Drizit	Diesel	Mix hydrocarbon degraders	Best results than commercial liquid containing surfactants	Cunningham et al. (2004)
Vermiculite	PAH	*Bacillus* and *Mucor*	Best results than free culture	Dan et al. (2006)
Ca alginate	Cd	*Streptomyces* and *Bacillus*	Less effective than culture type	Jeˇızeˇıquel and Lebeau (2008)
Alginate-lignin	Phenanthrene	*Phanerochaete chrysosporium*	Best results than free culture	Zhang et al. (2008)
Phosphorylated polyvinyl alcohol	Atrazine	*Agrobacterium radiobacter* and mix culture	Best results than free culture	Siripattanakul et al. (2008)
Alginate and biofilm on tezontle	Organophosphate pesticides (OP)	Bacterial consortium OP adopted	Best than suspension culture	Yañez-Ocampo et al. (2009)
Ca alginate	Diuron herbicide	*Delftia acidovorans* and *Arthrobacter*	One immobilized strain has better result than free culture	Bazot and Lebeau (2009)
Sodium alginate and diatomite	TPH	Hydrocarbon degrader mix consortium	Less effective than free inoculum and control plant	Farhana et al. (2012)

7.2.3 Rhizoremediation of Pollutants

Pollutant breakdown in the rhizosphere is improved due to the increase in the microbial activity than non-rhizosphere; this process is generally known as phytostimulation or plant assisted microbial degradation (Donnelly et al. 1994). Rhizodegradation involves the immobilization and removal of the pollutants which is strongly dependent on the rhizospheric processes including enhancement of microbial degradation. The number of PAH degraders and total microbial count were found higher in the rhizospheric region of ryegrass and clover than in the un-vegetated soil (Ma et al. 2010).

Symbiotic association is observed around the plant rhizosphere and their associated microbes due to the release of root exudates which provide necessary nutrients for the survival of microbes and in turn microbes responsible for improved soil conditions for plant growth. Especially, plants help in softening of soil due to aeration and water transport into the rhizosphere (Fig. 7.1). Plants also release allelopathic agents which suppress the growth of other plants in the same soil and protect them from competition, soil pathogens, toxins, and chemicals released from the unwanted plants.

In specific plant-microbe interaction, plant detects toxicants and secretes specific compounds in response to that stimuli and subsequently promotes the growth and activity of specific bacterial community (Fang et al. 2001). Due to this reason sometimes inhibition of inoculated bacterial culture was observed which interferes with the natural plant-microbe interaction and leads to the failure in enhanced degradation by bioaugmentation (Farhana et al. 2012) (Fig. 7.1). Nonspecific interaction occurs when the contaminants themselves may have similarities to the phytochemical which are released by the plant naturally (Phillips et al. 2012). These phytochemicals are used by the rhizospheric microbes as a primary carbon source, and co-metabolic process starts which slows down the pollutant degradation. Sometimes pollutants' degradative enzymes produced by microbes can also be produced by the plant itself. These secondary plant metabolites (SPMEs) are very diverse in nature, having structural similarities with organic pollutant which play an important role in the production of many degradative enzymes (Jha et al. 2015). Examples of pollutants analogy included PAH, their structures are similar to the root exudates morin (Donnelly et al. 1994), and pyrene resembles SPME confusarine released by plants (Singer et al. 2003). The degradation capacity increased in

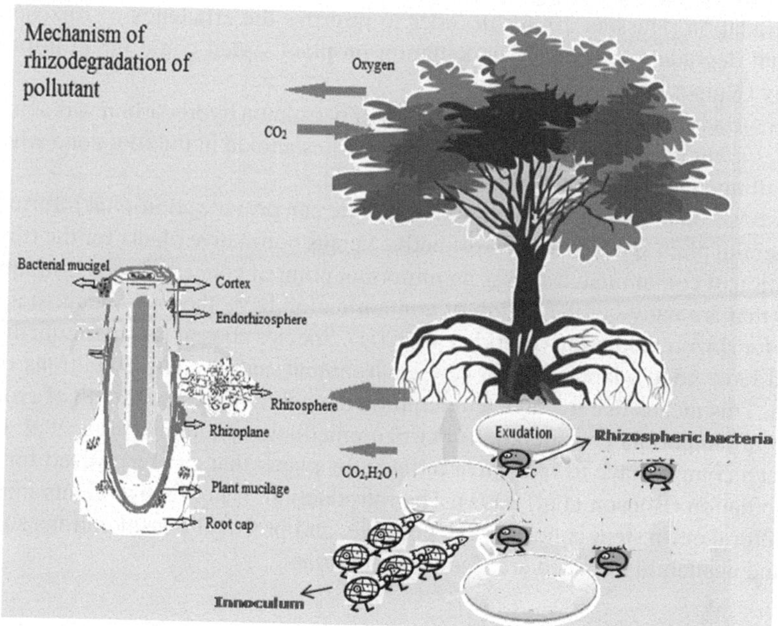

Fig. 7.1 Bacterial community attaches with rhizosphere due to mucilage, and negative interaction of rhizospheric and inoculated bacteria takes place due to competition for nutrients and niche

the presence of exudates acetate and alanine while decreased when malonate was produced by plant roots. It inferred that the degradation potential is associated with functional changes in microbial community.

Rhizospheric biodegradation is the smart technique which utilizes sunlight through plant to degrade pollutants with indigenous microbes. Due to limited methodological capabilities, there are many challenges which are difficult to explore; these includes which mechanisms influence microbial community composition in the rhizosphere, its type, degradative genes involved, and the effect of bioaugmentation in the rhizosphere. In order to assess the influence of plants on pollutant degradation, appropriate experimental design is an essential part of research study so that the conclusive results of pot experiments can be applied in the field plots or in scientific field studies (Mackova et al. 2006).

7.2.4 Plant Species Selection Criteria for Rhizodegradation

Plant root growth parameters are particularly important in the petroleum degradation. The large 0 roots penetrate deeply in the impermeable soil layers, and its root-associated microbial flora is exposed to more contaminants; similarly higher number of root tips provides more binding site for contaminants (Fig. 7.2). The monocot plant species have great potential of contaminant degradation due to its dense root systems which provide a high surface area to soil-microorganism interaction for biodegradation (Hussein 2006). In order to improve the efficiency of rhizospheric pollutant degradation, bacterial inoculation on plant's seed could be an important strategy (Kuiper et al. 2004).

The successful study of rhizodegradation of petroleum hydrocarbon was achieved with *Sesbania cannabina* plant; this plant tolerates anoxia in the root zone which is a prevailing condition of petroleum-polluted soil.

Observation of natural revegetation at the site can provide additional information on potential plant species. The use of native versus non-native plants for the rhizoremediation of contaminated sites is an important point of concern. In most situations, plants that are native to the region of contamination have shown to be most appropriate for rhizoremediation (Merkl et al. 2004). Species chosen for rhizoremediation should have good adaptability to new environment and climatic conditions of the region. This means that average temperature, annual rainfall, and length of growing season are important considerations in rhizoremediation planning (Frick et al. 1999) and each country have to recognize indigenous plants that can be utilized for phytoremediation (Robson et al. 2003). The introduction of non-native plants into any agricultural ecosystem is not always possible, and practical considerations such as cost and availability of seed are also very important.

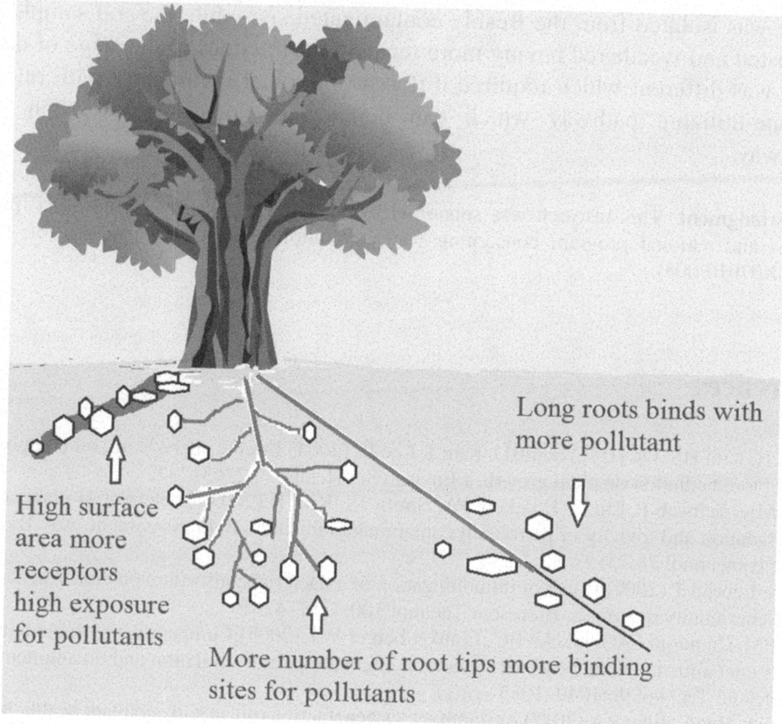

Fig. 7.2 Effect of root growth in petroleum rhizodegradation, morphology of roots plays an important role in binding of pollutants with roots which ultimately increases the pollutant degradation potential either by co-metabolic process or by pollutant uptake

7.3 Conclusion

In the bioremediation breakdown of the crude oil contaminant due to microbial activity enhanced in the presence of plant rhizosphere, the selection of native dominant plant species is required which can tolerate the exposed crude oil concentration. Bioaugmentation by using autochthonous microbes is of great interest because it is found to be best adopted with the contaminated environment. Inoculum addition or biostimulation techniques appeared to be effective in enhancing biodegradation of oil hydrocarbons in soil when used alone. However, in the presence of plant, bioaugmentation with free and sodium alginate-immobilized cultures was ineffective as sodium alginate degrades very quickly and plant growth required long time. One other reason was presence of root exudates which select bacterial species and their subsequent activity. Natural plant-microbe interaction seems to be responsible for higher hydrocarbon degradation directly and indirectly by promoting plant growth. The reason for ineffectiveness of bioaugmentation was that the inoculated culture was isolated from the freshly contaminated site while the soil sample was composted and weathered having more recalcitrant fraction, so the stage of degradation was different which required different microbial community with relevant substrate-utilizing pathway which can degrade petroleum hydrocarbon in a better way.

Acknowledgment This research was supported by the Education Ministry of China (project 308016) and national program concerning water pollution control and management, China (2009ZX07010-008).

References

Baek KH, Kim HS, Oh HM, Yoon BD, Kim J, Lee IS (2004) Effects of crude oil, oil components, and bioremediation on plant growth. J Environ Sci Health 39:2465–2472

Banks MK, Schwab P, Liu B, Kulakow PA, Smith JS, Kim R (2003) The effect of plants on the degradation and toxicity of petroleum contaminants in soil: a field assessment. Adv Biochem Eng Biotechnol 78:75–97

Bazot S, Lebeau T (2009) Effect of immobilization of a bacterial consortium on diuron dissipation and community dynamics. Bioresour Technol 100:4257–4261

Bento FM, Camargo FAO, Okeke BC, Frankenberger WT (2005) Comparative bioremediation of soils contaminated with diesel oil by natural attenuation, biostimulation and bioaugmentation. Bioresour Technol 96:1049–1055

Bhatti AA, Haq S, Bhat RA (2017) Actinomycetes benefaction role in soil and plant health. Microb Pathog 111:458–467

Burken JG (2004) Uptake and metabolism of organic compounds: green-liver model. In: McCutcheon SD, Schnoor JL (eds) Phytoremediation. Wiley, New York, pp 59–84

Cai Z, Zhou Q, Peng S, Kenan L (2010) Promoted biodegradation and microbiological effects of petroleum hydrocarbons by *Impatiens balsamina* L. with strong endurance. J Hazard Mater 183:731–737

Cunningham CJ, Ivshina IB, Lozinsky VI, Kuyukina MS, Philp JC (2004) Bioremediation of diesel-contaminated soil by microorganisms immobilised in polyvinyl alcohol. Int J Biodeterior Biodegrad 54:167–174

Dan S, Pei-jun L, Stagnitt F, Xian-he X (2006) Biodegradation of benzo[a]pyrene in soil by Mucor sp. SF06, and Bacillus sp. SB02 co-immobilized on vermiculite. J Environ Sci 18:204–1209

Das N, Chandran P (2011) Microbial degradation of petroleum hydrocarbon contaminants: an overview. Biotechnol Res Int 2011:941810. https://doi.org/10.4061/2011/941810

Dongmei G, Ying X, Maqbool F, Zhenhua X (2011) Bio-treatment of Hydrocarbon Polluted Soil with Agriculture Residues, Energy Procedia 5:1558–1562

Donnelly PK, Hegde RS, Fletcher JS (1994) Growth of PCB-degrading bacteria on compounds from photosynthetic plants. Chemosphere 28:981–988

Fang C, Radosevich M, Fuhrmann JJ (2001) Atrazine and phenanthrene degradation in grass rhizosphere soil. Soil Biol Biochem 33:671–678

Farhana M, Ying X, Zhao J, Wang Z, Gao D, Zhao Y-G et al (2012) Rhizodegradation of petroleum hydrocarbons by *Sesbania cannabina* in bioaugmented soil with free and immobilized consortium. J Hazard Mater 237-238:262–269

Frick C, Farrell R, Germida J (1999) Assessment of phytoremediation as an in-situ technique for cleaning oil-contaminated sites. Calgary Petroleum Technology Alliance of Canada (PTAC), Calgary

Gerardi M (2016) Wastewater bioaugmentation and biostimulation. In: Chapter 2: Bioaugmentation. DEStech, Lancaster
Gonzalez EC, Rojas-Avelizapa LI, Cruz-Camarillo R, Rojas-Avelizapa NG (2005) Chapter 1, effect of bacteria augmentation on aromatic and asphaltenic fraction removal in solid culture. In: Contaminated soils, sediments & water. Springer, Berlin, pp 1–11
Guiot SR, Tawfiki-Hajji K, Lépine F (2000) Immobilization strategies for bioaugmentation of anaerobic reactors treating phenolic compounds. Water Sci Technol 42:245–250
Herwijnen R, Joffe B, Ryngaert A, Hausner M, Springael D, Govers HA et al (2006) Effect of bioaugmentation and supplementary carbon sources on degradation of polycyclic aromatic hydrocarbons by a soil-derived culture. FEMS Microbiol Ecol 55:122–135
Hosokawa R, Nagai M, Morikawa M, Okuyama H (2009) Autochthonous bioaugmentation and its possible application to oil spills. World J Microbiol Biotechnol 25:1519–1528
Hussein EI (2006) In: investigation into the mechanism(s) which permit the high-rate degradation of PAHS….Thesis. Department of Biology, Georgia State University, Atlanta
ITRC (Interstate Technology and Regulatory Council) (2009) Phytotechnology technical and regulatory guidance and decision trees, revised. Phyto-3, Washington, DC
Je˘ıze˘ıquel K, Lebeau T (2008) Soil bioaugmentation by free and immobilized bacteria to reduce potentially phytoavailable cadmium. Bioresour Technol 99:690–698
Jha P, Panwar J, Jha PN (2015) Secondary plant metabolites and root exudates: guiding tools for polychlorinated biphenyl biodegradation. Int J Environ Sci Technol 12:789–802
Jørgensen KS, Puustinen J, Suortti AM (2000) Bioremediation of petroleum hydrocarbon-contaminated soil by composting in biopiles. Environ Pollut 107:245–254
Kastner M, Jammali MB, Mahro B (1998) Impact of inoculation protocols, salinity, and pH on the degradation of polycyclic aromatic hydrocarbons (PAHs) and survival of PAH-degrading bacteria introduced into soil. Appl Environ Microbiol 64:359–362
Kästner M, Miltner A. (2016) Application of compost for effective bioremediation of organic contaminants and pollutants in soil, Appl Microbiol Biotechnol 100:3433–3449
Kirk JL, Klironomos JN, Lee H, Trevors JT (2005) The effects of perennial ryegrass and alfalfa on microbial abundance and diversity in petroleum contaminated soil. Environ Pollut 133:455–465
Kuiper I, Lagendijk EL, Bloemberg GV, Lugtenberg BJJ (2004) Rhizoremediation: a beneficial plant-microbe interaction. Mol Plant Microbe Interact 17:6–15
Leahy JG, Colwell RR (1990) Microbial degradation of hydrocarbons in the environment. Microbiol Rev 54:305–315
Leenen EJTM, Dos Santos VAP, Grolle KCF, Tramper J, Wijffels R (1996) Characteristics of and selection criteria for support materials for cell immobilization in wastewater treatment. Water Res 30:2985–2996
Ma B, He Y, Chen H, Xu JM, Rengel Z (2010) Dissipation of polycyclic aromatic hydrocarbons (PAHs) in the rhizosphere: synthesis through meta-analysis. Environ Pollut 158:855–861
Mackova M, Dowling D, Macek T (2006) Phytoremediation rhizoremediation. In: Leigh MB (ed) Methods for rhizoremediation research: approaches to experimental design and microbial analysis. Springer, Berlin, pp 33–55
Martin-Gil J, Navas-Gracias L, Gomez-Sobrino E, Correa-Guimaraes A, Hernandez-Navarro S, Sanchez-Bascones M et al (2008) Composting and vermicomposting experiences in the treatment and bioconversion of asphaltenes from the prestige oil spill. Bioresour Technol 99:1821–1829
Merkl N, Schultze-Kraft R, Infante C (2004) Phytoremediation of petroleum-contaminated soils in the tropics-pre-selection of plant species from eastern Venezuela. J Appl Bot Food Qual 78:185–192
Mikkelsen A, Elgsaeter A (1995) Density distribution of calcium-induced alginate gels. A numerical study. Biopolymers 36:17–41
Mishra S, Jyot J, Kuhad RC, Lal B (2001) Evaluation of inoculum addition to stimulate in situ bioremediation of oily-sludge-contaminated soil. Appl Environ Microbiol 67:1675–1681

Nenadovic S, Nenadovic M, Kovacevic R, Matović L, Matović B, Jovanovic Z et al (2009) Influence of diatomite microstructure on its adsorption capacity for Pb (II). Sci Sinter 41:309–317

Odokuma LO, Dickson AA (2003) Bioremediation of a crude oil polluted tropical rain forest soil. Glob J Environ Sci 2:29–40

Partovinia A, Rasekh B (2018) Review of the immobilized microbial cell systems for bioremediation of petroleum hydrocarbons polluted environments. Crit Rev Environ Sci Technol 48:1–38

Peng S, Zhou Q, Cai Z, Zhang Z (2009) Phytoremediation of petroleum contaminated soils by Mirabilis Jalapa L. in a greenhouse plot experiment. J Hazard Mater 168:1490–1496

Phillips LA, Greer CW, Farrell RE, Germida JJ (2012) Plant root exudates impact the hydrocarbon degradation potential of a weathered-hydrocarbon contaminated soil. Appl Soil Ecol 52:56–64

Rahman KSM, Banat IM, Thahira J, Thayumanavan T, Lakshmanaperumalsamy P (2002) Bioremediation of gasoline contaminated soil by a bacterial consortium amended with poultry litter, coir pith and rhamnolipid biosurfactant. Bioresour Technol 81:25–32

Ramos JL, Duque E, Van Dillewjin P, Daniels C, Krell T, Espinosa-Urgel M et al (2010) Removal of hydrocarbons and other related chemicals via the rhizosphere of plants. In: Timmis KN (ed) Handbook of Hydrocarbon and Lipid Microbiology, pp 2575–2581

Robson D, Knight D, Farrell R (2003) Ability of cold-tolerant plants to grow in hydrocarbon-contaminated soil. Int J Phytoremediation 5:105–123

Singer AC, Crowley DE, Thompson IP (2003) Secondary plant metabolites in phytoremediation and biotransformation. Trends Biotechnol 21:123–130

Siripattanakul S, Wirojanagud W, McEvoy J, Eakalak K (2008) Effect of cell-to-matrix ratio in polyvinyl alcohol immobilized pure and mixed cultures on atrazine degradation. Wat Air Soil Pollut Focus 8:257–266

Tang J, Wang R, Niu X, Zhou Q (2010) Enhancement of soil petroleum remediation by using a combination of ryegrass (Lolium perenne) and different microorganisms. Soil Till Res 110:87–93

Thangarajan R, Adetutu EM, Moore BR, Ogunbanwo ST, Ball AS (2011) Comparison between different bio-treatments of a hydrocarbon contaminated soil from a landfill site. Afr J Biotechnol 10:15151–15162

Trevors JT, Kuikman P, Van Elsas JD (1994) Release of bacteria into soil: cell numbers and distribution. Review article. J Microbiol Methods 19:247–259

Trevors JT, van Elsas JD, Lee H, Wolters AC (1993) Survival of alginate-encapsulated Pseudomonas fluorescens cells in soil. Appl Microbiol Biotechnol 39:637–643

Ueno A, Ito Y, Yumoto I, Okuyama H (2007) Isolation and characterization of bacteria from soil contaminated with diesel oil and the possible use of these in autochthonous bioaugmentation. World J Microbiol Biotechnol 23:1739–1745

Van Gestel K, Mergaert J, Swings J, Coosemans J, Ryckeboer J (2003) Bioremediation of diesel oil-contaminated soil by composting with biowaste. Environ Pollut 125:361–368

Vogel TM, Walter MV (2001) Bioaugmentation. In: Hurst CJ, Crawford RL, Garland JL et al (eds) Manual of environmental microbiology. American Society for Microbiology, Washington, DC, pp 952–959

Wang ZY, Xu Y, Wang HY, Zhao J, Gao DM, Li FM, Xing B (2012) Biodegradation of crude oil in contaminated soils by free and immobilized microorganisms. Pedosphere 22:717–725

Wang ZY, Xu Y, Zhao J, Fengmin L, Dongmei G, Baoshan X (2011) Remediation of petroleum contaminated soils through composting and rhizosphere degradation. J Hazard Mater 190:677–685

Wilson NG, Bradley G (1996) Enhanced degradation of petrol (Slovene diesel) in an aqueous system by immobilized *Pseudomonas fluorescens*. J Appl Bacteriol 80:99–104

Yañez-Ocampo G, Sanchez-Salinas E, Jimenez-Tobon GA, Penninckx M, Ortiz-Hernández ML (2009) Removal of two organophosphate pesticides by a bacterial consortium immobilized in alginate or tezontle. J Hazard Mater 168:1554–1561

Yu KSH, Wong AHY, Yau KWY, Wong YS, Tam NFY (2005) Natural attenuation, biostimulation and bioaugmentation on biodegradation of polycyclic aromatic hydrocarbons (PAHs) in mangrove sediments. Mar Pollut Bull 51:1071–1077

Zawierucha I, Malina G (2011) Bioremediation of contaminated soils: effects of bioaugmentation and biostimulation on enhancing biodegradation of oil hydrocarbons. In: Singh A, Parmar N, Kuhad R (eds) Bioaugmentation, biostimulation and biocontrol. Soil biology: 108. Springer, Berlin, pp 187–201

Zhang K, Xu Y, Hua X, Han H, Wang J, Wang J et al (2008) An intensified degradation of phenanthrene with macroporous alginate–lignin beads immobilized *Phanerochaete chrysosporium*. Biochem Eng J 41:251–257

Chapter 8
Vermicomposting: An Eco-Friendly Approach for Recycling/Management of Organic Wastes

Ali Mohd Yatoo, Saiema Rasool, Shafat Ali, Sabhiya Majid, Muneeb U. Rehman, Md. Niamat Ali, Rafiqa Eachkoti, Shabhat Rasool, Shahzada Mudasir Rashid, and Sanah Farooq

A. M. Yatoo
Department of Environmental Science, University of Kashmir, Srinagar, Jammu and Kashmir, India

Centre of Research for Development, University of Kashmir, Srinagar, Jammu and Kashmir, India

S. Rasool (✉)
Forest Biotech Laboratory, Department of Forest Management, Faculty of Forestry, University of Putra Malaysia, Serdang, Selangor, Malaysia

S. Ali
Centre of Research for Development, University of Kashmir, Srinagar, Jammu and Kashmir, India

Department of Biochemistry, Govt. Medical College (GMC), Srinagar, Jammu and Kashmir, India

S. Majid · R. Eachkoti · S. Rasool · S. Farooq
Department of Biochemistry, Govt. Medical College (GMC), Srinagar, Jammu and Kashmir, India

M. U. Rehman
Department of Biochemistry, Govt. Medical College (GMC), Srinagar, Jammu and Kashmir, India

Department of Clinical Pharmacy, College of Pharmacy, King Saud University, Riyadh, Saudi Arabia

M. N. Ali
Centre of Research for Development, University of Kashmir, Srinagar, Jammu and Kashmir, India

S. M. Rashid
Division of Veterinary Biochemistry, Faculty of Veterinary Science and Animal Husbandry, SKUAST-Kashmir, Shuhama, Jammu and Kashmir, India

K. R. Hakeem et al. (eds.), *Bioremediation and Biotechnology*,
https://doi.org/10.1007/978-3-030-35691-0_8

8.1 Introduction

One of the burning problems faced by both developed and developing countries is the increasing quantity of organic waste. Because of speedy industrialization, urbanization, population explosion and modern agricultural practices, millions of tonnes of organic waste are produced per day globally. The most common practice used for the processing of this waste is either uncontrolled burning or dumping in landfills both of which leads to many environmental problems like release of greenhouse gases (CH_4, CO_2, N_2O) (Lee et al. 2009), leachate, ground- and surface water pollution (Mor et al. 2006), damage to natural beauty of that area, odour problems, become breeding grounds for mosquitoes and rats, destruction of fertility of soil, etc. Also an important organic resource is wasted.

Each human being generates some type of organic wastes which is having no value in the eyes of the owner and has to be thrown out. In order to maintain healthy and safe environment, management of this organic waste is important. The common organic wastes produced by modern societies are kitchen waste, agricultural waste, horticultural waste, vegetable waste, animal waste, sewage sludge, etc. (Gopalakrishnan 2005). Large amount of agricultural waste which is mainly destroyed by burning or disposed in landfills can be used as a soil amendment by following the simple recipe "What comes from soil must be returned to the soil" (Sequi 1990). Dumping of tonnes of organic wastes in landfills everyday creates both environmental and economic problems for the government to monitor and handle it for the safety of the environment. Although there are a number of physical and chemical methods which can be used for management of organic waste, they are either time-consuming or expensive (Zirbes et al. 2011). The problem to manage it has become more complicated because of mixing of biodegradable and non-biodegradable wastes at the source of generation. The main objective of organic waste management is to reduce environmental impacts caused by unplanned disposal of organic wastes (Pevey et al. 1985). One of the most promising techniques used for the sustainable management of organic wastes is vermicomposting (Singh et al. 2011). It is a process involving joint action of microorganisms and earthworms. Microorganisms help in biochemical degradation of organic waste, and earthworms help in fragmenting and making the substrate suitable for microorganisms. Earthworms make organic waste much more favourable for microbes by acting as mechanical grinders, thereby increasing its surface area, modifying its physicochemical status, decreasing its C:N ratio, etc. (Dominguez et al. 1997). They also act as an agent of aeration and turning (Ndegwa and Thompson 2001). Vermicomposting is like getting "gold from garbage" (Fig. 8.1). It is one such technique which can be used for sustainable management of organic waste, and at the same time, organic fertilizer can be produced which can be used for sustainable agriculture without having any negative effect on human health and soil ecosystem.

Fig. 8.1 Organic fertilizer from organic waste

8.2 Vermicomposting

Vermicomposting is a process involving joint action of microbes by which organic wastes are recycled and humus-like material known as vermicast is produced containing both macro- and micronutrients, beneficial microbes, enzymes and hormones required for proper growth of plants (Lee 1985; Bansal and Kapoor 2000; Jambhekar 1992). Vermicast is also known as black gold (Lim and Wu 2015; Patangray 2014) as it contains both macro- and micronutrients. Vermicomposting was started in Canada (1970), and the first experiment for organic waste management was done in Ontario (Canada) and is now processing more than 70 tonnes of refuse per week. Later on it was followed by the USA, Japan, Italy, the Philippines, Thailand, Brazil, India and other countries. Vermicomposting is a cost-effective, socially acceptable and eco-friendly technique for management of organic waste (Lim et al. 2016) using proper species of earthworm (Fig. 8.2). It is a very promising technique as it can give employment to millions of youth, can reduce the reliance on chemicals, can convert wastes into fertilizers and can make a motherland green and prosperous.

8.3 Species Used in Vermicomposting

There are around 8000 known earthworm species, but species which are appropriate for vermicomposting are only 7, and all belong to epigeic species such as *Eisenia foetida*, *Perionyx excavatus*, *Eudrilus eugeniae*, etc. *Eisenia foetida* (Fig. 8.3) is a well-known species and is used throughout the world for vermicomposting because of its tolerance to temperature (10–35 °C), disease resistance, fast multiplication, high speed of cocoon production, short life cycle and fast consumption of organic matter (Watanabe and Tsukamoto 1976; Hartenstein et al. 1979; Reinecke and Kriel 1981). Also it has the ability to double its population within 60 days and consume food equal to its body weight per day under appropriate conditions. It is commonly

Fig. 8.2 Vermicomposting of organic wastes

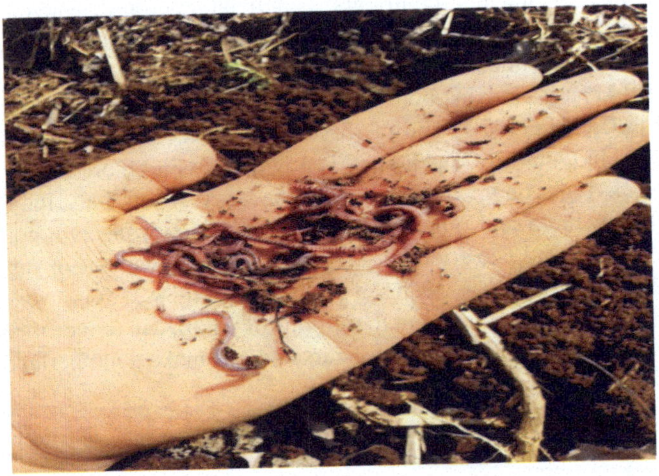

Fig. 8.3 *Eisenia foetida* species commonly used for vermicomposting

known as compost worm and is native to Europe. *Perionyx excavatus* and *Eudrilus eugeniae* are the best species to be used for organic solid waste management in tropical and subtropical regions (Kale 1998a, b). It has been confirmed that earthworm can process every type of organic waste like paper waste, kitchen waste, sewage sludge, animal waste, city refuse, industrial waste, etc. (Kale et al. 1982; Senapati and Dash 1982; Muyima et al. 1994; Edwards and Bohlen 1996; Ismail 2005, Saranraj and Stella 2012; Wu et al. 2014; Lim et al.2016).

8.4 Materials Used for Vermicomposting

In general every type of organic waste (Fig. 8.4) such as animal waste, agricultural waste, kitchen waste, etc. can be used as a material for vermicomposting. Most commonly cow dung, leaf litter and dried agricultural residues are key materials. Earthworms consume almost every type of organic waste and decrease the quantity by 50–60%. Each earthworm weighs about 0.4–0.6 g, eats squander proportions to its body weight and delivers cast equal to around 50% of the waste it devours a day. Most earthworms eat about half of their body weight of organic matter per day. But tiger worm can eat organic waste at a rate equal to its body weight per day. Under perfect conditions of temperature (20–25 °C) and moisture (50–60%), 1 tonne of organic waste can be transformed into vermicompost in just 30 days by about 5 kg of earthworms The main idea behind vermicomposting is to reduce organic waste pollution caused by inappropriate dumping of different types of organic wastes (Fig. 8.5). Fresh waste cannot be processed by worms; hence a partial decomposition of waste is required so that it becomes palatable for earthworms. To speed up the vermicomposting process, cow dung is usually used which acts as a basic food substrate for earthworms. Different types of wastes have been converted successfully by earthworms such as paper, coffee and tea waste, banana waste, sugarcane waste, agricultural waste, municipal waste, etc. Vermicomposting has also proven to be successful in recycling sewage sludge and solids from wastewater (Dominguez et al. 2000; Xing et al. 2012; Fu et al. 2015; Fernández-Gómez et al. 2015; Villar et al. 2016), brewery waste (Butt 1993), urban wastes, kitchen and cattle wastes (Allevi et al. 1987; Edwards and Burrows 1988; Elvira et al. 1995; Dominguez and Edwards 1996; Kaushik and Garg 2003) as well as other wastes like dead plants and mushroom waste (Edwards 1988), horse waste (Dharani et al. 2010; Brintha and Manimegala 2015), elephant dung (Elamparithi et al. 2017), etc. These wastes can

Fig. 8.4 Different types of organic wastes

Fig. 8.5 Dumping of organic wastes

lead to air, soil and water pollution if not managed properly. Mixing of these substrates with cow dung and other suitable organic waste can make them suitable for earthworms to feed upon and thereby decreasing the pollution level by using suitable species of earthworms. About 50–60% of waste found in municipalities and cities is organic in nature which has led to loss of organic resource because of lack of awareness and commitment to segregate materials and composting it. Some successful attempts have been made in some municipal areas to segregate the waste into organic and inorganic material and use it for vermicomposting e.g. Karnataka state (India) where a vermicomposting unit has been set up by Karnataka Compost Development Corporation for managing organic waste of the state. 100–200 tonnes of waste is processed, and the same is then sold to customers as organic fertilizer. Vermicomposting of human excreta (faeces) was studied by Bajsa et al. (2004), and he found that within 6 months human excreta can be converted into vermicompost by earthworms with safe pathogen quality, good physical texture and no bad odour. During vermicomposting of faeces (biosolids), the population of total coliform greatly decreased, and it was found that the population densities of coliform bacteria reduced by 98% as compared to those in fresh pig slurry when passed through the gut of earthworm species *Eisenia foetida*, *Eudrilus eugeniae* and *Eisenia andrei* (Monroy et al. 2008). It has also been reported that after 60 days of vermicomposting of faeces the number of coliform bacteria reduced from 39,000 to 0 MPN/g (Dominguez and Edwards 2004). Degradation and composting of "wastewater sludge" released from different industries such as sugarcane industry, potato and corn chip industry, paper pulp and cardboard industry, brewery and distillery industry, etc. by earthworms were studied by Kale (1998a, b), Kale et al. (1992), Seenappa et al. (1995), Gunthilingaraj and Ravignanam (1996), Lakshmi and Vizaylakshmi (2000).

8.5 Methods of Vermicomposting

There are different methods of vermicomposting but the most commonly used methods are pit method (Fig. 8.6) and windrows method (Fig. 8.7). Pit method is used for small-scale production of vermicompost, and for large-scale production, windrows method is used. Pit method is usually used to produce 5–10 tonnes of vermicompost per year and to meet personal requirements of a farmer, while windrows method is used for large-scale recycling of organic waste and production of 50–100 tonnes of vermicompost which can be used for commercial purposes. Both worms and vermicompost can be sold, worms at the rate of 150–200 rupees per kg and vermicompost at the rate of 10–15 rupees per kg.

Fig. 8.6 Small-scale production of vermicompost by pit method

Fig. 8.7 Large-scale production of vermicompost by windrows method

8.5.1 Pit Method

The steps used in pit method are:

1. Suitable-size pit is constructed over the surface of the ground. The pit size may vary depending upon the availability of raw materials. Generally the pit size is 3 × 2 × 1 m (L × W × D).
2. The first layer of pit is filled with paddy straw or rice husk for bedding.
3. The second layer is filled with cow dung (15–20 days old) to initiate microbial activity and serve as food substrate for earthworms.
4. In the third layer, earthworms are added up at the rate of 1 kg per 50 kg organic matter.
5. After adding earthworms every type of organic waste like vegetable peelings, rice husk, fruit waste, leaf litter, etc. is added to fill the pit up to the top.
6. The pit is finally covered with gunny bags to protect earthworms from being preyed upon by birds, rats, moles, etc. and to maintain adequate moisture (50–60%).
7. Water is to be sprayed twice or thrice in a week. After the end of 60 days, vermicompost will be ready and can be harvested for agricultural purposes. Vermicompost will be free floating having earthy smell and dark brown colour.

8.5.2 Windrows Method

As already stated it is used for large-scale production of vermicompost. Here organic wastes such as cow dung and agricultural waste is piled up in long rows (windrows). In order to improve oxygen content and porosity, these rows are turned. Usually windrow of appropriate size is made over the concrete surface, and the ideal height of the pile is 3–6 ft. and width 10–12 ft. The size of the pile may vary as per the availability of raw materials. To start a windrow, first organic waste and cow dung are spread on a concrete surface, and earthworms are added. After them manure is added every week to increase windrow depth. After the first windrow is made 3–4 ft. thick, the new windrow next to the first windrow is made. The earthworms will migrate from first to fresher feed. Then waste is added up to fill this pile, and this process continues till a number of windrows are made. Water will be sprayed as per the requirement, and after the end of 2–3 months, vermicompost will be ready from the first and subsequent windrows. After then it will be harvested and used for commercial purposes.

8.6 Factors Influencing Vermicomposting of Organic Wastes

There are a number of factors which effect processing of organic wastes during vermicomposting. Some of the important factors are mentioned below.

8.6.1 pH

Number, distribution and species of earthworm can be limited by pH of the substrate as worms are very responsive to pH (Edwards and Bohlen 1996; Chalasani et al. 1998). So for proper growth, development and action of earthworms, optimum pH is required. Earthworms can tolerate a pH range of 5–9, but the optimum pH for most of the composting worms is about 7.0 (Singh 1997; Narayan 2000; Pagaria and Totwat 2007; Suthar 2008). pH generally depends on the type of waste used for vermicomposting but can also decrease because of a number of chemical reactions occurring during decomposition of organic wastes.

8.6.2 Moisture

Adequate moisture is required for vermicomposting which is around 60–70%. As earthworms breathe through their skins, water should be sprayed twice or thrice a week so that proper moisture is maintained in the vermicomposting unit. It should not be more or less than the optimum range; otherwise it can be dangerous for growth and development of earthworms. When the moisture level is maintained, the time required for vermicomposting of organic waste is less and growth and development of worms are fast. The number and biomass of earthworms are also influenced by moisture (Olson 1928; El-Duweini and Ghabbour 1965; Wood 1974).

8.6.3 Temperature

Growth, metabolism, activity and reproduction of earthworms are significantly affected by temperature as reported by Evans et al. (1948). When temperature is ideal, vermicomposting is fast as worms feed faster on the substrate and reproduce faster. Compost worms can tolerate to a temperature range of 5–35 °C, but temperature beyond this level can be lethal, and worms will try to escape and if not possible will die quickly. The optimum temperature for most of the worms is 20–25 °C (Neuhauser et al. 1988), so maintenance of optimum temperature is essential for vermicomposting.

8.6.4 Food Substrate/Organic Waste

Earthworms can use almost every type of organic waste as their food substrate. Joshi (1997) reported that dairy waste, animal manure, biogas sludge, poultry waste and food industry waste can be recycled by vermicomposting. Under perfect conditions compost worms eat about half of their body weight per day. It is better to underfeed worms than to overfeed worms because if worms cannot eat organic waste quickly it will rot and produce bad odour and also oxygen level will decrease and worms may die.

8.6.5 Light

Earthworms hate light and are very susceptible to it. When exposed to sunlight, they go deep in to the soil or bin and if escape is not possible may get injured or killed. So vermicomposting should be done always in shady areas or provide any type of shelter.

8.6.6 Carbon and Nitrogen Ratio (C:N Ratio) "Greens and Browns"

For high-speed vermicomposting, carbon/nitrogen ratio should be maintained. If C:N ratio is high, the process of decomposition will be slow. To maintain C:N ratio to its desired level, nitrogenous substrate such as cow manure may have to be added. At the start of vermicomposting, the ideal C:N ratio should be below 30:1 and at the end should be decreased to 20:1 (Kavitha and Subramanian 2007). The microorganisms use carbon as a source of energy and nitrogen for protein synthesis. Thus adding a mixture of greens such as grass clippings and browns such as tree leaves will help to maintain ideal C:N ratio. The C:N ratio of different organic materials is shown in Table 8.1.

Table 8.1 Average carbon/nitrogen ratios of some organic substrates

S. no.	Organic material	C:N ratio
1.	Grass clippings	15–20:1
2.	Cow dung	20:1
3.	Sheep manure	10:1
4.	Food wastes	15–20:1
5.	Horse manure	25:1
6.	Vegetables	20–25:1
7.	Sewage sludge	16:1
8.	Fruit waste	30:1
9.	Paper	100:1
10.	Dry leaves	40–60:1

8.7 Advantages of Vermicompost

8.7.1 Organic Fertilizer

The most important advantage of vermicompost is that it is 100% organic. No harmful chemical is present in it and is not required to be mixed with anything. It is a complete fertilizer. Its addition can lead to organic enrichment to soil and can improve its physicochemical and biological properties (Ansari and Jaikishun 2011; Chauhan and Singh 2013).

8.7.2 More Nutritious

Vermicompost contains all essential nutrients in soluble forms of nitrogen, phosphorus, potassium, calcium, magnesium, etc. which are easily available to plants (Pathma and Sakthivel 2012; Orozco et al. 1996; Lim et al. 2015; Gupta et al. 2014b; Amanullah 2016). As compared to chemical fertilizers, vermicompost is not flushed from the soil easily as it contains worm mucus and remains attached to soil for longer period of time. Also vermicompost contains enzymes such as cellulase, amylase, lipase, etc. which help in releasing nutrients by breaking down of organic matter already present in soil and making them available to plants (Chaoui et al. 2003; Lunt and Jacobson 1994; Tiwari et al. 1989)

8.7.3 Beneficial Microorganisms

As the compost passes through the body of earthworms, it becomes enriched with microorganisms and contains bacteria, fungi and actinomycetes (Edwards 1983; Tomati et al. 1987). These microorganisms help to make plants more disease and pest resistant. As per reports some microorganisms such as the pseudomonads can induce resistance to plant diseases because they are antagonistic to plant pathogens and such effects were verified by experiments on suppression of diseases like verticillium wilt on strawberries and pythium and rhizoctonia on cucumbers and radishes under laboratory conditions (Chaoui et al. 2003).

8.7.4 Healthier Plants

Application of vermicompost makes the plants healthier and stronger. The chemical fertilizers on the other hand may increase yield of plants but do nothing for the health of plants as fertilizers feed the plant while vermicompost feeds the soil.

Continuous application of chemical fertilizers makes the plant susceptible to diseases. As per Singh et al. (2009), health and yield of wheat were better when grown on vermicompost as compared with chemical fertilizers, and it increases with continuous application of same amount of vermicompost.

8.7.5 *Plant Growth*

Vermicompost contains hormones which are very important for the growth of plants. These hormones encourage seed germination, growth and yield of plants. Microorganisms such as bacteria, actinomycetes, yeasts and fungi present in vermicompost produce plant growth hormones and plant growth regulators such as gibberellins, cytokinins, auxins, ascorbic acid, etc. (Frankenberger and Arshad 1995); vermicompost also adds beneficial microorganisms to the soil and supplies food for the existing microorganisms which enhances their biological properties and self-renewal capacity of soil fertility (Ouédraogo et al. 2001; Shiralipour et al. 1992).

8.7.6 *Water Retention*

Vermicompost has high porosity, aeration and water holding capacity (Edwards and Burrows 1988). It can hold up to nine times its own weight in water as it is a colloid. Thus during dry periods of time, it can make a big difference.

8.7.7 *Slow Nutrition Release*

Vermicompost releases nutrients very slowly, thus remaining available to plants for long period of time, whereas nutrients from inorganic fertilizers get released faster and plants are not able to absorb them with the speed they are released. The nutrients present in vermicompost also become available to plants in shorter period of time, while the conventional compost fails to release the required amount of important nutrients including NPK to plants in shorter time (Bonkowski and Schaefer 1997; Hammermeister et al. 2004; Subler et al. 1998).

8.7.8 *Rich in Humic Acids*

Vermicompost being rich in humic acids helps in the promotion of nutrient uptake and root growth of plants. This was confirmed by Canellas et al. (2002) that elongation of roots and lateral root formation in maize plants were enhanced by humic acids when isolated from vermicompost. This was also reported by Pramanik et al. (2007) that nutrient uptake by plants was increased by humic acids by stimulating root growth and by increasing proliferation of root hairs.

8.8 Nutrient Status of Vermicompost

The nutrient status of vermicompost usually depends upon the kind of organic waste and species of earthworm used in vermicomposting. In case of heterogenous waste, a wide range of nutrients will be available, while as in case of homogenous waste, only certain nutrients will be available. Besides containing other micronutrients, vermicompost is enriched with macronutrients which are present in soluble form and become available to plants within a month of application. Not only nutrients but also beneficial microorganisms such as bacteria, fungi, actinomycetes, plant growth promoters and other materials produced by microorganisms are also present in vermicompost (Joshi et al. 2015). The nutrient content present in vermicompost is five times more than what is present in potting soil mixtures. Ruz-Jerez et al. (1992) and Parkin and Berry (1994) conducted chemical analysis of vermicompost and found that as compared to good top soil vermicompost contains 5 times more nitrogen, 7 times more potassium and 1.5 times more calcium. As per Reinecke et al. (1992), phosphorus gets converted into plant-available form when it passes through the gut of earthworms. Therefore vermicomposting process is very important for agriculture as it makes phosphorus and other nutrients available to plants. The presence of essential nutrients and microorganisms in vermicompost is very important for the growth of productive and healthy plants. The vermicompost not only provides these but also has long-lasting effects as it increases water holding capacity of soil, modifies soil structure and improves soil stability and porosity (Ferreras et al. 2006). The presence of enzymes like chitinase, amylase, lipase, etc. in vermicompost can break down organic matter already present in soil and release nutrients, thereby making them accessible to plant (Chaoui et al. 2003; Tiwari et al. 1989).

Vermicompost on an average contains 1–2.5% nitrogen, 1.8–2.0% phosphorus and 1–1.5% potassium. It also contains other nutrients such as sodium, iron, zinc, calcium, magnesium, etc. as mentioned in Table 8.2.

8.9 Vermicompost as Plant Growth Promoter and Protector

8.9.1 Growth Promoter

Vermicompost is an exceptional plant growth promoter and protector as it contains both macro- and micronutrients and beneficial microorganisms and thereby can be used as a sustainable alternative to inorganic fertilizers (Sinha et al. 2009; Chauhan and Singh 2015). Continuous use of inorganic fertilizers over a long period of time

Table 8.2 Nutrient status of vermicompost

S. no.	Parameter	Nutrient content
1.	pH	6.8–7.5
2.	O.C (%)	15–20
3.	E.C ($mS \cdot cm^{-1}$)	0.18
4.	N (%)	1–2.5
5.	P (%)	1.8–2.0
6.	K (%)	1–1.5
7.	Ca (%)	0.17
8.	Mg (%)	0.06–0.3
9.	Zn (%)	0.005–0.11
10.	Na (%)	0.04–0.15
11.	Mn (%)	0.03–0.20
12	S (%)	0.3–0.5

restricts the overall growth and development of plants as it makes the soil acidic, decreases its water holding capacity, makes the plants susceptible to diseases (Ansari and Ismail 2001), etc. Vermicompost on the other hand is a perfect organic fertilizer for proper growth and yield of many plants (Lalitha et al. 2000; Amador et al. 2013; Gezahegn and Girum 2017) as it contains higher nutrient content as compared to other organic amendments (Joshi et al. 2015). Vermicompost not only improves physical and biochemical properties of soil, but also increases health-related secondary metabolites in plants (Goswami et al. 2017; Das et al. 2018). Besides containing plant growth protectors and growth-promoting materials produced by microbes, vermicompost also contains a significant quantity of micronutrients (Amanullah 2016); growth stimulators such as auxins, cytokinins, gibberellins and enzymes (Tejada and González 2009); and humic acids (Maji et al. 2017). All these compounds can lead to increase in the tiller number, leaf area and overall growth and yield (Joshi et al. 2015). Vermicompost also acts as a "slow release fertilizer," so nutrients remain accessible to plants for longer period of time, whereas nutrients from chemical fertilizers are released faster and depleted faster. When chemical fertilizers are applied to soil, a large portion of nitrogen is lost because of oxidation in the presence of sunlight. Suhane (2007) estimated that if 100 kg urea is applied in agricultural soil only 20–25 kg is accessible to plants while 40–50% is lost in air as ammonia and around 20–25 kg leaches and pollutes groundwater. Impact of vermicompost and inorganic fertilizers on strawberries was studied by Arancon et al. (2004) by applying them combined and separately. In dry shoot weight of strawberries, there was little difference when vermicompost at the rate of 10 tonnes per hectare and inorganic fertilizer at the rate of 85, 155 and 125 NPK kg per hectare, respectively, was applied. But yield and weight of commercial strawberries were highest in those treated with vermicompost after 220 days of transplanting. Similar types of results were also reported by other researchers like Ansari (2008) who observed higher yield of spinach, onion and potato. Dhanalakshmi et al. (2014) also observed higher number of leaves, branches, root and shoot length in the seeds of okra, chilli and brinjal grown in soil amended with vermicompost. Vermicompost is also known to have positive effects on other horticultural crops such as groundnut (Kumar et al. 2014), garlic (Suthar 2009), sweet corn and tomato (Lazcano et al. 2011; Abduli et al. 2013), banana, papaya (Reddy et al. 2014), etc.

8.9.2 Plant Protector

A great diversity of microorganisms such as fungi, bacteria, actinomycetes, etc. is present in vermicompost which makes it a perfect supplement for disease suppression in plants. As per Parle (1963), bacterial count in fresh vermicompost is around 32 million per gram, while it is around 6–9 million per gram in good soil. As vermicompost contains plant growth-promoting bacteria (PGPB), they promote growth directly by fixing nitrogen, by producing growth-related hormones such as 1-amino cyclopropane-1-carboxylate (ACC) deaminase (Glick 2014), and indirectly by producing siderophores, chitinase, cyanide, antibiotics and ß-1,3-glucanase which act as antagonists against pathogenic fungi (Han et al. 2005). It has been evidenced in the recent past that vermicompost has the ability to protect plants from diseases and pests either by inducing biological resistance or by suppressing or killing them (Anonymous 2001; Al-Dahmani et al. 2003; Gupta et al. 2014a; Mosa et al. 2015), as it contains some actinomycetes and antibodies, thereby increasing biological resistance of plants against pests and diseases. Edwards and Arancon (2004) reported that population of arthropods (buds, spider mite, aphids) and their damages to plants like tomato, cabbage and pepper decrease significantly when vermicompost was applied. Mean root disease in tomato was decreased from 82 to 18% and 89 to 26% in *Capsicum* when compost was applied to soil (Ayres et al. 2007).

8.10 Conclusion

It can be concluded that vermicomposting is the best way to recycle organic waste as it is an eco-friendly technique. As compared to other methods of waste management like incineration or waste disposal into landfills, vermicomposting causes no or less pollution and more benefits to environment and economy of the country. The end product of this process can be used as an organic fertilizer, thereby reducing the need for chemical fertilizers which are dangerous to the environment and human health. The quality of vermicompost and speed of recycling can be enhanced by adding different substrates like bone meal, egg shell, banana peel and certain microbial inoculants such as *Azotobacter*, *Azospirillum*, etc. besides maintaining proper temperature, pH, moisture, etc. As already mentioned it is like getting "gold from garbage". If vermicompost can be used as a substitute to inorganic fertilizer for organic food production, it will be a major move towards achieving economic, social and environmental sustainability throughout the globe. The popularity of organic food is growing throughout the world, so the demand for vermicompost will also be great in the future. Besides all these vermicomposting is actually a "one-time investment technology" because earthworms reproduce at a faster rate under optimum conditions and require no labour or energy input device. Therefore, vermicomposting of organic waste at the source of generation should be given first priority in dealing with this waste as it reduces transportation costs, chances of disease transmission and land space for dumping, decreases greenhouse gas emissions (CO_2, CH_4, N_2O), reduces surface and groundwater pollution and besides produces an organic fertilizer which can be used as a soil amendment.

Acknowledgement Special thanks are due to Research Centre, College of Pharmacy, King Saud University, Riyadh, and Deanship of Scientific Research, King Saud University, Kingdom of Saudi Arabia.

References

Abduli MA, Amiri L, Madadian E, Gitipour S, Sedighian S (2013) Efficiency of vermicompost on quantitative and qualitative growth of tomato plants. Int J Environ Res 7:467–472

Al-Dahmani JH, Abbasi PA, Miller SA, Hoitink HA (2003) Suppression of bacterial spot of tomato with foliar sprays of compost extracts under greenhouse and field conditions. Plant Dis 87:913–919

Allevi L, Citterio B, Ferrari A (1987) Vermicomposting of rabbit manure: modifications of microflora. Vermicomposting of rabbit manure: modifications of microflora. In: de Bertoldi M, Ferranti MP, L'Hennite P, Zucconi F (eds) Compost: production, quality and use. Elsevevier, Amsterdam, pp 115–126

Amador JA, Winiarski K, Sotomayor AD (2013) Earthworm communities along a forest coffee agroecosystem gradient: preliminary evidences supporting the habitat - dependent feeding hypothesis. Trop Ecol 54:365–374

Amanullah H (2016) Influence of organic and inorganic nitrogen on grain yield and yield components of hybrid rice in northwestern Pakistan. Rice Sci 23:326–333

Anonymous (2001) Vermicompost as Insect Repellent. Biocycle 1–19

Ansari A, Jaikishun S (2011) Vermicomposting of sugarcane bagasse and rice straw and its impact on the cultivation of Phaseolus vulgaris L. in Guyana, South America. J Agric Tech 7:225–234

Ansari AA (2008) Effect of vermicompost and vermiwash on the productivity of spinach (*Spinacia oleracea*), onion (*Allium cepa*) and potato (*Solanum tuberosum*). World J Agric Sci 4:554–557

Ansari AA, Ismail SA (2001) A case study on organic farming in Uttar Pradesh. J Soil Biol Ecol 27:25–27

Arancon NQ, Edwards CA, Bierman P, Welch C, Metzger JD (2004) Influences of vermicomposts on field strawberries: effects on growth and yields. Bioresour Technol 93:145–153

Ayres M, Ballard R, Wicks T, Barnett S, Ophel-Keller K (2007) Suppression of soil-borne plant diseases using compost. In: 3rd National Compost Research and Development Forum, Organized by COMPOST Australia, Murdoch University, Perth

Bajsa O, Nair J, Mathew K, Ho GE (2004) Pathogen die-off in vermicomposting process. Paper Presented at the International Conference on Small Water and Wastewater Treatment Systems, Perth

Bansal S, Kapoor KK (2000) Vermicomposting of crop residues and cattle dung with Eisenia foetida. Bioresour Technol 73:95–98

Bonkowski M, Schaefer M (1997) Interactions between earthworms and soil protozoa: a trophic component in the soil food web. Soil Biol Biochem 1:499–502

Brintha N, Manimegala G (2015) Vermicomposting of municipal solid waste using an earthworm Perionyx excavatus. Int J Mod Res Rev 3:711–715

Butt KR (1993) Utilization of solid paper mill sludge and spent brewery yeast as a feed for soil-dwelling earthworms. Bioresour Technol 44:105–107

Canellas LP, Olivares FL, Okorokova-Façanha AL, Façanha AR (2002) Humic acids isolated from earthworm compost enhance root elongation, lateral root emergence, and plasma membrane H+-ATPase activity in maize roots. Plant Physiol 130:1951–1957

Chalasani D, Krishna SR, Reddy AVS, Dutt C (1998) Vermiculture biotechnology for promoting sustainable agriculture. Asia Pacific J Rural Devel 8:105–117

Chaoui HI, Zibilske LM, Ohno T (2003) Effects of earthworm casts and compost on soil microbial activity and plant nutriant availability. Soil Biol Biochem 35:295–302

Chauhan HK, Singh K (2013) Effect of tertiary combinations of animal dung with agrowastes on the growth and development of earthworm *Eisenia fetida* during organic waste management. Int J Recyl Org Agric 2:11

Chauhan HK, Singh K (2015) Potancy of Vermiwash with Neem plant parts on the infestation of *Earias vittella* (Fabricius) and productivity of okra (*Abelmoschus esculentus*) (L.) Moench. Asian J Res Pharm Sci 5:36–40

Das S, Charan TK, Mukherjee S, Seal S, Sah RK, Duary B, Kim K, Bhattacharya SS (2018) Impact of edaphic factors and nutrient management on the hepatoprotective efficiency of Carlinoside purified from pigeon pea leaves: an evaluation of UGT1A1 activity in hepatitis induced organelles. Environ Res 161:512–523

Dhanalakshmi V, Remia KM, Shanmugapriyan R, Shanthi K (2014) Impact of addition of vermicompost on vegetable plant growth. Int Res J Biol Sci 312:56–61

Dharani S, Manimegala G, Gunasekaran G, Ananthakrishnaswamy S, Sarojini S (2010) Vermicomposting of fly ash: a potential hazardous waste converted into nutritive compost by earthworm (*Eudrilus eugeniae*). Int J Global Environ 1:384–397

Dominguez J, Edwards C, Subler S (1997) Comparison of vermicomposting and composting. Biocycle 38:57–59

Dominguez J, Edwards CA (2004) Vermicomposting organic wastes: soil zoology for sustainable development in the 21st century. MikhaTl, Cairo, Egypt, pp 369–395

Dominguez J, Edwards CA, Webster M (2000) Vermicomposting of sewage sludge: effect of bulking materials on the growth and reproduction of the earthworm *Eisenia andrei*. Pedobiologia 44:24–32

Edwards CA (1983) Utilization of earthworm composts as plant growth media. In: International Symposium on Agricultural and Environmental Prospects in Earthworm. Rome, Italy, pp 57–62

Edwards CA, Arancon NQ (2004) Vermicompost suppress plant pests and diseases attacks. Biocycle 45:51–54

Edwards CA and Bohlen PJ (1996). Biology and ecology of earthworm. (3rd edn.), Chapman & Hall, London. 426

Edwards CA, Burrows I (1988) The potential of earthworm composts as plant growth media. In: Edwards CA, Neuhauser E (eds) Earthworms in waste and environmental management. SPB Academic Press, The Hague, pp 21–32

Elamparithi V, Manimegala G, Brintha N, Gunasekaran G (2017) Screening of microbial population from vermicompost of and *Eisenia fetida* and *Lampito mauritii*. Int J Zool Appli Sci 2:115–123

El-Duweini AK, Ghabbour SI (1965) Population density and biomass of earthworms in different types of Egyptian soils. J Appl Ecol 2:271–287

Elvira C, Dominguez J, Sampedro L, Mato S (1995) Vermicomposting for the paper pulp industry. Bio Cycle USA 62–63

Evans AC, Guild WJML (1948) Studies on the relationships between earthworms and soil fertility. 4th on the life-cycles of some British Lumbricidae. Ann Appl Bio 35:471–484

Ferreras L, Gomez E, Toresani S, Firpo I, Rotondo R (2006) Effect of organic amendments on some physical, chemical and biological properties in a horticultural soil. Bioresour Technol 97:635–640

Fernández-Gómez MJ, Nogales R, Plante A, Plaza C, Fernández JM (2015) Application of a set of complementary techniques to understand how varying the proportion of two wastes affects humic acids produced by vermicomposting. Waste Manag 35:81–88

Frankenberger WT, Arshad M (1995) Phytohormones in soils: microbial production and function. Marcel Dekker, New York

Fu X, Huang K, Chen X, Li F, Cui G (2015) Feasibility of vermistabilization for fresh pelletized dewatered sludge with earthworms Bimastus parvus. Bioresour Technol 175:646–650

Gezahegn D, Girum T (2017) Growth and reproductive performance of *Eisenia fetida* in three varieties of flower (rose, carnation and hypericum) leftovers. J Entomol Nematol 31:29–35

Glick BR (2014) Bacteria with ACC deaminase can promote plant growth and help to feed the world. Microbiol Res 169:30–39

Gopalakrishnan N (2005) Village wealth from urban waste. In: Jeyaraaj R, Jayraaj IA (eds) Proceedings of the UGC sponsored national level workshop on vermitechnology transfer to NSS programme officers. Rohini Press, Coimbatore, pp 29–31

Goswami L, Nath A, Sutradhar S, Bhattacharya SS, Kalamdhad A, Vellingiri K, Kim K (2017) Application of drum compost and vermicompost to improve soil health, growth, and yield parameters for tomato and cabbage plants. J Environ Manag 200:243–252

Gunthilingaraj K, Ravignanam T (1996) Vermicomposting of sericulture wastes. Madras Agricul J 83:455–457

Gupta R, Yadav A, Garg VK (2014a) Influence of vermicompost application in potting media on growth and flowering of marigold crop. Int J Recycl Org Waste Agri 31:1–7

Gupta S, Kushwah T, Yadav S (2014b) Role of earthworms in promoting sustainable agriculture in India. Int J Curr Microbiol Appl Sci 37:49–460

Hammermeister AM, Warman PR, Jeliakova EA, Martin RC (2004) Nutrient supply and lettuce growth in response to vermicomposted and composted cattle manure. J Biores Technol 2:54–65

Han J, Sun L, Dong X, Cai Z, Sun X, Yang H et al (2005) Characterization of a novel plant growth-promoting bacteria strain Delftia tsuruhatensis HR4 both as a diazotroph and a potential bio-control agent against various plant pathogens. Syst Appl Microbiol 28:66–76

Hartenstein R, Neuhauser EF, Kaplan DL (1979) Reproductive potential of the earthworm Eisenia foetida. Oecologia 43:329–340

ICAR Research Complex for NEH Regionz Umiam—793 103, Meghalaya

Ismail SA (2005) The earthworm book. Other India Press, Mapusa, p 101

Jambhekar HA (1992) Use of earthworm as a potential source to decompose organic waste. In Proc. Natl. Sem. On organic farming, MPKV, College of Agriculture, Pune, pp 52–53

Joshi R, Singh J, Vig AP (2015) Vermicompost as an effective organic fertilizer and biocontrol agent: effect on growth, yield and quality of plants. Rev Environ Sci Biotechnol 14:137–159

Joshi SN (1997) Worm composting. Inora News Lett 1(2)

Kale RD (1998a) Earthworms: nature's gift for utilization of organic wastes. St. Lucie Press, New York

Kale RD (1998b) Earthworm Cinderella of organic farming. Prism Book, Bangalore. 88

Kale RD, Bano K, Krishnamoorthy RV (1982) Potential of Perionyx excavatus for utilizing organic wastes. Pedobiologia 23:419–425

Kale RD, Mallesh BC, Kubra B, Bhagyaraj DJ (1992) Influence of vermicompost application on available micronutrients and selected microbial populations in paddy field. Soil Biol Biochem 24:1317–1320

Kaushik P, Garg VK (2003) Vermicomposting of mixed solid textile sludge and cow dung with the epigeic earthworms *Eisenia fetida*. Bioresour Technol 90:311–316

Kavitha R, Subramanian P (2007) Bioactive compost-a value added compost with microbial inoculants and organic additives. J Appl Sci 7:2514–2518

Kumar DS, Kumar PS, Kumar VU, Anbuganapathi G (2014) Influence of biofertilizer mixed flower waste vermicompost on the growth, yield and quality of groundnut (Arachis hypogea). World Appl Sci J 31:1715–1721

Lakshmi BL, Vizaylakshmi GS (2000) Vermicomposting of sugar factory filter filter pressmud using African earthworms species (Eudrillus eugeniae). J Pollut Res 19:481–483

Lalitha R, Fathima K, Ismail SA (2000) Impact of biopesticides and microbial fertilizers on productivity and growth of *Abelmoschus esculentus*. Vasundh Earth 1(2):4–9
Lazcano C, Revilla P, Malvar RA, Domínguez J (2011) Yield and fruit quality of four sweet corn hybrids (*Zea mays*) under conventional and integrated fertilization with vermicompost. J Sci Food Agri 91:1244–1253
Lee DH, Behera SK, Kim JW, Park HS (2009) Methane production potential of leachate generated from Korean food waste recycling facilities: a lab-scale study. Waste Manag 29:876–882
Lee KE (1985) Earthworms: their ecology and relationship with soil and land use. Academic Press, Sydney.:411
Lim SL, Lee LH, Wu TY (2016) Sustainability of using composting and vermicomposting technologies for organic solid waste biotransformation: recent overview, greenhouse gases emissions and economic analysis. J Clean Prod 111:262–278
Lim SL, Wu TY (2015) Determination of maturity in the vermicompost produced from palm oil mill effluent using spectroscopy, structural characterization and thermogravimetric analysis. Ecol Eng 84:515–519
Lim SL, Wu TY, Lim PN, Shak KP (2015) The use of vermicompost in organic farming: overview, effects on soil and economics. J Sci Food Agri 95:1143–1156
Lunt HA, Jacobson HG (1994) The chemical composition of earthworm casts. Soil Sc 58:367–375
Maji D, Misra P, Singh S, Kalra A (2017) Humic acid rich vermicompost promotes plant growth by improving microbial community structure of soil as well as root nodulation and mycorrhizal colonization in the roots of *Pisum sativum*. Appl Soil Ecol 110:97–108
Monroy F, Aira M, Domínguez J (2008) Changes in density of nematodes, protozoa and total coliforms after transit through the gut of four epigeic earthworms (Oligochaeta). Appl Soil Ecol 39:127–132
Mor S, Ravindra K, Dahiya R, Chandra A (2006) Leachate characterization and assessment of groundwater pollution near municipal solid waste landfill site. Environ Monit Asses 118:435–456
Mosa WF, Paszt LS, Frac M, Trzciński P (2015) The role of biofertilization in improving apple productivity-a review. Adv Appl Microbiol 1:21–27
Muyima NY, Reinecke AJ, VIljoen-Reinecke SA (1994) Moisture requirements of Dendrobaena Veneta (Oligochaeta), a candidate for vermicomposting. Soil Biol Biochem 26:973–976
Narayan J (2000) Vermicomposting of biodegradable wastes collected from Kuvempu University campus using local and exotic species of earthworm. In Proceedings of a national conference on industry and. environment:417–419
Ndegwa PM, Thompson SA (2001) Integrating composting and vermicomposting in the treatment and bioconversion of biosolids. Bioresour Technol 76:107–112
Neuhauser EF, Loehr RC, Malecki MR (1988) The potential of earthworms for managing sewage sludge. In: Edwards CA, Neuhauser EF (eds) Earthworms in waste and environmental management. SPB, The Hague, pp 9–20
Olson HW (1928) The earthworms of Ohio. Ohio Biol Surv Bull 17:47–90
Orozco FH, Cegarra J, Trujillo LM, Roig A (1996) Vermicomposting of coffee pulp using the earthworm Eisenia fetida: effects on C and N contents and the availability of nutrients. Biol Fertil Soils 1:162–166
Ouédraogo E, Mando A, Zombré NP (2001) Use of compost to improve soil properties and crop productivity under low input agricultural system in West Africa. Agric Ecosyst Environ 1:259–266
Pagaria P, Totwat KL (2007) Effects of press mud and spent wash in integration with phosphogypsum on metallic cation build up in the calcareous sodic soils. J Ind Soc Soil Sci 55:52–57
Parkin TB, Berry EC (1994) Nitrogen transformations associated with earthworm casts. Soil Biol Biochem 26:1233–1238
Parle JN (1963) A microbiological study of earthworm casts. J Gen Microbiol 31:13–23
Patangray AJ (2014) Vermicompost: beneficial tool for sustainable farming. Asian J Multidisciplinary Stud 2:254–257

Pathma J, Sakthivel N (2012) Microbial diversity of vermicompost bacteria that exhibit useful agricultural traits and waste management potential. Springer Plus 1:26
Pevey HS, Rowe DR, Tchabonoglous G (1985) Environmental engineering. McGraw Hill International, New Delhi
Pramanik P, Ghosh GK, Ghosal PK, Banik P (2007) Changes in organic–C, N, P and K and enzyme activities in vermicompost of biodegradable organic wastes under liming and microbial inoculants. Bioresour Technol 1:2485–2494
Reddy YTN, Kurian RM, Ganeshamurthy AN, Pannersalvam P, Prasad SR (2014) Effect of organic practices on growth, fruit yield, quality and soil health of papaya cv. Arka Prabhat Indian Hort J 41:9–13
Reinecke A, Viljoen SV, Saayman R (1992) The suitability of Eudrilus eugeniae, Perionyx excavatus and Eisenia fetida (Oligochaeta) for vermicomposting in southern Africa in terms of their temperature requirements. Soil Biol Biochem 24:1295–1307
Reinecke AJ, Kriel JR (1981) Influence of temperature on the reproduction of the earthworm Eisenia fetida (Oligochaeta). S Afr J Zool 16:96–100
Ruz-Jerez BE, Ball PR, Tillman RW (1992) Laboratory assessment of nutrient release from a pasture soil receiving grass or clover residues, in the presence or absence of Lumbricus rubellus or Eisenia fetida. Soil Biol Biochem 24:1529–1534
Saranraj P, Stella D (2012) Vermicomposting and its importance in improvement of soil nutrients and agricultural crops. Novus Nat Sci Res 1:14–23
Seenappa SN, Rao J, Kale R (1995) Conversion of distillery wastes into organic manure by earthworm Eudrilus eugeniae. J IAEM 22:244–246
Senapati BK, Dash MC (1982) Earthworms as waste conditioners. Indust Eng J 11:53–58
Sequi P (1990) The role of agriculture in nutrient cycling. Alma Mater Studiorum 3:155–162
Shiralipour A, McConnell DB, Smith WH (1992) Uses and benefits of MSW compost: a review and an assessment. Biomass Bioenergy 3267–79
Singh J (1997) Habitat preferences of selected Indian earthworm species and their efficiency in reduction of organic material. Soil Biol Biochem 29:585–588
Singh PK, Rajiv S, Sunil HK, Ravindra S Sunita K (2009) Studies on earthworms vermicompost as a sustainable alternative to chemical fertilizers for production of wheat crops. Collaborative research on vermiculture studies, College of Horticulture, Noorsarai, Bihar, India and Griffith University, Brisbane, Australia
Singh R, Embrandiri A, Ibrahim M, Esa N (2011) Management of biomass residues generated from palm oil mill: vermicomposting a sustainable option. Resour Conserv Recycl 55:423–434
Sinha RK, Herat S, Bharambe G, Patil S, Bapat PD, Chauhan K et al (2009) Vermiculture biotechnology: the emerging cost-effective and sustainable technology of the 21st century for multiple uses from waste and land management to safe and sustained food production. Environ Res J 3:41–110
Subler S, Edwards C, Metzger J (1998) Comparing vermicomposts and composts. Biocycle 39:63–66
Suhane RK (2007) Vermicompost (In Hindi); Pub. of Rajendra Agriculture University, Pusa, p 88
Suthar S (2008) Microbial and decomposition efficiencies of monoculture and polyculture vermireactors based on epigeic and anecic earthworms. World J Microbial Technol 24:1471–1479
Suthar S (2009) Vermicomposting of vegetable-market solid waste using *Eisenia fetida*: impact of bulking material on earthworm growth and decomposition rate. Ecol Eng 35:914–920
Tejada M, González JL (2009) Application of two vermicomposts on a rice crop: effects on soil biological properties and rice quality and yield. Agron J 101:336–344
Tiwari SC, Tiwari BK, Mishra RR (1989) Microbial populations, enzyme activities and nitrogen-phosphorus- potassium enrichment in earthworm casts and in surrounding soil of a pineapple plantation. J Biol Fertil Soils 8:178–182
Tomati U, Grappelli A, Galli EV (1987) The presence of growth regulators in earthworm-worked wastes. In on earthworms. In: Proceedings of international symposium on earthworms. Selected Symposia and Monographs, Unione Zoologica Italiana, vol 2, pp 423–435

Villar I, Alves D, Pérez-Díaz D, Mato S (2016) Changes in microbial dynamics during vermicomposting of fresh and composted sewage sludge. Waste Manag 48:409–417
Watanabe H, Tsukamoto J (1976) Seasonal change in size and stage structure of Lumbricid Eisenia foetida population in a field compost and its practical application as the decomposer of organic waste matter. Rev Ecol Biol Sol 13:141–146
Wood TG (1974) The distribution of earthworms (*Megascolecidae*) in relation to soils, vegetation and altitude on the slopes of Mt. Kosciusko, Australia. J Anim Ecol 43:87–106
Wu TY, Lim SL, Lim PN, Shak KPY (2014) Biotransformation of biodegradable solid wastes into organic fertilizers using composting or/ and vermicomposting. Chem Eng Trans 39:1579–1584
Xing M, Li X, Yang J, Huang Z, Lu Y (2012) Changes in the chemical characteristics of water-extracted organic matter from vermicomposting of sewage sludge and cow dung. J Hazard Mater 205:24–31
Zirbes L, Renard Q, Dufey J, Tu KP, Duyet HN, Lebailly P et al (2011) Valorisation of water hyacinth in vermicomposting using epigeic earthworm Perionyx excavatus in Central Vietnam. Biotechnol Agron Soc Environ 15:85–93

Chapter 9
Bio-fertilizers: Eco-Friendly Approach for Plant and Soil Environment

Umair Riaz, Shahzada Munawar Mehdi, Shazia Iqbal, Hafiza Iqra Khalid, Ayesha Abdul Qadir, Wajiha Anum, Munir Ahmad, and Ghulam Murtaza

9.1 Introduction

The soil is a living entity because of the presence of a multitude microflora including actinomycetes (Bhatti et al. 2017), algae, bacteria, and fungi (Khanday et al. 2016; Bhat et al. 2017; Sofi et al. 2017). According to an estimate, about 1×10^8 microorganisms exist in 1 g of soil. Majority of these microorganisms are beneficial for agriculture. Some of the organisms are harmful; however, they are very low in number. It has been reported that only 5–7% of soil microorganisms are harmful (Chowdhury and Mukherjee 2006). Soil degradation is the major limitation in achieving higher crop yields in the developing world, especially among farmers with poor resources (Khosro and Yousef 2012). The extensive and imbalanced utilization of pesticides and chemical fertilizers to enhance the crop production has resulted in various social, environmental, and economic concerns (Santos et al. 2012). Chemical fertilizers are technically based materials which consist of known amounts of macro- and micronutrients. The injudicious application of these fertilizers no doubt has improved the crop yield especially in developing countries but

U. Riaz (✉)
Soil and Water Testing Laboratory for Research, Bahawalpur, Pakistan

S. M. Mehdi
Rapid Soil Fertility, Survey & Soil Testing Institute-Punjab, Lahore, Pakistan

S. Iqbal · H. I. Khalid · A. A. Qadir · G. Murtaza
Institute of Soil and Environmental Sciences, University of Agriculture, Faisalabad, Pakistan

W. Anum
Department of Agronomy, Regional Agricultural Research Institute (RARI), Bahawalpur, Pakistan

M. Ahmad
Soil Science Department, College of Food and Agricultural Sciences, King Saud University, Riyadh, Saudi Arabia

K. R. Hakeem et al. (eds.), *Bioremediation and Biotechnology*,
https://doi.org/10.1007/978-3-030-35691-0_9

also has induced adverse effects on the ecosystem including the contamination of atmosphere and soil and groundwater and increased disease attack by weakening the plant's roots (Chun-Li et al. 2014). Hence, new fertilization strategies with lower cost, more efficiency, and eco-friendly properties are required.

Bio-fertilizers can potentially participate for sustainable agriculture and environment. Recently, the efforts have been made to develop nutrient-rich fertilizer with high quality, called bio-fertilizer, to certify bio-safety. Bio-fertilizer has been known as a substitute for commercial inorganic fertilizer in order to upsurge crop yield by increasing soil fertility in sustainable agriculture. These potential biological fertilizers are eco-friendly as they keep the environment safe and also act as cost-effective agricultural inputs (Khosro and Yousef 2012; Adesemoye and Kloepper 2009).

Bio-fertilizers have arisen as a promising strategy for better nutrient supply in agriculture in recent years. Our whole agriculture is dependent on microbial activities in many ways. A great potential appears for making the use of microbes in enhancing crop yield (Bloemberg et al. 2000).

The term "bio-fertilizer" is defined as "materials consisting of live or cells of effective strains of phosphate-solubilizing, nitrogen-fixing, or cellulolytic microorganism used for seed, soil, or composting area application, for increasing microbial number and to hasten the microbial process which supplements the nutrients that can be simply acquired by plants". The application of bio-fertilizers as soil or seed inoculation multiplies and participates in nutrient cycling and then increases crop productivity (Adesemoye and Kloepper 2009).

9.2 Difference Between Bio-fertilizers, PGPR, and Organic Fertilizers

Though a big difference exists among bio-fertilizer and organic fertilizer, bio-fertilizers have been termed as the organic fertilizer earlier. Bio-fertilizers are microbial inoculants comprising of live cells of microbes like algae, bacteria, and fungi, separately or in combination, which may benefit the crop by increasing productivity, while the organic fertilizers are obtained from or consist of plant sources (green manure) or animal sources (animal manure). Plant growth-promoting rhizobacteria (PGPRs) are microorganisms which make the association with a host plant and enhance the growth of their host (Vessey 2003). However, all the PGPRs cannot be termed as bio-fertilizers. For instance, the bacteria that improve plant growth through the control of harmful organism are termed as biopesticides, but they are not bio-fertilizers. However, some PGPR can improve the growth of plants by working as both biopesticides and bio-fertilizer. For example, *Burkholderia cepacia* strains can stimulate the growth of maize via siderophore production under the low iron condition and also possess biocontrol ability to *Fusarium* sp. (Bevivino et al. 1998). Bio-fertilizers duty comprise of a living cell which enhances the plant growth through enhanced nutrient availability.

9.3 History of Bio-fertilizer

The application of bio-fertilizers in agriculture has begun a long time ago. The acquaintance about microbial inoculum application and its benefits passes from generation to generation in the long history of farmers. The concept of bio-fertilizer emerges from the production of compost on a small scale (Khosro and Yousef 2012; Halim 2009). In this process, microbial culture hastens the decomposition process of agricultural by-products and organic residues and gives healthy crops to harvest (Halim 2009). Beneficial bacterial inoculation with plants can be happening for centuries. Though bacteria were not discovered until 1683, when Von Leeuwenhoek noticed microscopic "animals," the utilization of these bacteria for plant growth stimulation in agriculture has been done since ancient times. Theophrastus (372–287 BC) proposed different soil mixing for the remediation of soil defects (Vessey 2003). From this practice, farmers noticed that application of soil collected from legumes boosted the crop yield, while the application of soil taken from non-legume crops did not affect the crop yield. In the last decades of the nineteenth century, the practice of seed mixing with "naturally inoculated" soil became an endorsed technique of legume inoculation in the USA (Nobbe and Hiltner 1986). In the 1930s, *Bacillus megaterium* was used on a large scale for phosphate solubilization in Eastern Europe. In the 1930s and 1940s, inoculation of legumes with associative, nonsymbiotic, rhizospheric bacteria, like *Azotobacter*, was done on a large scale in Russia (Amutha 2011). Bio-fertilizer's commercial history started with the launch of "Nitragin," a laboratory *rhizobia* culture, by Nobbe and Hiltner in 1895 (Kribacho 2010). In the USA, *Rhizobium* inoculant was first prepared and marketed by the private sector in the 1930s (Smith 1992).

After *rhizobia*, *Azotobacter* was discovered followed by blue-green algae (Kribacho 2010). Vesicular-arbuscular mycorrhizae (VAM) and *Azospirillum* are discovered recently (Rana and Ramesh 2013). In the late 1960s in India, the production of rhizobial inoculum was firstly commenced at IARI, New Delhi, in 1956 (Amutha 2011). In Malaysia, production of microbial inoculants on an industrial scale began at the end of the 1940s. Picking up was started in the 1970s by taking legumes-*Bradyrhizobium* inoculation as a guide. The Malaysian Rubber Board (MRB), a government research institute, has conducted research on young rubber trees in the large plantation by the application of *Rhizobium* inoculums. Bio-fertilizers are generally made as inoculants (carrier based), having active microorganisms (Vessey 2003).

9.4 Mechanisms of Action of Bio-fertilizers

Bio-fertilizers have attracted a significant attention of the researchers in last few years due to their role in improving crop yields, reducing the chemical fertilizers cost, and being less detrimental to the environment (Khan et al. 2010). Bio-fertilizers can stimulate the plant growth either through direct or indirect mechanisms.

Direct mechanism affects the activity of plant growth directly; however these direct ways differ between strains and species. These mechanisms include nitrogen fixation, phosphate solubilization, phytohormones production (auxin, cytokinins, ethylene, gibberellic acid, and abscisic acid), and increasing iron availability through siderophore production. Direct improvement of nutrient uptake has been testified due to increasing influxes of specific ions at the root surface when biofertilizers were applied (Bertrand et al. 2000). Several rhizobacterial genera, e.g., *Agrobacterium*, *Azospirillum*, *Paenibacillus polymyxa*, *Pseudomonas*, and *Erwinia*, are known to produce auxins. *Bacillus* and *Rhizobium* were also found to produce auxin at a different temperature and pH (Ju et al. 2018; Ansari et al. 2013). For instance, many bacteria have established iron uptake systems through siderophore production (DalCorso et al. 2013; Saha et al. 2013; Kundan et al. 2015). Iron is not easily accessible for the plant uptake as it exists as very low-soluble ferric ions (Ganz 2013; Saha et al. 2013; Kundan et al. 2015). Therefore, the microbial siderophores scavenge the iron from minerals by Fe^{3+}complex formation, which is soluble and is taken up by active transport mechanisms. This mechanism is active only under low iron solubility (Saha et al. 2013; Kundan et al. 2015).

Indirect mechanisms refer to the inhibition of the functioning of pathogenic organisms of plants. Indirect mechanisms comprise the production of degrading enzymes, ACC deaminase, induced systemic resistance, antibiotics, competition, hydrogen cyanide, quorum quenching, and siderophore production (Balogh et al. 2010; Frampton et al. 2012).

9.4.1 Nitrogen Fixation

Fixation of atmospheric nitrogen into useable nitrogen that is then converted to ammonia is called nitrogen fixation. Biological nitrogen fixation usually occurs at slight temperatures by nitrogen-fixing microorganisms (Bakulin et al. 2007). Rhizobial bacteria lead to root nodule formation by initiating a series of reactions (Gage 2004). In the root nodule, the bacteria do not contain a cell wall (bacteroid). They fix atmospheric nitrogen by the action of an enzyme called nitrogenase enzyme and then produce ammonia (Olanrewaju et al. 2017). Figure 9.1 shows the nitrogen fixation mechanism.

This biological fixation occurs in a nitrogenase complex, which is a complex enzyme. The Nitrogenase complex is explained as a metalloenzyme consisting of two components: (1) dinitrogenase, which consists of a metal cofactor, and (2) dinitrogenase reductase, which is an iron protein. Dinitrogenase reductase supplies high reducing power electrons, while dinitrogenase uses these electrons to reduce N_2 to NH_3. This process utilizes a large amount of energy, necessitating 16 ATP moles for 1 mole nitrogen reduction. For more ATP production, microbial carbon is allocated to oxidative phosphorylation, rather than storing energy in the form of glycogen through the synthesis of glycogen synthesis. An oxygen-sensitive gene, nitrogenase gene (nif), is required for this process. The nif genes also activate

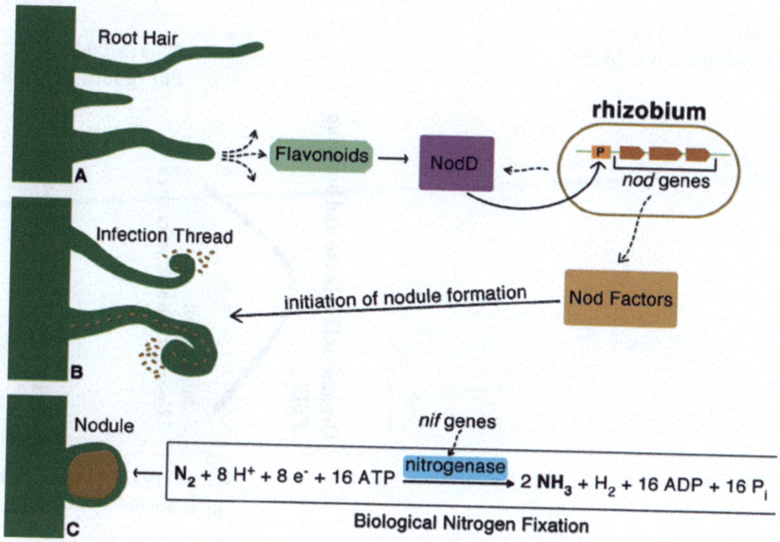

Fig. 9.1 Mechanism of biological nitrogen fixation (*Source*: www.Googleimages.com)

molybdenum, iron, protein, and many other regulatory genes. This gene prevents oxygen from inhibiting the nitrogen fixation and simultaneously supplying enough oxygen for bacteroide respiration inside the nodule. To bind free oxygen, bacterial hemoglobin is introduced (Kundan et al. 2015).

9.4.2 Phosphate Solubilization

Figure 9.2 shows the mechanism of phosphate solubilization.

The main mechanism of phosphate solubilization involves the use of chemicals such as organic acids, siderophores, hydroxyl ion, carbon dioxide, and protons (Rodríguez and Fraga 1999). Organic acids with hydroxyl and carboxyl ions either reduce the pH or make chelates with cations and release the phosphates in a plant-available form (Khosro 2012; Sharma et al. 2017). Organic acid converts tricalcium phosphate to dibasic and monobasic phosphates, and this process boosts phosphorus bioavailability. The type and amounts of organic acid vary with different organisms. Aliphatic acids are more efficient in P solubilization comparative to fumaric acid, citric acids, and phenolics. Tri- and dicarboxylic acids are more efficient compared to aromatic acids and monocarboxylates (Mahdi et al. 2010a). Gaseous "O_2/CO_2" exchange, the release of proton and bicarbonate, lowered pH of the medium (Sharma et al. 2017). Thus, phosphorus availability and rhizosphere pH are inversely related (Olanrewaju et al. 2017).

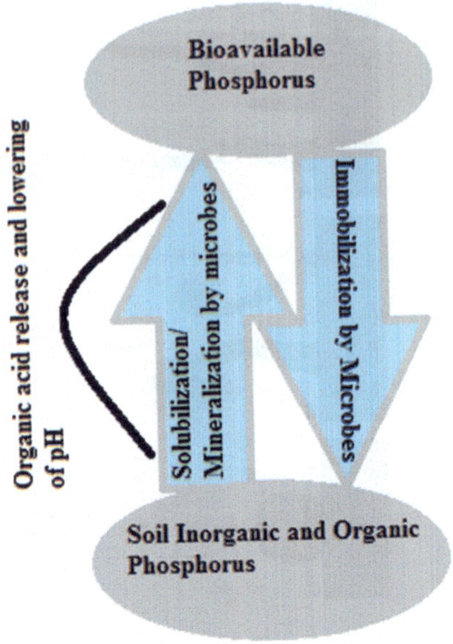

Fig. 9.2 Mechanism of phosphate solubilization by microbes

9.4.3 Zinc Solubilization

The zinc-solubilizing bio-fertilizers acts by secreting organic acids. These organic acids replace the zinc on insoluble chelated compounds and make it accessible for plant uptake (Mahdi et al. 2010b).

9.4.4 Potassium Solubilization

The potassium-solubilizing bio-fertilizers containing potassium-solubilizing microorganisms solubilize silicates through organic acid production and release in the rhizosphere. These organic acids provide H^+ ions and activate hydrolysis. Organic acids such as hydroxyl, carboxylic acids, oxalic acid, citric acid, and keto acids promote the removal of silicates from the cationic complexes into a free or dissolved state. This breakdown of potassium silicate complex also released potassium in the plant-available form (Ju et al. 2018). Figure 9.3 shows the mechanism of potassium and silicate solubilization.

9.4.5 Silicate Solubilization

Some microbial metabolisms produce several organic acids. These organic acids have a double role in weathering of silicate minerals. Organic acids provide H^+ ions and activate hydrolysis. Organic acids such as hydroxyl, carboxylic acids, oxalic

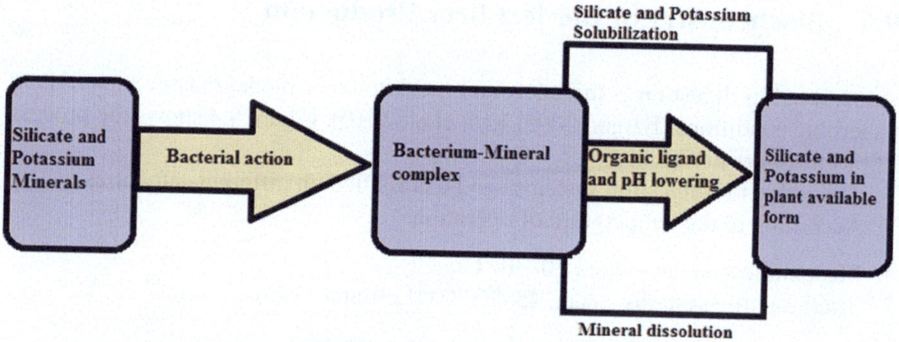

Fig. 9.3 Schematic diagram of silicate and potassium solubilization

acid, citric acid, and keto acids promote the removal of silicates from the cationic complexes into free or dissolved state and then help in silicate retention in the dissolved state in a medium (Rana and Ramesh 2013).

9.4.6 *Sulfur Oxidation*

Plants uptake the sulfur in the form of sulfates. Sulfur-oxidizing microbes oxidize the sulfur to sulfates (Ju et al. 2018).

9.5 Production of Bio-fertilizer

Several factors are needed to be considered in the production of bio-fertilizers including growth profile of microbes, formulation of inoculum, types, and optimum conditions of the organism. The inoculum formulation, application method, and product storage are all critical for the accomplishment of a biological product. Generally, six stages are involved in the production of a bio-fertilizer, i.e., (1) selection of active microbes, (2) isolation of target microbes, (3) carrier material selection, (4) selection of propagation method, (5) phenotype testing, and (6) large scale tests. In the first step, the selection of either nitrogen fixer or organic acid bacteria is made, and the isolation of target microbes is done. Next, the isolated organism is streaked on petri dishes. Selection of right carrier material is of critical importance. For powder bio-fertilizer production, peat or tapioca flour is the best carrier material. Microbial culture from petri dishes is transferred into small flasks. In case of large-scale bio-fertilizer production, it is transferred into the fermenter. At the last stage, large-scale testing in a different environment is performed, and its limitations and effectiveness are analyzed (Khosro and Yousef 2012).

9.6 Biochemistry of Bio-fertilizer Production

Anaerobic bio-digestion is the microbial breakdown of biodegradable materials in anaerobic conditions (Ezigbo 2005; Kim et al. 2010). Figure 9.4 shows the process of bio-fertilizer production.

Anaerobic bio-digestion systems can be classified on different categories.

According to the temperature of operation:

1. Mesophilic systems (i.e., 20–40 °C).
2. Thermophilic systems (i.e., 45–70 °C) (Lettinga 1995).

According to the total suspended solid concentration:

1. Dry systems (between 20 and 40% of total solids).
2. Wet systems (dry matter content of approximately 10%) (Braber 1995).

According to the number of stages considered:

1. Single stage.
2. Multistage processes (Vandevivere et al. 2003).

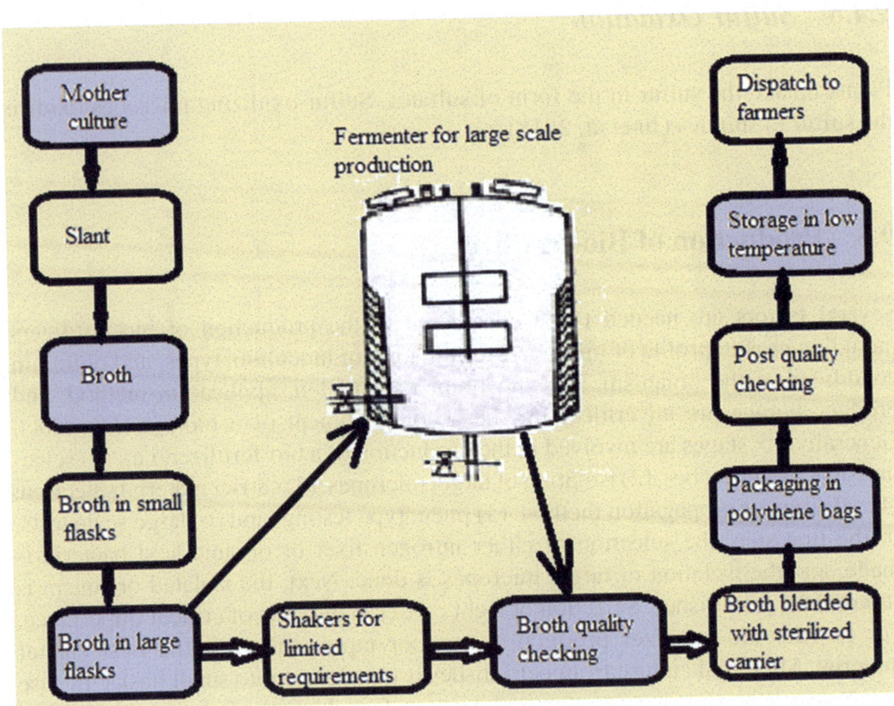

Fig. 9.4 Schematic diagram of bio-fertilizer production

Three biochemical steps are involved in bio-fertilizer preparation that consist of breaking down of complex substances into simpler ones in anaerobic digestion process. Four main stages and three major bacterial groups can be considered in order to simplify the AD process.

9.6.1 Hydrolysis

Hydrolysis is the first step in anaerobic digestion process, in which complex compounds are passed through the cell membrane and then hydrolyzed to monomer compounds (long-chain fatty acids, amino acids, and sugars) through the controlled extracellular enzymes actions, emitted by fermentative bacteria (Ponsá et al. 2008). It is a rate-limiting step for this process. Many groups of anaerobic bacteria take part in this step such as clostridia and bactericides. Some facultative bacteria also take part in this process, e.g., streptococci, etc. (Christy et al. 2014). This is an important step because microorganisms release enzymes to break down large molecules into smaller ones as they cannot use large molecules directly as their food. Extracellular enzymes "cut" the larger compounds into smaller molecules that the microorganism then engulfs and use as nutrition and energy source. Different types of extracellular enzymes are secreted by microorganisms to complete biodegradation and break down a variety of organic materials. Some microorganisms are specific, and they secrete specific enzymes for a specific function. For example, saccharolytic microorganisms secrete enzymes that biodegrade only different sugars; proteolytic microorganism biodegrades only proteins. For biodegradation of proteins, sugars, and fats, different enzymes are secreted (Schnurer and Jarvis 2009). Table 9.1 shows some extracellular enzymes. The rate of hydrolysis reaction varies with the nature of the substrate. Protein decomposition rate is usually faster than cellulose and hemicellulose transformation (Schnurer and Jarvis 2009).

Table 9.1 Some important enzymes, their substrates, and breakdown products (Schnurer and Jarvis 2009)

Enzymes	Substrate	Breakdown products
Cellulase	Cellulose	Cellobiose and glucose
Proteinase	Proteins	Amino acids
Amylase	Starch	Glucose
Lipase	Fats	Glycerol and fatty acids
Hemicellulase	Hemicellulose	Sugars, such as mannose, glucose, xylose, and arabinose
Pectinase	Pectin	Sugars, such as galactose, polygalacturonic, and arabinose acid

9.6.2 A Fermentative Step (Acidogenesis)

In this step, the organic compounds formed in the hydrolytic phase are converted into short-chain volatile fatty acids (VFAs) such as acetic acids, butyric acids, alcohols, carbon dioxide, and hydrogen. Hydrogen is formed as an intermediate product, and it affects the composition of the final product. If hydrogen partial pressure is too low, it would increase the concentration of reduced compounds. Usually, fatty acids, simple sugars, and amino acids are changed into alcohols and organic acids during this phase (Chandra et al. 2012; Gerardi 2003).

9.6.3 Acetogenesis

In this step, acetic acid, hydrogen, and carbon dioxide are produced by the degradation of fatty acids, aromatic compounds, and alcohols (Al Seadi et al. 2008). These acetic acid, hydrogen, and carbon dioxide are used as substrates by the microorganisms active in this phase and carried out anaerobic oxidation (Aslanzadeh 2014). The collaboration of anaerobic oxidation microorganisms is with the methane-forming microorganisms and with the next group. This type of collaboration is dependent on hydrogen partial pressure present in the system (Schnurer and Jarvis 2009; Chandra et al. 2012). When products are transformed into methane, some are converted into volatile fatty acids, alcohols, and methanogenic substrates. Volatile fatty acids with more than 1 unit carbon chain are oxidized to hydrogen and acetate (Al Seadi et al.2008). During the production of hydrogen, protons act as the final electron acceptors, and symbiotic relationship interspecies hydrogen transference happens. Partial pressure plays an important role in this process. Oxidation reactions occur only at low hydrogen partial pressure, explaining the importance of collaboration with the methanogens since they will incessantly utilize hydrogen, to produce methane (Chandra et al. 2012).

9.6.4 Methanogenesis

It is the final, critical (Al Seadi et al. 2008), and rate-limiting biochemical step of the whole anaerobic digestion process. In this step, carbon dioxide and methane are produced by the use of intermediate products through the action of methanogenic bacteria under stern anaerobic conditions (Aslanzadeh 2014).

9.7 Bio-fertilizer Classification

Bio-fertilizers are categorized on the basis of microorganisms' type. Table 9.2 displays the organization of bio-fertilizers.

Table 9.2 Classification of bio-fertilizers

Bio-fertilizer groups		Examples
Nitrogen-fixing bio-fertilizers	Free-living	*Azotobacter, Anabaena, Acetobacter, Beijerinckia, Clostridium, Klebsiella, Nostoc*
	Symbiotic	*Rhizobium (legume), Frankia (non- legume), Anabaena azollae*
	Associative symbiotic	*Azospirillum* sp.
	Fungi	*Penicillium* sp., *Aspergillus awamori*
Phosphate-solubilizing bio-fertilizers	Bacteria	*Bacillus* sp., *Pseudomonas* sp., *Phosphaticum, Burkholderia, Micrococcus, Rhizobium, Agrobacterium, Achromobacter, Aerobacter, Flavobacterium, Erwinia*
	Fungi	*Aspergillus awamori, Penicillium*
Phosphate-mobilizing bio-fertilizers	Arbuscular mycorrhizal fungi	*Glomus, Gigaspora, Scutellospora* sp., *Acaulospora* sp.
	Ectomycorrhiza	*Laccaria* sp., *Pisolithus* sp., *Boletus* sp., *Amanita* sp.,
	Ericoid mycorrhiza	*Pezizella ericae*
Potassium-solubilizing bio-fertilizers		*Bacillus* sp., *Aspergillus niger*
Silicate-solubilizing bio-fertilizers		*Bacillus* sp., *Bacillus circulans, Bacillus mucilaginous*
Zinc-solubilizing bio-fertilizers		*Bacillus* sp., *Pseudomonas* sp., *Acinetobacter, Enterobacter, Flavobacterium, Serratia, Gluconacetobacter, Burkholderia, Saccharomyces* sp.
Sulfur-oxidizing bio-fertilizers		*Thiobacillus* sp.
Organic matter decomposer bio-fertilizers	Cellulolytic	*Cellulomonas, Trichoderma*
	Lignolytic	*Arthrobacter, Agaricus*
Plant growth-promoting rhizobacteria ((PGPR)		*Pseudomonas* sp.

9.7.1 Nitrogen Fixing Bio-fertilizers

Nitrogen is an important macronutrient for crop growth purpose. It is present in the atmosphere in a free state. The part of this nitrogen bargains its entry into the soil by fixation that is performed by a special group of microorganisms. This process is called biological nitrogen fixation, and microorganisms that perform this function are called nitrogen fixer or nitrogen-fixing microorganisms. In this process, the nitrogen is converted into a form that is plant usable (Gothwal et al. 2007). Nitrogen fixer microorganisms are used as bio-fertilizer which is able to fix atmospheric nitrogen to meet plants' need of nitrogen. They are grouped into symbionts such as *Azolla*, *Frankia*, and *Rhizobium*; free-living, *Azospirillum* and *Azotobacter*; and the blue-green algae (Gupta 2004). Some species of nitrogen-fixing microorganisms are shown in Fig. 9.5. Though many genera of nitrogen-fixing microorganisms are reported, only *Azospirillum* and *Azotobacter* have been verified to improve the yield of legumes and cereals under field condition. *Rhizobium* spp., which can fix the atmospheric nitrogen and are mainly associated with legumes, were the first recognized bio-fertilizer and have been commercially used for legumes for more than 100 years (Kannaiyan 2002).

El-Komy (2005) confirmed the advantageous effect of *Bacillus megaterium* and *Azospirillum lipoferum* co-inoculation for improving wheat plant nutrition of nitrogen and phosphorus. The bacterial mixture inoculation gave more balanced nutrition to plants. Improvement in nitrogen and phosphorus uptake by root was the chief mechanism of plants-bacterial interaction.

9.7.2 *Phosphate-Solubilizing Bio-fertilizer*

Phosphorus is classified as organic P and inorganic P in soil. However, only a little of total P (0.1% or 1 ppm) is available for plants due to low solubility and high soil P-adsorbing capacities. Plants absorb P as anions of phosphate (HPO_4^{-2} or $H_2PO_4^-$) from the soil solution, but these phosphate anions are reactive and become inaccessible for plants. When P fertilizers are applied in soil, they often become intricate due to the complex formation with aluminum and iron in low pH soils (Dorahy et al.

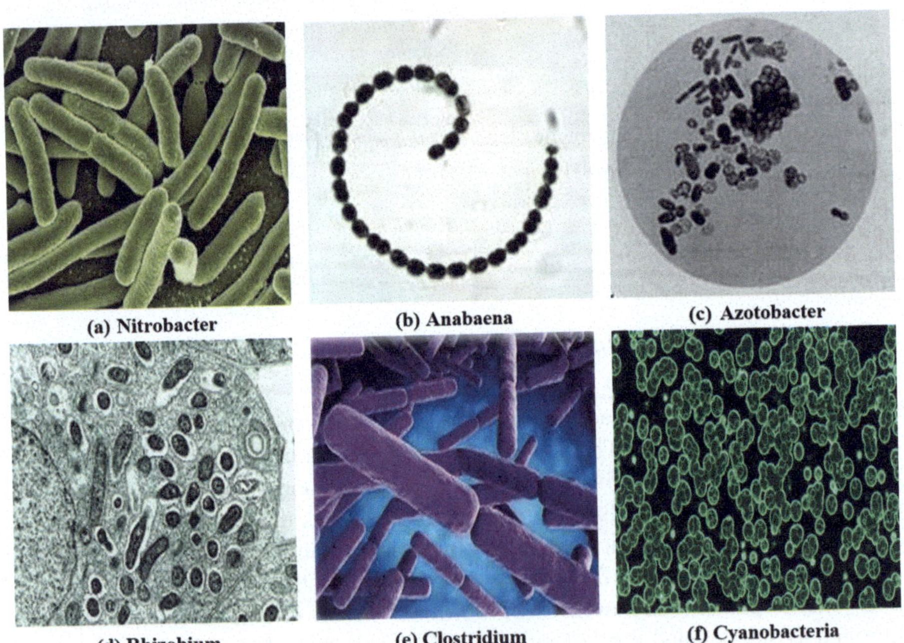

Fig. 9.5 Nitrogen-fixing microorganisms

2005), fixation with calcium and magnesium in high pH soil, and precipitation (Mittal et al. 2008). The overall P utilization efficiency is lower than optimum P utilization efficiency in Pakistani soils (Vance 2001).

Phosphate-solubilizing bio-fertilizers (PSB) contain microorganisms that solubilize the fixed phosphate and make it bioavailable. Many soil fungi and bacteria have the competency to transform insoluble phosphates into soluble forms. This process is accomplished by the excretion of organic acids in the rhizosphere by these organisms. These organic acids decline the soil pH and cause the dissolution of phosphate complexes and make them available to plants (Gupta 2004). Several bacterial species have been found with phosphate-solubilizing ability. These solubilize inorganic phosphate compounds, such as rock phosphate, hydroxyapatite, dicalcium phosphate, and tricalcium phosphate. The more common genera of soil bacteria are *Bacillus* and *Pseudomonas* and fungi. Among other bacterial genera, *Burkholderia*, *Micrococcus*, *Rhizobium*, *Agrobacterium*, *Achromobacter*, *Aerobacter*, *Flavobacterium*, and *Erwinia* are P solubilizer (Subbarao 1988). *Arthrobotrys oligospora*, a nematode fungus, also can solubilize the rock phosphate (Duponnois et al. 2006). The fungus is less effective compared to bacteria in phosphorus solubilization (Alam et al. 2002). Phosphate-solubilizing bacteria exist in large numbers in plant and in the rhizosphere. These bacteria are both aerobic and anaerobic, but aerobic strains are usually found in submerged soils (Raghu and Macrae 2000). Examples include *Bacillus* spp., *Pseudomonas* sp., and *Aspergillus* sp. (Ju et al. 2018) (Fig. 9.6).

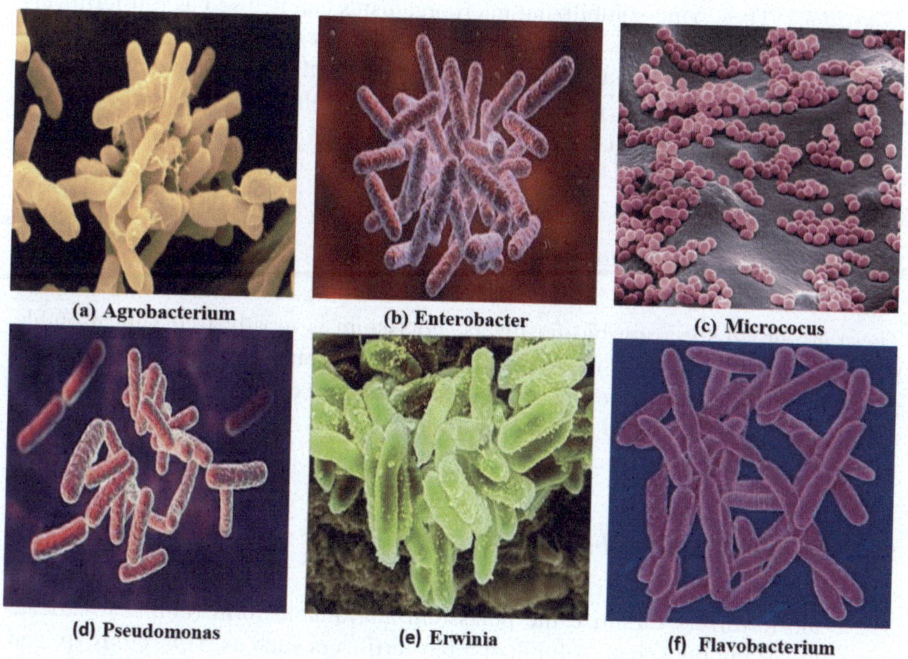

Fig. 9.6 Phosphorus-solubilizing microorganisms

9.7.3 Phosphate-Mobilizing Bio-fertilizers

Phosphate-mobilizing bio-fertilizers work by foraging the soil phosphates and mobilizing the insoluble phosphorus compounds in the soil. Phosphate-solubilizing bio-fertilizer is broad spectrum and also mobilizes the phosphate sometimes (Chang and Yang 2009). Examples are mycorrhiza (Ju et al. 2018).

Mycorrhizae form a symbiotic association with plants. In this association, the fungal partner is penetrated in the root cell and fulfills its carbon necessities from the plant, and in return, the plant is helped by surplus nutrient supply especially phosphorus, copper, calcium, zinc, etc. (Sadhana 2014).

9.7.4 Zinc-Solubilizing Bio-fertilizer

Many nitrogen fixers and phosphorus solubilizer are well accepted as bio-fertilizers nowadays (Subba 2001), but these provide only macronutrients. Soils are also deficient in many micronutrients. The most important of which is zinc because of its low availability. Out of the total, about 75% of applied zinc gets fixed (residual and crystalline iron oxide-bound zinc), and only 1–4%, of totally applied zinc, is used by the plants. Zinc gets fixed by either forming complex by organic ligand or by means of chemisorptions (Alloway 2008). This fixed zinc can be made available by the action of microorganisms such as *Saccharomyces* sp., *B. subtilis*, and *Thiobacillus thiooxidans*. These zinc-solubilizing microorganisms can be used as bio-fertilizers (Raj 2007). In a study, it was recommended that *Bacillus* sp. can be used for increasing zinc availability either alone or in combination with zinc compounds such as zinc carbonate, zinc sulfide, and zinc oxide that are insoluble and cheaper than zinc sulfate (Mahdi et al. 2010b).

9.7.5 Potassium-Solubilizing Bio-fertilizer

These are broad-spectrum bio-fertilizers. Potassium is mostly found in insoluble silicate mineral compounds in the soil. These mineral compounds are unavailable to plants. Only through weathering or solubilization process, these minerals are made accessible for plant uptake (Ju et al. 2018).

9.7.6 Potassium-Mobilizing Bio-fertilizer

These bio-fertilizers mobilize the potassium-unavailable form (bound to silicate minerals). Many phosphate-solubilizing bio-fertilizers such as *Aspergillus* sp. and *Bacillus* sp. carried out phosphate solubilization as well as potassium mobilization (Ju et al. 2018).

9.7.7 Silicate-Solubilizing Bio-fertilizers

Silicate is found in soils as silicate minerals that are unavailable. Many microbes produce several organic acids for converting silicon into an available form (Rana and Ramesh 2013).

9.7.8 Sulfur-Oxidizing Bio-fertilizer

The *Thiobacillus* sp. is a good example of the sulfur-oxidizing microorganism (Ju et al. 2018); commercial bio-fertilizer: Sulfogreen, Sulphomex.

9.7.9 Plant Growth-Promoting Bio-fertilizer (PGPB)

Plant growth-promoting bio-fertilizers are crop specific bio-fertilizers. They produce anti-metabolites and hormones and improve root growth and hasten the process of organic matter decomposition. This decomposition process helps in mineralization and increases the bioavailability of nutrients (Bhattacharyya and Jha 2012). Examples are *Pseudomonas* spp.

9.7.10 Liquid Bio-fertilizers

Liquid bio-fertilizers are usually defined as a "suspensions having agriculturally useful microorganisms." *It is more advantageous than* the carrier inoculants. Liquid bio-fertilizers consisting of microorganisms, such as phosphobacteria *Rhizobium* and *Azospirillum*, are now been used effectively for horticulture crops, vegetables, pulses, sugarcane, rice, millets, and cotton. The reasons behind the increasing use of liquid bio-fertilizers over conventional carrier-based bio-fertilizers are higher competition potentials with native population, quick and easy quality control protocols, longer shelf life (12–24 months), higher populations can be sustained, properties remained unchanged during storage up to 45 °C, more tolerant to temperature, typical fermented smell helps in its easy identification, easy to produce and use for farmers, no contamination, very high enzymatic activity in the meantime contamination is zero, high potential for export, can compete in the global market because of organic crop production, improved soil and seeds survival, their dosages are ten times less than carrier-based powder bio-fertilizers, and cuts the chemical fertilizer use by 15–40% (Rana and Ramesh 2013).

Commercial bio-fertilizer, chitosan concentrate, bass liquid potash, *Azospirillum* bio-fertilizer, liquid consortia bio-fertilizer, potash-mobilizing bio-fertilizer, phosphate-solubilizing bio-fertilizer, etc.

9.7.11 Composting

Compost is used in agriculture as well as in landscaping, as a fertilizer and soil conditioner. Compost is a decomposed remnant of organic matter in the presence of oxygen. This compost making process is called composting. Composting is a biological decomposition of organic waste material in the presence of oxygen at an elevated temperature, carried out by active microorganisms which break down the cellulolytic material. Factors that affect this process include pH, temperature, particle size, oxygen levels, nutrient levels, number, and species of microorganisms (Riaz et al. 2018). Compost is advantageous over chemical fertilizers because of its many useful functions that include means of land reclamation, controls of soil erosion, provides nutrients and support to crops by serving as an absorbent, porous, growing medium and retains soluble mineral and moisture, protects against chemical fertilizers by acting as a buffer, and causes easier till of heavy soils (Somani n.d. www.agriinfo.in).

9.8 Characteristics of Some Microbes Used as Bio-fertilizers

9.8.1 *Rhizobium*

It has its place in the family *Rhizobiaceae* and forms symbiotic relations (Mahdi et al. 2010a). *Rhizobium* is known to fix atmospheric nitrogen in legumes (Gupta 2004). Rhizobia are special bacteria that live either in the soil or in nodules, formed on the roots especially legumes. The *Rhizobium* colony is whitish, slightly transparent, fast growing, water-soaked, and shiny in nature (Somani n.d. www.agriinfo.in). They can fix nitrogen at the rate of 50–100 kg ha^{-1} with only legumes. It is associated with pulse legumes, red-gram, chickpea, pea, black gram, and lentil; oilseed legumes, groundnut and soybean; and forage legumes, lucerne and berseem. It inhabits on the roots of the legumes and forms root nodules (tumor-like growths), which act as ammonia production factories (Mahdi et al. 2010a).

9.8.2 *Azotobacter*

Azotobacter belongs to family *Azotobacteriaceae*, is Gram-negative, and is a free-living, aerobic soil-dwelling, heterotrophic nitrogen-fixing bacterium, used as a bio-fertilizer in most crops (Mahdi et al. 2010a). They range from 2 to 10 μm long and 1 to 2 μm wide in size (Somani n.d. www.agriinfo.in). *Azotobacter* are present in neutral and alkaline soils. Most commonly occurring species of *Azotobacter* is *A. chroococcum* in arable soils (Rana and Ramesh 2013). Other reported species are *A. beijerinckii*, *A. insignis*, *A. macrocytogenes*, and *A. vinelandii* (Subba 2001). *A. chroococcum* is capable of fixing N_2 (2–15 mg N_2 fixed/g of carbon) in culture media. The proliferation of *Azotobacter* is limited by a lack of organic matter (Rana and Ramesh 2013). The *Azotobacter* number hardly exceeds 105 g^{-1} of soil because of the presence of antagonistic microorganisms and lack of organic matter in the soil (Subba 2001). A plant requires nitrogen for its growth, and *Azotobacter* performs nonsymbiotic nitrogen fixation (Somani n.d. www.agriinfo.in). *Azotobacter* is reported in many crops such as sugarcane, rice, bajra, maize, and vegetables (Arun 2007).

9.8.3 *Azospirillum*

Azospirillum belongs to family *Spirilaceae* (Mahdi et al. 2010a) and is a Gram-negative, heterotrophic, motile bacterium and is associated with roots of monocots (Somani n.d. www.agriinfo.in). It lives inside plant roots and does not form root nodules (Rana and Ramesh 2013). *A. brasilense* and *A. lipoferum* are most widely distributed and most beneficial species of this genus. Other species are *A. amazonense*, *A. halopraeferens*, and *A. brasilense* (Mahdi et al. 2010a). The organism multiplies under both aerobic and anaerobic environment. It stimulates the phytohormone production, drought tolerance, and disease resistance. It can fix the substantial amount of nitrogen (20–40 kg N/ha) in non-leguminous plants' rhizosphere such as oilseeds, cereals, cotton, millets, etc. (Rana and Ramesh 2013). The *Azospirillum* forms symbiotic association with C4 plants because they grow and fix nitrogen on salts of organic acids such as aspartic and malic acid (Arun 2007). Thus, it is recommended mainly for maize, sugarcane, sorghum, pearl millet, etc. (Mahdi et al. 2010a).

9.8.4 *Acetobacter*

It is an endotrophic, symbiotic bacteria with the ability of atmospheric nitrogen fixation. It is capable of living inside the sugar plant tissues. It needs high sugar levels that are available in sugarcane tissues. Usage of Acetobacter on a large scale increases crop production (Somani n.d. www.agriinfo.in).

9.8.5 *Beijerinckia*

Beijerinckia is an aerobic, nonsymbiotic free-living, and slow-growing bacteria. The *Beijerinckia* colonies are wrinkled, round, flat, and raised in shape. These microorganisms reside in the rhizosphere of crops and fix the atmospheric nitrogen in acid soil (pH 3.0–4.0). It is commonly used for monocots and applied at 250 g per 10 kg of seeds (Somani n.d. www.agriinfo.in).

9.8.6 *Azolla*

Azolla (*Azolla pinnata*) is an aquatic weed found in shallow ditches, tank, idle pond, and channels. It is found floating on the water surface through small and closely overlapped scale-like leaves and through hanging roots deep in the water. *Azolla* is usually associated with rice cultivation in many countries, for example, the Philippines, Vietnam, Thailand, and China. Azolla bio-fertilizers increased the yield of rice in many experiments. It is known to contribute to 40–60 kg N/ha per rice crop (Rana and Ramesh 2013). Azolla also forms a symbiotic association with blue-green algae (*Anabaena azollae*). They both are applied as co-inoculation. This symbiotic association of *Azolla pinnata* and *Anabaena azollae* is termed as *AZOLLA-ANABAENA COMPLEX*. In this association, blue-green algae fix atmospheric nitrogen for *Azolla*, and *Azolla* provides food and shelter to the algae in return. This ability gives this association a great potential as bio-fertilizer for the agricultural field. It can serve as an alternative fertilizer to chemical nitrogenous fertilizers. It is reported that Azolla application increased rice yields by 0.5–2 t/ha in a field trial (Gupta 2004).

9.8.7 *Cyanobacteria*

Cyanobacteria are symbiotic, free-living, aquatic, and one-celled to many-celled and are red, brown, or purple in color. They cannot live in acidic conditions (Rana and Ramesh 2013). They form a symbiotic association with ferns, fungi, liverworts, and plants, but the most common symbiotic association is formed by *Anabaena azollae* with the ability of nitrogen fixation (Mahdi et al. 2010a). This is only used in paddy fields. BGA bio-fertilizers are applied by a broadcast method in standing water as an algal mass in a paddy field after 1 week of transplantation (Somani n.d. www.agriinfo.in).

9.8.8 *Mycorrhizae*

Mycorrhiza signifies “fungus roots.” Among fungi, arbuscular mycorrhizal (AM) fungi are more abundant and account for 5–50% of soil microbe’s biomass. Out of 150 species of fungi in class *Zygomycetes*, order *Glomales*, an insignificant magnitude is assumed to be *mycorrhizal*. Only six genera of fungi produce *arbuscular mycorrhizal fungi* (AMF). Four genera, *Acaulospora*, *Gigaspora*, *Entrophospora*, and *Scutellospora*, form spores, similar to zygospores. Two genera (*Glomus* and *Sclerocytis*) yield only chlamydospores. Arbuscular mycorrhizal fungi (AMF) form a symbiotic relationship with host plants at the root system, first evolved 400 million years ago (Sawers et al. 2008) (Table 9.3).

Table 9.3 Different crops recommended bio-fertilizer and their application method

Crops	Bio-fertilizer	Method of application
Chickpea, pea, groundnut, soybean, beans, lentil	*Rhizobium*	Seed treatment
Rice	*Azospirillum*	-do-
Oilseeds	*Azotobacter*	-do-
Maize and sorghum	*Azospirillum*	-do-
Tobacco	*Azotobacter*	-do-
Rubber, coconuts	*Azotobacter*	-do-
Fruit plants	*Azotobacter*	-do-
Leguminous plants/trees	*Rhizobium*	-do-

Singh et al. (2015)

9.9 Application Methods

Mainly three types of bio-fertilizer application:

(a) Seed treatment or seed inoculation.
(b) Seedling root dip.
(c) Soil application.

9.10 Advantages of Bio-fertilizer over Chemical Fertilizer

Industrially formulated materials, which comprise known quantities of macro- (nitrogen, phosphorus, potassium) and micronutrients or combination of two or more of these nutrients, are called chemical fertilizers. The practice of chemical fertilizers may lead to air and groundwater pollution as a result of eutrophication of water bodies (Youssef and Eissa 2014). According to Chun-Li et al. (2014), soil acidification as well as atmospheric and groundwater contamination increases due to heavy use of chemical fertilizers and pesticides. These heavy doses reduce immunity of plant roots and make them prone to unwanted diseases. In this scenario, use of nutrient-rich high-quality fertilizers such as bio-fertilizer is a safe and healthy approach to pledge bio-safety.

Bio-fertilizer has been recognized as a competitive option to chemical fertilizer to increase soil fertility and crop production in sustainable farming. These eco-friendly and cost-effective inputs help the farmer to increase productivity of soil in a sustainable way Bio-fertilizers are able to fix nitrogen, solubilize and mobilize phosphate, and promote rhizobacteria (Bhat et al. 2010). The effectiveness of bio-fertilizer depends on selective microorganism that may be useful for the soil suitable packaging for a longer shelf life and adaptable to environment and user (Brar et al. 2012). Microorganisms are not applied directly to the field; instead these are settled

Table 9.4 Replacement of chemical fertilizer by bio-fertilizer

Sr. no.	Bio-fertilizer	Substitutes/ha per year
1	*Rhizobium*	108.6–217.3 kg of urea
2	*Azolla*	20–40 kg urea/10 mg
3	*Azospirillum*	60 kg urea in maize
4	BGA	54–65 kg urea
5	*Frankia*	195 kg urea

Bhowmik and Das (2018)

on some material. This material not only makes the application easier but also increases shelf life and facilitates rapid growth (Mahdi et al. 2010a).

Chemical fertilizers alters the metabolic activities that may be due to drop in osmotic potential. The chemical fertilizers releases more salt ions to the growth media, thus the osmotic pressure outside the embryo organs increases, and, consequently, water is osmotically bound, and thus salt concentration enhances, and water accessibility decreases for the embryo germination (Rafiq et al. 2010).

The bio-fertilizers act as a soil conditioner, and the conditioning property increased organic matter contents to the soil which in turn improves soil structure, prevents oil erosion as well as desertification and increases oil and water retention capacity (Swathi 2010). A functional relationship developed within rhizospheric microorganisms, and due to this, holistic system plant flourishes and grows fruitfully (Ju et al. 2018).

The high cost of chemical fertilizer and unavailability at the time of application further aggravate the economic conditions of farmers. Bio-fertilizer practice considers not only economical but also environment-friendly. Similar to chemical fertilizers, bio-fertilizers increase the soil fertility, crop production, and productivity without causing environment problems (Yadav and Sarkar 2019). Human, plants, and the environment are protected to pollution as well as save wages through bio-fertilization. Additionally, it upgrades soil biota and reduces the use of synthesis fertilizers (Jalilian et al. 2012) (Table 9.4).

Preventive Measures in the Use of Bio-Fertilizer:

1. Bio-fertilizers should not be blended with nitrogen fertilizers.
2. Bio-fertilizers should not be applied with fungicides.
3. Bio-fertilizers should not be exposed to sunlight directly.
4. Bio-fertilizers should always be stockpiled at room temperature, not below 0 and above 35 °C.
5. Used solution should not be kept overnight (Hari and Perumal 2010) (Table 9.5).

Table 9.5 Benefits and drawbacks of diverse application methods

Method	Benefits	Drawbacks	Reference
Carrier-based inoculation	Readily available Readily prepared Inexpensive	Carrier can contaminate the inoculants by unwanted microbes such as peat. No uniformity on the carrier. Short-term storage ability	Smith (1992); Brockwell (1980); Brockwell (1977); Bezdicek et al. (1978)
Seed coating and covering	Easy application No need of specific machine Practiced by farmers in case of pesticide application to seeds	Application of pesticides to the seeds. Sticking agents harmful to bacteria. Flexibility in seeding is less	Brockwell (1977); Brockwell (1980); Bashan and Carrillo (1996); Bashan and Holguin (1997); Bashan and Levanony (1990)
Pelleting	Easy application Preferred by farmers Adaptable in seeding and application Lime pellets can be used for acid soils	Less moisture hindered the survival of bacteria. Need special machinery to prepare thus expensive	Brockwell (1977); Bezdicek et al. (1978); Bashan and Levanony (1990); Bordeleau and Prevost (1981)
Direct soil application	Injection in the root zone is possible Easy and simple	Expose to sun Dehydration problems Require more volume	Brockwell (1977)
Root dipping	Require nursery Simple and easy	Liquid media and bacterial cells needed in large quantity Easily contaminate from environment	Brockwell (1977); Bordeleau and Prevost (1981); Bashan and Levanony (1990)

References

Adesemoye AO, Kloepper JW (2009) Plant–microbes interactions in enhanced fertilizer-use efficiency. Appl Microbiol Biotechnol 85:1–12

Al Seadi T, Ruiz D, Prassl H, Kottner M, Finsterwaldes T, VolkeS JR (2008) Handbook of biogas. University of Southern Denmark, Esbjerg

Alam S, Khalil S, Ayub N, Rashid M (2002) In vitro solubilization of inorganic phosphate by phosphate solubilizing microorganism (PSM) from maize rhizosphere. Int J Agric Biol 4:454–458

Alloway BJ (2008) Zinc in soils and crop nutrition. International Zinc Association, Brussels

Amutha AI (2011) The growth kinetics of Arachis Hypogaea L. Var. TMV-7 under the inoculation of biofertilizers with reference to physiological and biochemical studies. PHD thesis. Department of Botany and Research Centre Scott Christian College (Autonomous). Manonmaniam Sundaranar University, Nagercoil

Ansari MW, Trivedi DK, Sahoo RK (2013) A critical review on fungi mediated plant responses with special emphasis to piriformospora indica on improved production and protection of crops. Plant Physiol Biochem 70:403–410
Arun KS (2007) Bio-fertilizers for sustainable agriculture. Mechanism of P-solubilization, 6th edn. Agribios, Jodhpur, pp 196–197
Aslanzadeh S (2014) Pretreatment of cellulosic waste and high rate biogas production. Doctoral thesis on resource recovery. University of Borås, Borås, pp 1–50
Bakulin MK, Grudtsyna AS, Pletneva AY (2007) Biological fixation of nitrogen and growth of bacteria of the genus Azotobacter in liquid media in the presence of perfluorocarbons. Appl Microbiol Biotechnol 43:399–402
Balogh B, JonesJB IF, Momol M (2010) Phage therapy for plant disease control. Curr Pharm Biotechnol 11:48–57
Bashan Y, Carrillo A (1996) Bacterial inoculants for sustainable agriculture. In New horizons in agriculture: agroecology and sustainable development; J. Pérez-Moreno and R. Ferrera-Cerrato(eds.), pp. 125–155, Proceedings of the 2nd international symposium on agroecology, sustainable agriculture and education. San Luis Potosi, Mexico, 16-18.11.1994. Published by Colegio de Postgraduados en ciencias agricolas, Montecillo, Mexico
Bashan Y, Levanony H (1990) Current status of Azospirillum inoculation technology: Azospirillum as a challenge for agriculture. Can J Microbiol 36:591–608
Bashan Y, Holguin G (1997) Azospirillum-plant relationships: environmental and physiological advances (1990–1996). Can J Microbiol 43:103–121
Bertrand H, Plassard C, Pinochet X, TraineB NP (2000) Stimulation of the ionic transport system in Brassica napus by a plant growth-promoting rhizobacterium. Can J Microbiol 46:229–236
Bevivino A, Sarrocco S, Dalmastri C, Tabacchioni S, Cantale C, Chiarini L (1998) Characterization of a free-living maize-rhizosphere population of Burkholderia cepacia: eject of seed treatment on disease suppression and growth promotion of maize. FEMS Microbiol Ecol 27:225–237
Bezdicek DF, Evans DW, Abeda B, Witters RE (1978) Evaluation of peat and granular inoculum for soybean yield and N2 fixation under irrigation. Agron 70:865–868
Bhat M, Yadav S, Ali T, Bangroo S (2010) Combined effects of rhizobium and vesicular arbuscular fungi on green gram (*Vigna radiata* L.) under temperate conditions. Indian J Ecol 37:157–161
Bhat RA, Dervash MA, Mehmood MA, Bhat MS, Rashid A, Bhat JIA, Singh DV, Lone R (2017) Mycorrhizae: a sustainable industry for plant and soil environment. In: Varma A et al (eds) Mycorrhiza-nutrient uptake, biocontrol, Ecorestoration. Springer International, Berlin, pp 473–502
Bhattacharyya PN, Jha DK (2012) Plant growth-promoting rhizobacteria (PGPR): emergence in agriculture. World J Microbiotechnol 28:1327–1350
Bhatti AA, Haq S, Bhat RA (2017) Actinomycetes benefaction role in soil and plant health. Microb Pathog 111:458–467
Bhowmik SN, Das A (2018) Biofertilizers: a sustainable approach for pulse production. In: Legumes for soil health and sustainable management. Springer, Singapore, pp 445–485
Bloemberg GV, Wijfijes AHM, GEM L, Stuurman N, Lugtenberg BJJ (2000) Simultaneous imaging of Pseudomonas fluorescens WCS 3655 populations expressing three different autofluorescent proteins in rhizosphere: new perspective for studying microbial communities. Mol Plant Microb Int 13:1170–1176
Bordeleau LM, Prevost D (1981) Quality of commercial legume inoculants in Canada. In Proceeding of the 8th North American Rhizobium Conference, K.W. Clark and J.H.G. Stephens (eds.), pp. 562–565, University of Manitoba, Winnipeg, Canada
Braber K (1995) Anaerobic digestion of municipal solid waste: a modern waste disposal option on the verge of breakthrough. Biomass Bioenergy 9:365–376
Brar SK, Sarma SJ, Chaabouni E (2012) Shelf-life of biofertilizers: an accord between formulations and genetics. J Biofertil Biopestic 3:109
Brockwell J (1977) Application of legume seed inoculants. In A Treatise on Dinitrogen Fixation, ed. R. W. F. Hardy & A. H. Gibson. John Wiley and Sons, New York, Section IV, pp. 277–310

Brockwell J (1980) Experiments with crop and pasture legumes—principles and practice. In Methods for Evaluating Biological Nitrogen Fixation. Ed F J Bergersen, pp 417–488, Wiley, Chichester, U.K.

Chandra R, Takeuchi H, Hasegawa T (2012) Methane production from lignocellulosic agricultural crop wastes: a review in context to second generation of biofuel production. Renew Sustain Energy Rev 16:1462–1476

Chang CH, Yang SS (2009) Thermotolerant phosphate solubilizing microbes for multifunctional bio-fertilizer preparation. Biorese Technol 100:1648–1658

Chowdhury A, Mukherjee P (2006) Biofertilizers. Newslett, ENVIS Centre Environ Biotechnol. 9:2–7

Christy PM, Gopinath LR, Divya D (2014) A review on anaerobic decomposition and enhancement of biogas production through enzymes and microorganisms. Renew Sust Energ Rev 34:167–173

Chun-Li W, Shiuan-Yuh C, Chiu-Chung Y(2014) Present situation and future perspective of bio-fertilizer for environmentally friendly agriculture. Annu Rep 1–5

DalCorso G, Manara A, Furini A (2013) An overview of heavy metal challenge in plants: from roots to shoots. Metallomics 5:1117–1132

Dorahy CG, Rochester IJ, Blair GJ (2005) Response of field-grown cotton (*Gossypium hirsutum* L.) to phosphorus fertilisation on alkaline soils in eastern Australia. Soil Res 42:913–920

Duponnois R, Kisa M, Plenchette C (2006) Phosphate solubilizing potential of the nemato fungus Arthrobotrys oligospora. J Plant Nutr Soil Sci 169:280–282

El-Komy HMA (2005) Co-immobilization of A. lipoferum and B. megaterium for plant nutrition. Food Technol Biotech 43:19–27

Ezigbo U (2005) Studies on the production of biogas from droppings and cow dung. Unpublished B.Sc. Thesis. Department of Botany, University of Jos, pp 110–26

Frampton RA, Pitman AR, Fineran PC (2012) Advances in bacteriophage-mediated control of plant pathogens. Int J Microbiol 326–452

Gage DJ (2004) Infection and invasion of roots by symbiotic, nitrogen- fixing rhizobia during nodulation of temperate legumes. Microbiol Mol Biol Rev 68:280–300

Ganz T (2013) Systemic iron homeostasis. Physiol Rev 93:1721–1741

Gerardi MH (2003) The microbiology of anaerobic digesters. Wiley, Hoboken, pp 89–92

Gothwal RK, Nigam VK, Mohan MK, Sasmal D, Ghosh P (2007) Screening of nitrogen fixers from rhizospheric bacterial isolates associated with important desert plants. Appl Ecol Environ Res 6:101–109

Gupta AK (2004) The complete technology book on biofertilizer and organic farming. National Institute of Industrial Research Press, Delhi, pp 242–253

Halim NA (2009) Effects of using enhanced bio-fertilizer containing N-fixer bacteria on patchouli growth. Thesis. Faculty of Chemical & Natural Resources Engineering, University Malaysia, Pahang. 145, 42, pp 913–920

Hari M, Perumal K (2010) Booklet on bio-fertilizer (phosphabacteria). Shri Annm Murugapa Chettiar Research Centre, Chennai, pp 1–6

Jalilian J, Modarres-Sanavy SAM, Saberali SF, Sadat-Asilan K (2012) Effects of the combination of beneficial microbes and nitrogen on sunflower seed yields and seed quality traits under different irrigation regimes. Field Crop Res 127:26–34

Ju I, Wj B, Md S, Ia O, Oj E (2018) A review: biofertilizer—a key player in enhancing soil fertility and crop productivity. J Microbiol Biotechnol Rep 2(2)

Kannaiyan S (2002) Biofertilizers for sustainable crop production. Biotechnology of biofertilizers. Narosa, New Delhi, p 377

Khan RU, Rashid A, Khan MS, Ozturk E (2010) Impact of humic acid and chemical fertilizer application on growth and grain yield of rainfed wheat (*Triticum aestivum* L.). Pak J Agric Res 1:23

Khanday M, Bhat RA, Haq S, Dervash MA, Bhatti AA, Nissa M, Mir MR (2016) Arbuscular mycorrhizal fungi boon for plant nutrition and soil health. In: Hakeem KR, Akhtar J, Sabir M

(eds) Soil science: agricultural and environmental prospectives. Springer International, Berlin, pp 317–332
Khosro M (2012) Phosphorus solubilizing bacteria: occurrence, mechanisms and their role in crop production. Resour Environ 2:80–85
Khosro M, Yousef S (2012) Bacterial bio-fertilizers for sustainable crop production: a review. APRN J Agric Biol Sci 7:237–308
Kim MD, Song M, Jo M, Shin SG, Khim JH, Hwang S (2010) Growth condition and bacterial community for maximum hydrolysis of suspended organic materials in anaerobic digestion of food waste-recycling wastewater. Appl Microbiol Biotechnol 85:1611–1618
Kribacho (2010) Fertilizer ratios, Krishak and Bharati Cooperative Ltd. J Sci 5:7–12
Kundan R, Pant G, Jadon N, Agrawal PK (2015) Plant growth promoting rhizobacteria: mechanism and current prospective. J Fertil Pestic 6:1–9
Lettinga G (1995) Anaerobic digestion and wastewater treatment systems. Antonie Van Leeuwenhoek 67:3–28
Mahdi SS, Hassan GI, Samoon SA, Rather HA, Dar SA, Zehra B (2010a) Bio-fertilizers in organic agriculture. J Phytology 2:42–54
Mahdi SS, Dar SA, Ahmad S, Hassan GI (2010b) Zinc availability—a major issue in agriculture. Res J Agric Sci 3:78–79
Mittal V, Singh O, Nayyar H, Kaur J, Tewari R (2008) Stimulatory effect of phosphate-solubilizing fungal strains (Aspergillus awamori and Penicillium citrinum) on the yield of chickpea (Cicer arietinum L. cv. GPF2). Soil Biol Biochemi 40:718–727
Nobbe F, Hiltner L (1986) Inoculation of the soil for cultivating leguminous plants. US Patent. 570813
Olanrewaju OS, Glick BR, Babalola OO (2017) Mechanisms of action of plant growth promoting bacteria. World J Microbiol Biotechnol 33:197
Ponsá S, Ferrer I, Vázquez F, Font X (2008) Optimization of the hydrolytic–acidogenic anaerobic digestion stage (55 C) of sewage sludge: influence of pH and solid content. Water Res 42:3972–3980
Rafiq MA, Ali A, Malik MA, Hussain M (2010) Effect of fertilizer levels and plant densities on yield and protein contents of autumn planted maize. Pak J Agric Sci 47:201–208
Raghu K, Macrae IC (2000) Occurrence of phosphate-dissolving microorganisms in the rhizosphere of rice plants and in submerged soils. J Appl Bacteriol 29:582–586
Riaz U, Murtaza G, Farooq M (2018) Influence of different sewage sludges and composts on growth, yield, and trace elements accumulation in rice and wheat. Land Degrad Dev 29: 1343–1352
Raj SA (2007) Bio-fertilizers for micronutrients. Biofert Newslet (July), pp 8–10
Rana R, Ramesh KP (2013) Biofertilizers and their role in agriculture. Pop Kheti 1:56–61
Rodríguez H, Fraga R (1999) Phosphate solubilizing bacteria and their role in plant growth promotion. Biotechnol Adv 17:319–339
Sadhana B (2014) Arbuscular mycorrhizal fungi (AMF) as a biofertilizer—a review. Int J Curr Microbiol App Sci 3:384–400
Saha R, Saha N, Donofrio RS, Bestervelt LL (2013) Microbial siderophores: a mini review. J Basic Microbiol 53:303–317
Santos VB, Araujo SF, Leite LF (2012) Soil microbial biomass and organic matter fractions during transition from conventional to organic farming systems. Geoderma 170:227–231
Sawers RJH, Gutjahr C, Paszkowski U (2008) Cereal mycorrhiza: an ancient symbiosis in modern agriculture. Trends Plant Sci 13:93–97
Schnurer A, Jarvis A (2009) Microbiological handbook for biogas plant. Swedish Waste Management, Swedish Gas Centre, Malmö, pp 1–74
Sharma S, Kumar V, Tripathi RB (2017) Isolation of phosphate solubilizing microorganism (PSMs) from soil. J Microbiol Biotechnol Res 1:90–95

Singh M, Kumar N, Kumar S, Lal M (2015) Effect of co-inoculation of B. japonicum, psb and a fungi on microbial biomass carbon, nutrient uptake and yield of soybean (*Glycine max* L. Merril), pp 14–18

Smith R (1992) Legume inoculant formulation and application. Can J Microbiol 38:485–492

Sofi NA, Bhat RA, Rashid A, Mir NA, Mir SA, Lone R (2017) Rhizosphere mycorrhizae communities an input for organic agriculture. In: Varma A et al (eds) Mycorrhiza-nutrient uptake, biocontrol, ecorestoration. Springer International, Berlin, pp 387–413

Somani LL (n.d.) Biofertilizer: commercial production technology & quality control. www.agri-info.in

Subba RNS (2001) An appraisal of biofertilizers in India. In: Kannaiyan S (ed) The biotechnology of biofertilizers. Narosa, New Delhi. (in press)

Subbarao NS (1988) Phosphate solubilizing microorganism. In: biofertilizer in agriculture and forestry. Regional Biofert. Dev. Centre, Hissar, pp 133–142

Swathi V (2010) The use and benefits of bio-fertilizer and biochar on agricultural soils" unpublished B.Sc. Doctoral dissertation, Thesis, Department of Chemical and Biological Engineering, Chalmers University of Technology, Goteborg, Sweden

Vance CP (2001) Symbiotic nitrogen fixation and phosphorus acquisition. Plant nutrition in a world of declining renewable resources. Plant Physiol 127:390–397

Vandevivere P, De Baere L, Verstraete W (2003) Types of anaerobic digesters for solid wastes. In: Biomethanization of the organic fraction of municipal solid wastes. IWA, London

Vessey JK (2003) Plant growth promoting Rhizobacteria as bio-fertilizers. J Plant Soil 225:571–586

Yadav KK, Sarkar S (2019) Biofertilizers, impact on soil fertility and crop productivity under sustainable agriculture. Environ Ecol 37:89–93

Youssef MMA, Eissa MFM (2014) Biofertilizers and their role in management of plant parasitic nematodes. A review. J Biotechnol Pharm Res 5:1–6

Chapter 10
Phytoremediation of Heavy Metals: An Eco-Friendly and Sustainable Approach

Shamsul Haq, Asma Absar Bhatti, Zubair Ahmad Dar, and Suhail Ahmad Bhat

10.1 Introduction

"Environmental pollution has become a severe public health concern because it a major source of health risk and causes several serious diseases throughout the world" (Briggs 2003). The most serious concern of environmental pollution is the presence of toxic metals. The severity of toxic metals on humans has been known since ages, and its exposure continues and is increasing in many areas. Heavy metals severely affect human beings and even cause death (Jarup 2003). The effect of some of the toxic metals on human beings is shown in Table 10.1. Industrialization has increased the heavy metal pollution, and concentration of these heavy metals is higher in industrial areas (Suvaryan et al. 2011; Adesuyi et al. 2015; Jiao et al. 2015).

Naturally "heavy metals are found in the earth's crust" (Jadia and Fulekar 2008; Ismail et al. 2013), with density more than 5 g cm^{-3} (Alloway and Ayres 1997) and an atomic number greater than 20 (Jadia and Fulekar 2008). As, Ni, Hg, Cd, Cr, Pb and Zn are the most common toxic metals in soil and water bodies. "Trace elements are metals whose percentage in rock composition does do not exceed 0.1%" (Ovko and Romic 2011). "Heavy metals in the soil occur naturally from the weathering of parent materials as traces (<1000 mg kg^{-1}) and are not toxic" (Wuana and Okieimen 2011; Parizanganeh et al. 2012). "Anthropogenic sources such as mining, smelting, electroplating, energy and fuel production, power transmission, intensive agriculture, sludge dumping, and melting operations, are the main contributor to heavy metal pollution" (Ismail et al. 2013; Dembitsky 2003; Igwe and Abia 2006; Ali et al. 2013).

S. Haq (✉) · A. A. Bhatti · Z. A. Dar
Sher-e-Kashmir University of Agricultural Science and Technology, Jammu, Jammu and Kashmir, India

S. A. Bhat
Pondicherry University, Puducherry, India

K. R. Hakeem et al. (eds.), *Bioremediation and Biotechnology*,
https://doi.org/10.1007/978-3-030-35691-0_10

Table 10.1 Toxic effect of some heavy metals on human beings (Dixit et al. 2015)

Toxic metal	Effect
Silver	The tissues become grey or bluish grey and cause problems in breathing, irritation in throat and pain in stomach
Arsenic	Oxidative phosphorylation and ATP synthesis are affected
Barium	"Cause cardiac arrhythmias, respiratory failure, gastrointestinaldysfunction, muscle twitching and elevated blood pressure"
Cadmium	"Carcinogenic, mutagenic, endocrine disruptor, lung damage and fragile bones, affects calcium regulation in biological systems"
Chromium	"Hair loss"
Copper	"Brain and kidney damage, elevated levels result in liver cirrhosis andchronic anaemia and stomach and intestine irritation"
Mercury	"Autoimmune diseases, depression, drowsiness, fatigue, hair loss,insomnia, loss of memory, restlessness, disturbance of vision, tremors, temper outbursts, brain damage, lung and kidney failure"
Nickel	"Allergic skin diseases such as itching, cancer of the lungs, nose, sinuses, throat through continuous inhalation, immunotoxic, neurotoxic, genotoxic, affects fertility, hair loss"
Lead	"Excess exposure in children causes impaired development, reduced intelligence, short-term memory loss, disabilities in learning and coordination problems, risk of cardiovascular disease"
Selenium	"Affects endocrine function, impairment of natural killer cell activity, hepatotoxicity and gastrointestinal disturbances"
Zinc	"Dizziness, fatigue", etc.

"Heavy metals in the soil from anthropogenic sources tend to be more mobile, hence bioavailable than pedogenic, or lithogenic ones" (Wuana and Okieimen 2011; Kaasalainen and Yli-Halla 2003).

10.2 Phytoremediation and Mechanisms

The practice of utilizing green plants to release, transfer, stabilize and degrade the toxic pollutants from the soil is known as phytoremediation (Elekes 2014; Paz-Ferreiro et al. 2014). "Phytoremediation is a natural technology with great potential" (Banarjee 2018; Bhat et al. 2018). Several plant roots can absorb and immobilize metal pollutants, whereas other plant species have the ability to break down or accumulate organic pollutants. The term phytoremediation consists of two words: phyto derived from the Greek means plant, and remedium derived from Latin means able to cure or restore (Vamerali et al. 2010). Chaney (1983) was the first to introduce such thought, and later various plant species were developed which have the capability to remove toxic metals from the polluted environments. Phytoremediation is used to remediate a variety of organic (Cluis 2004) and inorganic contaminants (Vamerali et al. 2010) (Fig. 10.1).

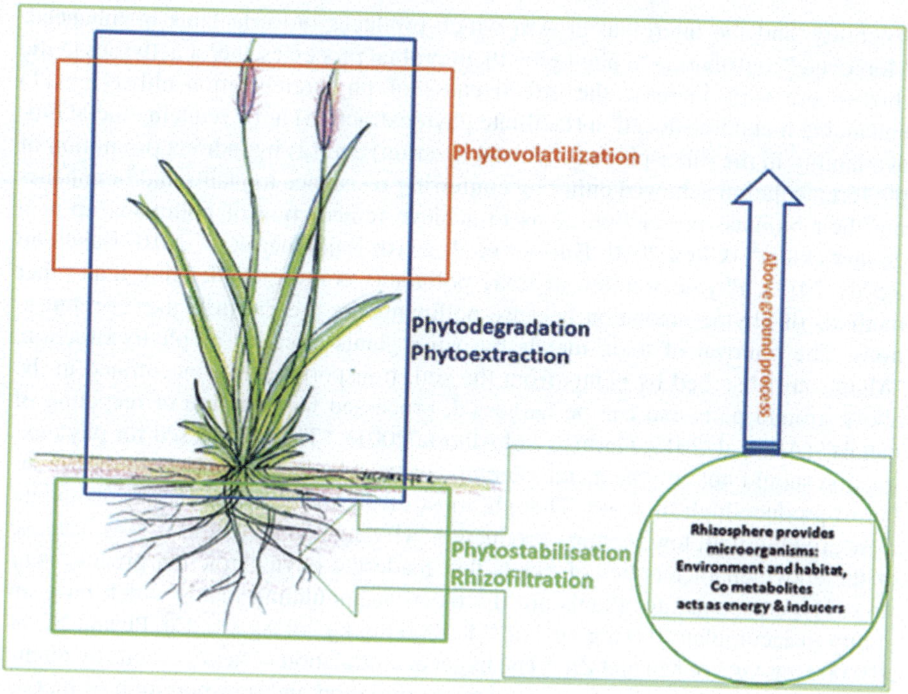

Fig. 10.1 Process of phytoremediation

The technique of phytoremediation is effective whereever contaminants are low to medium in concentration; at higher concentration there is a reduction in plant and microbial growth. "Mechanisms involved in the uptake, translocation, and storage of micronutrients are the same involved to translocate and storage heavy metals" (Subhashini and Swamy 2013). The various methods of phytoremediation are as follows.

10.2.1 Phytoextraction

The absorption of metals from soil by metal accumulators in their harvestable parts is called phytoextraction. Phytoremediation of metals has gained tremendous interest against the microbial remediation from the last few decades (Kramer 2005; Pilon-Smits 2005; Doty 2008). Various strategies of phytoremediation can simultaneously be adapted at a given time for the remediation of a particular contaminant. "Heavy metals exist in colloidal, ionic, particulate and dissolved phases, with higher affinity for humic acids, organic clays and oxides coated with organic matter" (Connell and Miller 1984). Metal's bioavailability in the rhizosphere is greatly affected by plants as well as microbial activities (Ma et al. 2011; Miransari 2011; Aafi et al. 2012). Lipophilic compounds in plant exudates or lysates increase water solubility, and the microbial growth which produces biosurfactants is enhanced. Moreover, "contributing to plant growth, microbial processes and/or activities in the rhizosphere soils increase the effectiveness of phytoremediation either by: (1) enhancing metal translocation (facilitate phytoextraction) or by reducing metal bio-availability in the rhizosphere (phytostabilization) and (2) by indirect promotion of phytoremediation achieved either by conferring resistance to plants and/or enhancing their biomass production so as to achieve remediation of pollutants up to a greater extent" (Glick 2010; Kuffner et al. 2010; Rajkumar et al. 2010; Babu and Reddy 2011). Phytoextraction of toxic pollutants is more challenging than other methods of phytoremediation because pollutants are present in higher concentrations. The removal of toxic metals by using plants is called as phytoextraction. "Metals are absorbed by plants from the soil, transported and concentrated in the above ground parts and can be harvested, processed for dumping or recycling of metals" (Ali et al. 2013; Garbisu and Alkorta 2001). "The plants used for phytoextraction should not only be metal tolerant, but must be fast growing with the potential to produce high biomass. Though, most of the metal-accumulating plants are slow growing with low biomass production" (Evangelou et al. 2007). It is because of the above characteristics of plants that made the phytoextraction process very slow. Hyperaccumulator plants are the metal-accumulating plants which have an ability to accumulate 100 mg kg^{-1} of Cd; 1000 mg $kg^{-1}$of As, Co, Cu, Pb and Ni; or >10,000 mg kg^{-1} of Mn and Zn. "The hyper accumulation of heavy metals by plants depends upon numerous steps, including assimilation and transportation of metals across the membranes of root cells, loading of metals into xylem and translocation to the shoots and sequestration and detoxification of metals within plant tissues" (Yang et al. 2005). The ideal site for metal detoxification is the epidermis, trichomes and cuticle (Rascio and Navari-Izzo 2011). "Cation Diffusion Facilitator (CDF) family members like metal transporter proteins there in the tonoplast are over expressed in Zn and Ni hyper accumulators and these transporters are also reported to be concerned in Ni accumulation by Ni hyperaccumulators" (Gustin et al. 2009; Hammond et al. 2006; Persans et al. 2001; Rascio and Navari-Izzo 2011). "About 400 plants have been known as hyper accumulators which comprises only <0.2% of higher plants" (McGrath and Zhao 2003). However, "these species have an essentially low ability to absorb metals but can accumulate higher concentrations of metals if grown in the soils treated with chemical amendments to increase metal phytoavailability and plant uptake" (Meers et al. 2005). It is reported that plant families like Asteraceae, Brassicaceae, Euphorbiaceae, Fabaceae, Flacourtiaceae and Violaceae accumulate heavy metals in high concentrations (Kumar et al. 1995). Some species of family Brassicaceae are considered as a potential contender for phytoextraction because of its property to scavenge the heavy metals. Pb, Cd, Zn and Ni are the most common metals that the family Brassicaceae scavenges. It is found that "one-third of the concentration of zinc in the tissues of *B. juncea* is present, thus more capable to remediate Zn than *Thlaspi caerulescens* which is commonly known zinc hyperaccumulator and the reason behind this fact is that the production of biomass 10-times greater in *B. juncea* than *T. caerulescens*" (Ebbs and Kochian 1997). The ability of various *Brassica* species to resist and accumulate toxic metals has been verified by experimental study (Kumar et al. 1995). Indian mustard (*B. juncea*) is found to be the chief plant for the remediation of heavy metals, viz. Cd, Cr-IV, 137Cs, Cu, Ni, Pb, U and Zn, from the soil (Jiang et al. 2000). Sunflower (*H. annuus*) and several other plant species remediate radionuclides from the polluted soil. Some limited plant species perform the function of phytoextraction though they have only few desired traits, but genetic modified plants possess important characters for phytoextraction. Phytoextraction is mostly categorized into "(1) chelate-assisted phytoextraction or induced phytoextraction—this approach involves the use of artificial chelates in order to improve the mobility of heavy metal ions so that they become amenable to plants; (2) continuous phytoextraction—this approach involves the natural capability of plants to uptake and scavenge the toxicants" (Salt et al. 1997; Ayyappan et al. 2016).

10.2.2 *Rhizofiltration*

Rhizofiltration is "a type of phytoremediation that uses plant roots to absorb, concentrate and precipitate contaminants present in the soil through the plant root system into the harvestable parts of the roots and above-ground shoots" (Verma et al. 2006). In this process, the plants after raising them hydroponically are transplanted into metal-contaminated water. The roots and shoots of these plants absorb and concentrate toxic metals in them which are then harvested for safe disposal (Padmavathiamma and Li 2007). Rhizofiltration is mostly used for those metals which are retained within the roots like Pb, Cd, Cu, Ni, Zn and Cr (US Environmental Protection Agency 2000). Sunflower has the maximum capability to remediate Pb from water amongst other species like Indian mustard, tobacco, rye, spinach and corn. The bioaccumulation coefficient of Indian mustard is 563 for lead and also has the ability to effectively remove Pb from water with concentration between 4 mg/L and 500 mg/L (Raskin and Ensley 2000; US Environmental Protection Agency 2000). In the rhizosphere the pH changes because several chemical ooze out from the roots which precipitate metals on the root surfaces. "Once roots get saturated with these toxicants, either only the roots or the entire plants are harvested for more processing" (Zhu et al. 1999a). The greatest advantages of rhizofiltration are its ability to utilize terrestrial as well as aquatic plants for in situ or ex situ remediation strategies, and also the contaminants are not transported to shoots. Therefore, plant species that are not hyperaccumulators can also be used in the process of phytoremediation. Terrestrial plants are an ideal selection for rhizofiltration as they have strong and deep root system, which increases the root area (Raskin and Ensley 2000). If the contamination is high in water, then this method is not possible because for the absorption of contaminants by plant roots the contaminants must be in the solution form. For efficient remediation of pollutants, plants should have the property to neutralize their effect of toxic metals and scavenge the same. This can be achieved only when the plants have able and rapidly growing roots. The other essential criteria for effective remediation of toxic metals from an area are the low maintenance cost, easy to handle and plant resistance towards to toxic metals. Aquatic plants like water hyacinth (Zhu et al. 1999b), pennywort (Dierberg et al. 1987) and duckweed (Mo et al. 1989) have the enhanced ability to remediate metals from water, but because of their lesser and young root systems, their potential for rhizofiltration becomes very limited (Dushenkov et al. 1995). Zhu et al. (1999c) Reported that "water hyacinth is a potential macrophyte for the deletion of trace elements from waste streams". Various evidences demonstrate that "terrestrial plants with dense and fibrous root systems are suitable for this technique as they have greater metal-absorbing powers, and the chief examples include Sunflower (*Helianthus annuus* L.) and Indian mustard (*Brassica juncea* Czern.). Indian mustard is known to eradicate a wide concentration of Pb (4–500 mg l^{-1})" (Raskin and Ensley 2000). Metals like Cd, Zn, Cu, Ni and Cr are also removed by these terrestrial plants (Dushenkov et al. 1995) from hydroponic solutions. Blastofiltration (blasto means "seedling" in Greek) is the method in which heavy metals are removed from water with the help of young seedlings water and is considered as advanced technique for treating polluted water. In this technique, "there is a remarkable improvement in surface to volume ratio that typically occurs after germination and some germinating seedlings also adsorb huge quantities of toxic metal ions; this is why young seedlings are suitable for restoring water quality" (Salt et al. 1997). It has been found that blastofiltration is more active and inexpensive than rhizofiltration, but the only benefit of rhizofiltration over blastofiltration is that the rhizofiltration can be used both in situ and ex situ as well.

10.2.3 Phytostabilization

Phytostabilization is the process by which plants reduce the environmental contaminants reduced by stabilizing them. Phytostabilization is the way of achievement of decontamination of toxic metals from the soil. Several plant species accumulate various metals, absorb, adsorb them on the root surfaces and subsequently precipitate them in the root zone. Soil contaminants are immobilized by certain plant species by absorbing and accumulating them in roots, adsorbing them on the surface of roots or precipitating them in the root zone. The phytostabilization uses "abilities of exudates of several plant roots for decrease in the bio-availability of toxic substances as the main reason of phytoremediation to prevent the migration of metals in the environment" (Cheraghi et al. 2011). The extensive root system and a low translocation of metals from roots to shoots are the prerequisite conditions for the plant species to be used for phytostabilization. Phytostabilization is a "plant-based remediation technique that is aimed at reducing the risk of metal pollutants by stabilizing them through formation of a vegetative cap at the plant rhizosphere, where sequestration (binding and sorption) processes immobilize metals so as to make them unavailable for livestock, wildlife and human exposure" (Munshower 1994; Cunningham et al. 1995; Wong et al. 2003). In contrast to other phytoremediation techniques, the major aim of phytostabilization is to stabilize the toxic contaminants and not to remove them from a site, thus decreasing the threat to the environment and human health. Furthermore, phytostabilization is considered to be more advantageous than other techniques because it is cost-effective and very easy to execute (Berti and Cunningham 2000). Generally, "this technique is used to remediate soils contaminated with Zn, As, Cr, Cd, Pb and Cu". In this technology, the hazardous waste dumping is not necessary (US Environmental Protection Agency 2000), and surface and groundwater resources are preserved due to immobilization of toxic metals. This technique inhibits the formation of toxic leachate and prevents soil erosion by decreasing the percolation of water in the soil. There are certain criteria for plants to be followed in order to be used for this technology, which are as follows: (1) low translocation of toxic metals from root to shoot system, (2) should possess rapid growth rate and must be resistant to heavy metals and (3) should be economical. Basically, "this technique is not only applicable at sites with high organic load and porosity but is also efficient for a wide range of surface contamination sites" (Berti and Cunningham 2000). The drawbacks of this technique is that "it is not applicable to those areas which are heavily contaminated because such conditions become an obstacle in plant growth and development" (Berti and Cunningham 2000).

10.2.4 Mechanisms of Phytostabilization

Toxic metals are "adsorbed and precipitated into less soluble forms like carbonates and sulphides, metal complexes with organic compounds and accumulation in root tissues in the rhizosphere" (Mendez and Maier 2008; Wong et al. 2003). Plants in contaminated soil promote microbial populations mostly heterotrophic, which enhance growth rate of plants and cause stabilization of toxic metals. The plant species which keep the metals away from their shoot system are best suited for phytostabilization. However, monitoring of metals in shoots is important although they restrict the metals in their roots only (Mendez and Maier 2008). "Arsenic is best accumulated by *Cynodon dactylon* and thus a promising candidate for phytostabilization" (Leung et al. 2007). A very important role is played by mycorrhizae in stabilizing the metals, and a few mycorrhizae like ericoid and ectomycorrhizal fungi stabilize metals in the rhizosphere (Meharg 2003). "Hyphae of Mycorrhizal fungi have polyphosphate which can bind heavy metals up to saturation and greater than 60% metals are reported to be retained in apoplast cell walls" (Bucking and Heyser 1999; Yang et al. 2005; Khanday et al. 2016; Bhat et al. 2017; Sofi et al. 2017). There are certain plant species which make metals less available by detoxifying them in the roots by the release of organic acids (Brunner et al. 2008; Qin et al. 2007). "Another process for the detoxification of metals is immobilization of metals within fine roots through binding with pectins in the cell walls and to the negatively charged cytoplasm-membrane surfaces owing to their strong electrochemical potential" (Rengel and Zhang 2003). Valence of metals can be reduced by several plant species by the release of redox enzymes and thereby transforming the toxic metals into other forms which are very less toxic (Ali et al. 2013). The best studied example of this strategy adapted by plants is the conversion of more toxic Cr^{+4} to less toxic Cr^{+3} (Bluskov et al. 2005). Using phytostabilization as a technique, extremely promising results have been obtained from chromium and lead stabilization in soils. Cr^{6+} which is highly toxic is converted into Cr^{3+} which is not that toxic by plants which are deep rooted (James 1996). The plants which "can live in metal contaminated soils without affecting growth and retain low concentrations of metals in aerial parts, although concentration of metals is very high in the roots, is known as metal excluder plants" (Kramer 2010; Wei et al. 2005). It has been found that there are certain plants which exclude metals from aerial parts. These include "Ni-excluders such as *Silene vulgaris, Zea mays*, Cu excluder *Hyparrhenia hirta* and Co excluder *Armeria maritima*" (Brewin et al. 2003; Seregin et al. 2003). However, "excluder plants can grow in metal-contaminated soils without affecting their growth and keeping metal concentration in aerial parts at minimum levels" (Wei et al. 2005). There are different approaches which plants utilize to exclude metals, which include mycorrhizae, cell walls and plasma membranes (Hall 2002). "Mycorrhizae in general adopt the same mechanisms as those are adopted by higher plants like binding to extracellular materials or sequestration in the vacuolar compartment" (Hall 2002; Tam 1995). The various hypotheses, which explain the mechanism of metal exclusion, are (1) cell wall metal binding, (2) exudation of metal-chelating ligands and (3) formation of redox and pH barriers at the plasma membrane (Taylor 1987). "There are conflicting reports about the role of the cell wall in metal tolerance of plants" (Hall 2002). There are some researchers which are of the opinion that a very small role is played by the cell wall cell, while others are of the opinion that the accumulated heavy metals in the cell wall remain protein or silicate bound (Bringezu et al. 1999). "Metal tolerant plants select for homeostasis to maintain the high concentration of metals due to their inability to tolerate reactive oxygen species or free radicals" (Dietz et al. 1999; Sharma and Dietz 2009; Panda et al. 2003). The extracellular chelation of Al with citrate and malate activates the tolerance of metals in wheat (Delhaize and Ryan 1995) and Al-resistant *Arabidopsis* discharge of organic acids from roots (Larsen et al. 1998). The soils which cannot be remediated by phytoextraction rapidly, for those soils phytostabilization is very good choice for degradation of contaminants.

10.2.5 Phytovolatilization

In phytovolatilization, the toxic metals like Hg, Se and As are transformed into less toxic and volatile forms into the atmosphere (Malik and Biswas 2012; Marques et al. 2009). Groundwater, soil, sediments and sludges are usually remediated by this technology. Previously "only microorganisms were known to play this role" (Karlson and Frankenberger 1989), but it has been revealed recently that plants (*B. juncea, B. napus*) also hold an excellent property to carry out the process of phytostabilization (Terry et al. 1992). There are some aquatic plants, viz. *Azolla*, rabbit foot grass, rice and pickle weed, are the best volatilizer (Zayed et al. 2000). "Even though this remediation approach has added benefits of minimal site disturbance, less erosion and no need to arrange of contaminated plant material, it is still considered as the most controversial of all phytoremediation technologies as discharge of mercury into the environment is likely to be recycled by precipitation and then redeposit back into the ecosystem" (Henry 2000). It has been revealed that "*Brassica juncea* volatilizes Se into the atmosphere through assimilation of Se from the soil into organic seleno-amino acids, selenocysteine and seleno-methionine" (Banuelos et al. 1993; Banuelos and Meek 1990; Terry et al. 2000). "A gene responsible for reducing mercuric ion into elemental mercury through enzyme mercury reductase has been introduced into *Arabidopsis thaliana* which finally volatilizes large amounts of Hg into the atmosphere" (Rugh et al. 1996). To make a transgenic plant which shows excellent mercury volatilization, a bacterial Hg ion reductase gene was recently incorporated. Further, "it has also been reported that bacterial organomecurial lyase (merB) and mercuric reductase (merA) genes were incorporated into model plants such as *A. thaliana* and *N. tabacum*; the resulting transgenic plants have the potential to absorb elemental mercury (II) as well as methyl mercury from the soil and convert it into a volatile form (Hg0)" (Heaton et al. 1998). "The plantlets of transgenic yellow poplar (*Liriodendron tulipifera*) plantlets were produced which exhibited resistance to and grew well in normally toxic levels of ionic mercury, and the transgenic plantlets volatilized about 10-times more elemental mercury than non-transgenic plantlets" (Rugh et al. 1998). Presently by means of this technology, "tritium (3H), a radioactive isotope of hydrogen, is decayed to stable helium with a half-life of about 12 years as reported Dushenkov et al. (1995)".

10.2.6 Advantages and Limitations of Phytoremediation

The various advantages using this technique for remediating metal contaminants are as follows:

1. Low cost.
2. Environment-friendly.
3. A good range of obnoxious metals are remediated.
4. A very attractive technique.

On the other hand, phytoremediation has certain limitations. It takes years to clean the site as it is a very lengthy process. Furthermore, it is only used to clean the subsurface layer of the soil. The various advantages and disadvantages of phytoremediation and various mechanisms are shown in Table 10.2.

10.2.7 Plant Selection Criteria for Phytoremediation

Root depth, soil contaminants, soil and regional climate are the basis on which plant species are selected to carry out the process of phytoremediation. The depth of soil is directly impacted by the root (US Environmental Protection Agency 2001), and

Table 10.2 Advantages and limitations of phytoremediation (Elekes 2014; Kumar et al. 1995; Ashraf et al. 2013; Newman et al. 1997)

Mechanisms	Advantages	Limitation
Phytoextraction	• Less expensive • Permanent removal of toxic metals from the environment • Substantial decrease in disposal of waste material • Recycling of contaminates	• "Metal hyperaccumulators are generally slow growing with a small biomass and shallow root systems" • Plant biomass must be harvested and removed • Show phytotoxic effect
Phytostabilization	• No requirement of hazardous and biomass disposal • "Very effective when rapid immobilization is needed to preserve ground- and surface waters" • Soil erosion is reduced and soil water is decreased • It is cost-efficient • Revegetation enhances the stability of ecosystem	• Formation of leachate is prevented • Higher doze of fertilizers is used • Phytostabilization is an interim measure • "Phytostabilization causes stabilization by decreasing the amount of water moving through the soil"
Phytovolatilization	• Mercuric ion is converted into less toxic one • "Contaminants or metabolites released to the atmosphere might be subject to more effective or rapid natural degradation processes such as photodegradation"	• "Hazardous metabolite may be released into the atmosphere which may accumulate in vegetation" • In plant tissues metabolites are found in low levels
Rhizofiltration	• "Both terrestrial and aquatic plants are used for either in situ or ex situ" • Hyperaccumulators may be used other than accumulators • An ex situ system can be used anywhere	• Adjusting the pH constantly • Plants are first grown in a greenhouse • Harvesting and plant disposal are done periodically • "The chemical speciation and interaction of all species in the influent have to be understood and accounted for"

"it varies among different types of plants, and vary significantly for one species depending on local conditions such soil structure, depth of a hard pan, soil fertility, cropping pressure, or other conditions" (Pivetz 2001). "The cleaning depths are approximately <3 ft. for grasses, <10 ft. for shrubs and <20 ft. for deep rooting trees" (Sharma and Reddy 2004). It has been reported that grasses have fast growth, huge biomass, strong resistance and ability to decontaminate various soil types as compared to trees and shrubs (Shu et al. 2002; Elekes 2014). "They are pioneers and usually are adapted to adverse conditions such as low soil nutrient content, stress environment and shallow soils" (Malik et al. 2010; Xia et al. 1999; Ye et al. 2000; Sinha et al. 2013). Leaching, runoff and erosion are reduced by the large surface area of their fibrous roots and thus offer advantages for phytoremediation (Garba et al. 2012). The erosion is prevented by extensive canopy of shrubs and trees. In addition, high nutrient is provided by shrubs and trees to the grass, while they lower the water stress and improve soil physical properties (Hamzah and Priyadarshini 2014; Tiedemann and Klemmedson 2004). Developed root system of grasses stabilizes the soils and reduces erosion, while nitrogen is added by legumes (Kidd et al. 2009; Sanchez et al. 2001; Carvalho et al. 2013).

In a relatively short time, grasses develop a large biomass and are metal-tolerant biosystems, which accumulate high concentration of toxic metals (Elekes 2014). However, "the shorter growing period of the seasonal flowering plants is a better option in phytoremediation over perennial plants, as it can be harvested yearly or seasonally, and the area can be replanted with subsequent seasonal flowering plants" (Sinha et al. 2013). "It is better to use plant species adapted to the climatic and soil conditions of the area to be de-polluted" (Elekes 2014; Pivetz 2001; Tordoff et al. 2000). The native plant species are favoured because they are tolerant to stress conditions, have low maintenance cost and are environmental and human friendly than non-native or genetically engineered species (Compton et al. 2003). However, "particular non-native plant may work best remediation of specific contaminant and can be safely used under circumstances where the possibility of invasive behaviour has been eliminated" (USEPA 2000) (Table 10.3).

Table 10.3 Main characteristics of phytoremediation processes (Subhashini and Swamy 2013; Ali et al. 2013; Vamerali et al. 2010; Elekes 2014; Alkorta et al. 2004; Pulford and Watson 2003; Mendez and Maier 2008; Tangahu et al. 2011; Bhat et al. 2018)

Process	Pollutants	Media	Criteria for selection
Phytoextraction	Organic and inorganic	• Soils • Sediments • Water • Sludges	• "Metal tolerance to high concentrations" • "Capability of higher accumulation of metals" • "fast growth" • "Easy harvesting and disposal" • "Translocation factor should be high" • "Should be easily managed" • "Pathogen and pest resistance"
Phytostabilization	Heavy metals and chlorinated solvents	• Soil • Sediments • Sludges	• Extended root system should develop • Translocation of metals to shoots should be very low • Should have a capability to retain pollutants in their roots or rhizosphere
Phytovolatilization	Chlorinated solvents and inorganic compounds	• Groundwater • Soil • Sediments • Sludge	
Rhizofiltration	Toxic metals and organic compounds	• Surface • Waters • Wastewaters	• Should be metal resistant • Should have very high surfaces for adsorption • Hypoxia tolerant • Preference to terrestrial plants because they have long and fibrous roots

10.3 Conclusion

Heavy metal pollution of soil and water poses a severe threat to the environment as well as human health. The increase in human health risk has led researchers to focus on eco-friendly and risk-free technology. Phytoremediation is a conventional and environment-friendly technique to depollute the toxic substances present in the natural environment. Exploring deep into the understanding of the mechanism of it would certainly increase our knowledge level, thus allowing us to choose the suitable process of phytoremediation and the appropriate species for the remediation of polluted environment. Phytoremediation has shown promising results in cleaning up the various pollutants in the various environment, be it soil or water. The new phytoremediation technology should be encouraged to safeguard our environment, and much work should be carried in this cheap method of pollution control.

References

Aafi NE, Brhada F, Dary M, Maltouf AF, Pajuelo E (2012) Rhizostabilization of metals in soils using *Lupinus luteus* inoculated with the metal resistant rhizobacterium *Serratia* sp. MSMC 541. Int J Phytoremediation 14:261–274

Adesuyi AA, Hjoku KL, Akinola MO (2015) Assessment of heavy metals pollution in soil and vegetation around selected industries in Lagos state, Nigeria. J Geosci Env Prot 3:11–19

Ali H, Khan E, Sajad MA (2013) Phytoremediation of heavy metals—concepts and applications. Chemosphere 91:869–881

Alkorta I, Hernandez-Allica J, Becerril JM, Amezaga I, Albizu I, Garbisu C (2004) Recent findings on the phytoremediation of soils contaminated with environmentally toxic heavy metals and metalloids such as zinc, cadmium, lead, and arsenic. Rev Environ Sci Biotechnol 3:71–90

Alloway BJ, Ayres DC (1997) Chemical principles of environmental pollution, vol 168. Blackie Academic, London

Ashraf MA, Maah MJ, Yusoff I (2013) Evaluation of natural phytoremediation process occurring at extin mining catchment. Chiang Mai J Sci 40:198–213

Ayyappan D, Sathiyaraj G, Ravindran KC (2016) Phytoextraction of heavy metals by Sesuvium portulacastrum L. a salt marsh halophyte from tannery effluent. Int J Phytoremediation 18(5):453–459

Babu AG, Reddy S (2011) Dual inoculation of *Arbuscular mycorrhizal* and phosphate solubilizing fungi contributes in sustainable maintenance of plant health in fly ash ponds. Water Air Soil Poll 219:3–10

Banarjee P (2018) Phytoremediation: using natural strength for curing nature. Acta Sci Agric 2(2):44–153

Banuelos G, Meek D (1990) Accumulation of selenium in plants grown on selenium-treated soil. J Environ Qual 19:772–777

Banuelos G, Cardon G, Mackey B, Ben-Asher J, Wu L, Beuselinck P, Akohoue S, Zambrzuski S (1993) Boron and selenium removal in boron-laden soils by four sprinkler irrigated plant species. J Environ Qual 22:786–792

Berti WR, Cunningham SD (2000) Phytostabilization of metals. In: Raskin I, Ensley BD (eds) Phytoremediation of toxic metals: using plants to clean-up the environment. Wiley, New York, pp 71–88

Bhat RA, Dervash MA, Mehmood MA, Bhat MS, Rashid A, Bhat JIA, Singh DV, Lone R (2017) Mycorrhizae: a sustainable industry for plant and soil environment. In: Varma A et al (eds) Mycorrhiza-nutrient uptake, biocontrol, ecorestoration. Springer International, Berlin, pp 473–502

Bhat RA, Dervash MA, Qadri H, Mushtaq N, Dar GH (2018) Macrophytes, the natural cleaners of toxic heavy metal (THM) pollution from aquatic ecosystems. In: Environmental contamination and remediation. Cambridge Scholars, Cambridge, pp 189–209

Bluskov S, Arocena J, Omotoso O, Young J (2005) Uptake, distribution, and speciation of chromium in *Brassica juncea*. Int J Phytoremediation 7:153–165

Brewin L, Mehra A, Lynch P, Farago M (2003) Mechanisms of copper tolerance by *Armeria maritima* in Dolfrwynog bog, North Wales-initial studies. Environ Geochem Health 25:147–156

Briggs D (2003) Environmental pollution and the global burden disease. Br Med Bull 68:1–24

Bringezu K, Lichtenberger O, Leopold I, Neumann D (1999) Heavy metal tolerance of *Silene vulgaris*. J Plant Physiol 154:536–546

Brunner I, Luster J, Günthardt-Goerg MS, Frey B (2008) Heavy metal accumulation and phytostabilisation potential of tree fine roots in a contaminated soil. Environ Pollut 152:559–568

Bucking H, Heyser W (1999) Elemental composition and function of polyphosphates in *Ectomycorrhizal fungi*—an X-ray microanalytical study. Mycol Res 103:31–39

Carvalho A, Nabais C, Roiloa SR, Rodriguez-Echeverria S (2013) Revegetation of abandoned copper mines: the role of seed banks and soil amendments. Web Ecol 13:69–77

Cheraghi M, Lorestani B, Khorasani N, Yousefi N, Karami M (2011) Findings on the phytoextraction and phytostabilization of soils contaminated with heavy metals. Biol Trace Elem Res 144:1133–1141

Cluis C (2004) Junk-greedy greens: phytoremediation as a new option for soil decontamination. BioTeach J 2:61–67

Compton HR, Prince GR, Fredericks SC, Gussman CD (2003) Phytoremediation of dissolved phase organic compounds: optimal site considerations relative to field case studies. Remediation 13:21–37

Connell DW, Miller GJ (1984) Chemistry and ecotoxicology of pollution, vol 444. Wiley, New York

Cunningham SD, Berti WR, Huang JWW (1995) Phytoremediation of contaminated soils. Trends Biotechnol 13:393–397

Delhaize E, Ryan PR (1995) Aluminum toxicity and tolerance in plants. Plant Physiol 107:315

Dembitsky V (2003) Natural occurrence of Arseno compounds in plants, lichens, Fungi, algal species, and microorganisms. Plant Sci 165:1177–1192

Dierberg FE, DeBusk TA, Goulet NA (1987) In: Reddy KR, Smith WH (eds) Aquatic plants for water treatment and resource recovery. Magnolia, Orlando, pp 497–504

Dietz KJ, Baier M, Kramer U (1999) Free radicals and reactive oxygen species as mediators of heavy metal toxicity in plants, heavy metal stress in plants. Springer, Berlin, pp 73–97

Dixit R, Wasiulah MD, Pandiyan K, Singh UB, Sahu A, Shukla R, Singh BP, Rai JP, Sharma PK, Lade H, Paul D (2015) Bioremediation of heavy metals from soil and aquatic environment: an overview of principles and criteria of fundamental processes. Sustainability 7:2189–2212

Doty SL (2008) Enhancing phytoremediation through the use of transgenics and endophytes. New Phytol 179:318–333

Dushenkov V, Kumar PB, Motto H, Raskin I (1995) Rhizofiltration: the use of plants to remove heavy metals from aqueous streams. Environ Sci Technol 29:1239–1245

Ebbs SD, Kochian LV (1997) Toxicity of zinc and copper to *Brassica* species: implications for phytoremediation. J Environ Qual 26:776–781

Elekes CC (2014) Eco-technological solutions for the remediation of polluted soil and heavy metal recovery. In: Hernández-Soriano MC (ed) Environmental risk assessment of soil contamination. In Tech, Rijeka, pp 309–335

Evangelou MW, Ebel M, Schaeffer A (2007) Chelate assisted phytoextraction of heavy metals from soil. Effect, mechanism, toxicity, and fate of chelating agents. Chemosphere 68:989–1003

Garba ST, Osemeahon AS, Humphrey M, Barminas JT (2012) Ethylenediaminetetraacetic acid (EDTA)-assisted phytoremediation of heavy metal contaminated soil by *Eleusine indica* L. Gearth. J Environ Chem Ecotoxicol 4:103–109

Garbisu C, Alkorta I (2001) Phytoextraction: a cost-effective plant-based technology for the removal of metals from the environment. Bioresour Technol 77:229–236
Glick BR (2010) Using soil bacteria to facilitate phytoremediation. Biotechnol Adv 28:367–374
Gustin JL, Loureiro ME, Kim D, Tikhonova M, Salt DE (2009) MTP1-dependent Zn sequestration into\shoot vacuoles suggests dual roles in Zn tolerance and accumulation in Zn-hyperaccumulating plants. Plant J 57:1116–1127
Hall J (2002) Cellular mechanisms for heavy metal detoxification and tolerance. J Exp Bot 53:1–11
Hammond JP, Bowen HC, White PJ, Mills V, Pyke KA, Baker AJ, Whiting SN, May ST, Broadley MR (2006) A comparison of the *Thlaspi caerulescens* and *Thlaspi arvense* shoot transcriptomes. New Phytol 170:239–260
Hamzah A, Priyadarshini R (2014) Identification of wild grass as remediator plant on artisanal gold mine tailing. Plant Sci Int 1:33–40
Henry (2000) An overview of phytoremediation of lead and mercury. NNEMS Report, Washington, DC, pp 3–9
Igwe JC, Abia AA (2006) A bioseparation process for removing heavy metals from waste water using biosorbents. Afr J Biotechnol 5:1167–1179
Ismail S, Khan F, Zafar Iqbal M (2013) Phytoremediation: assessing tolerance of tree species against heavy metal (PB and CD) toxicity. Pak J Bot 45:2181–2186
Jadia CD, Fulekar MH (2008) Phytoremediation: the application of vermicompost to remove zinc, cadmium, copper, nickel and Lead by sunflower plant. Environ Eng Manag J 7:547–558
James BR (1996) Peer reviewed: the challenge of remediating chromium-contaminated soil. Environ Sci Technol 30:248–251
Jarup L (2003) Hazards of heavy metal contamination. Br Med Bull 68:167–182
Jiang W, Liu D, Hou W (2000) Hyperaccumulation of lead by roots, hypocotyls, and shoots of *Brassica juncea*. Biol Planta 43(4):603–606
Jiao X, Teng Y, Zhan Y, Wu J, Lin X (2015) Soil heavy metal pollution and risk assessment in Shenyang industrial district, northeast China. PLoS One 10(5):1–9
Kaasalainen M, Yli-Halla M (2003) Use of sequential extraction to assess metal partitioning in soils. Environ Pollut 126:225–233
Karlson U, Frankenberger WT (1989) Accelerated rates of selenium volatilization from California soils. Soil Sci Soc Am J 53:749–753
Khanday M, Bhat RA, Haq S, Dervash MA, Bhatti AA, Nissa M, Mir MR (2016) Arbuscular mycorrhizal fungi boon for plant nutrition and soil health. In: Hakeem KR, Akhtar J, Sabir M (eds) Soil science: agricultural and environmental prospectives. Springer International, Berlin, pp 317–332
Kidd P, Barcelo J, Bernal MP, Navari-Izzo F, Poschenrieder C, Shilev S, Clemente R, Monterroso C (2009) Trace element behaviour at the root-soil Interface: implications in phytoremediation. Environ Exp Bot 67:243–259
Kramer (2005) Phytoremediation: novel approaches to cleaning up polluted soils. Curr Opin Biotechnol 2:133–141
Kramer U (2010) Metal hyperaccumulation in plants. Annu Rev Plant Biol 61:517–534
Kuffner M, De Maria S, Puschenreiter M, Fallmann K, Wieshammer G, Gorfer M (2010) Culturable bacteria from Zn and cd accumulating *Salix caprea* with differential effects on plant growth and heavy metal availability. J Appl Microbiol 108:1471–1484
Kumar PBA, Dushenkov V, Motto H, Raskin I (1995) Phytoextraction: the use of plants to remove heavy metals from soils. Environ Sci Technol 29:1232–1238
Larsen PB, DegenhardtJ TCY, Stenzler LM, Howell SH, Kochian LV (1998) Aluminum-resistant Arabidopsis mutants that exhibit altered patterns of aluminum accumulation and organic acid release from roots. Plant Physiol 117:9–17
Leung H, Ye Z, Wong M (2007) Survival strategies of plants associated with arbuscular *Mycorrhizal fungi* on toxic mine tailings. Chemosphere 66:905–915
Ma Y, Prasad MNV, Rajkumar M, Freitas H (2011) Plant growth promoting rhizobacteria and endophytes accelerate phytoremediation of metalliferous soils. Biotechnol Adv 29:248–258

Malik N, Biswas A (2012) Role of higher plants in remediation of metal contaminated sites. Sci. Rev Chem Commun 2:141–146

Malik RN, Husain SZ, Nazir I (2010) Heavy metal contamination and accumulation in soil and wild plant species from industrial area of Islamabad, Pakistan. Pak J Bot 42:291–301

Marques A, Rangel AOSS, Castro PML (2009) Remediation of heavy metal contaminated soils: phytoremediation as a potentially promising clean-up technology. Crit Rev Environ Sci Technol 39:622–654

McGrath SP, Zhao FJ (2003) Phytoextraction of metals and metalloids from contaminated soils. Curr Opin Biotechnol 14:277–282

Meers E, Ruttens A, Hopgood M, Samson D, Tack F (2005) Comparison of EDTA and EDDS as potential soil amendments for enhanced phytoextraction of heavy metals. Chemosphere 58:1011–1022

Meharg AA (2003) The mechanistic basis of interactions between mycorrhizal associations and toxic metal cations. Mycol Res 107:1253–1265

Mendez MO, Maier RM (2008) Phytostabilization of mine tailings in arid and semiarid environments-an emerging remediation technology. Environ Health Perspect 116:278–283

Miransari M (2011) Hyperaccumulators, arbuscular mycorrhizal, fungi and stress of heavy metals. Biotechnol Adv 29:645–653

Mo SC, Choi DS, Robinson JW (1989) Uptake of mercury from aqueous solution by duckweed: the effect of pH, copper, and humic acid. J Environ Health Sci 24:135–146

Munshower FF (1994) Practical handbook of disturbed land revegetation. Lewis, Boca Raton

Newman LA, Strand SE, Choe N, Duffy J, Ekuan G, Ruszaj M, Shurtleff BB, Wilmoth J, Heilman P, Gordon MP (1997) Uptake and biotransformation of trichloroethylene by hybrid poplars. Environ Sci Technol 31:1062–1067

Ovko M, Romic M (2011) Soil contamination by trace metals: geochemical behavior as an element of risk assessment. In: Dar IA (ed) Earth and environmental sciences. InTech, Rijeka, pp 437–456

Padmavathiamma PK, Li LY (2007) Phytoremediation technology, hyper-accumulation metals in plants. Water Air Soil Pollut 184:105–126

Panda S, Chaudhury I, Khan M (2003) Heavy metals induce lipid peroxidation and affect antioxidants in wheat leaves. Biol Plant 46:289–294

Parizanganeh AH, Bijnavand V, Zamani AA, Hajabolfath A (2012) Concentration, distribution and comparison of total and bioavailable heavy metals in top soils of Bonab district in Zanjan Province. Open J Soil Sci 2:123–132

Paz-Ferreiro J, Lu H, Fu S, Mendez A, Gasco (2014) Use of phytoremediation and biochar to remediate heavy metal polluted soils: a review. Solid Earth 5:65–75

Persans MW, Nieman K, Salt DE (2001) Functional activity and role of cation-efflux family members in Ni hyperaccumulation in *Thlaspi goesingense*. Proc Natl Acad Sci 98:9995–1000

Pilon-Smits E (2005) Phytoremediation. Ann Rev Plant Biol 56:15–39

Pivetz P (2001) Phytoremediation of contaminated soil and ground water at hazardous waste sites. EPA/540/S-01/500. United States Environmental Protection Agency (EPA), Washington, DC, p 36

Pulford ID, Watson C (2003) Phytoremediation of heavy metal-contaminated land by trees—a review. Environ Int 29:529–540

Qin R, Hirano Y, Brunner I (2007) Exudation of organic acid anions from poplar roots after exposure to Al, cu and Zn. Tree Physiol 27:313–320

Rajkumar M, Ae N, Prasad MNV, Freitas H (2010) Potential of siderophore-producing bacteria for improving heavy metal phytoextraction. Trends Biotechnol 28:142–149

Rascio N, Navari-Izzo F (2011) Heavy metal hyperaccumulating plants: how and why do they do it and what makes them so interesting. Plant Sci 180:169–181

Raskin I, Ensley BD (2000) Phytoremediation of toxic metals: using plants to clean up the environment. Wiley, New York

Rengel Z, Zhang WH (2003) Role of dynamics of intracellular calcium in aluminium-toxicity syndrome. New Phytol 159:295–314

Rugh CL, Wilde HD, Stack NM, Thompson DM, Summers AO, Meagher RB (1996) Mercuric ion reduction and resistance in transgenic *Arabidopsis thaliana* plants expressing a modified bacterial merA gene. Proc Natl Acad Sci 93:3182–3187

Rugh CL, Senecoff JF, Meagher RB, Merkle SA (1998) Development of transgenic yellow poplar for mercury phytoremediation. Nat Biotechnol 16:925–928

Salt DE, Pickering IJ, Prince RC, Gleba D, Dushenkov S, Smith RD, Raskin I (1997) Metal accumulation by aquacultured seedlings of Indian mustard. Environ Sci Technol 31(6):1636–1644

Sanchez JE, Willson TC, Kizilkaya K, Parker E, Harwood RR (2001) Enhancing the mineralizable nitrogen pool through substrate diversity in long term cropping systems. Soil Sci Soc Am J 65:1442–1447

Seregin I, Kozhevnikova A, Kazyumina E, Ivanov V (2003) Nickel toxicity and distribution in maize roots. Russ J Plant Physiol 50:711–717

Sharma SS, Dietz KJ (2009) The relationship between metal toxicity and cellular redox imbalance. Trends Plant Sci 14:43–50

Sharma HD, Reddy KR (2004) Geoenvironmental engineering: site remediation, waste containment and emerging waste management technologies. Wiley, New York

Shu WS, Xia HP, Zhang ZQ (2002) Use of vetiver and three other grasses for revegetation of Pb/Zn mine tailings: field experiment. Int J Phytoremediation 4:47–57

Sinha S, Mishra RK, Sinam G, Mallick S, Gupta AK (2013) Comparative evaluation of metal phytoremediation potential of trees, grasses and flowering plants from tannery wastewater contaminated soil in relation with physico-chemical properties. Soil Sediment Contam Int J 22:958–983

Sofi NA, Bhat RA, Rashid A, Mir NA, Mir SA, Lone R (2017) Rhizosphere mycorrhizae communities an input for organic agriculture. In: Varma A et al (eds) Mycorrhiza-nutrient uptake, biocontrol, ecorestoration. Springer International, Berlin, pp 387–413

Subhashini V, Swamy AVVS (2013) Phytoremediation of Pb and Ni contaminated soils using *Catharanthus roseus* (L.). Univers J Environ Res Technol 3:465–472

Suvaryan Y, Sargsyan V, Sargsyan A (2011) The problem of heavy metal pollution in The Republic of Armenia: overview and strategies of balancing socioeconomic and ecological development. In: Simeonov L, Kochubovski M, Simeonova B (eds) Environmental heavy metal pollution and effects on child mental development: Risk and prevention strategies, ICSAE IOP Publishing IOP Conf. Series: Earth and Environmental Science. Springer Science and Business Media, Dordrecht, pp 309–315. https://doi.org/10.1088/1755-1315/142/1/012023

Tam PC (1995) Heavy metal tolerance by ectomycorrhizal fungi and metal amelioration by *Pisolithus tinctorius*. Mycorrhiza 5:181–187

Tangahu BV, Abdullah SRS, Basri H, Idris M, Anuar N, Mukhlisin M (2011) A review on heavy metals(As, Pb, and Hg) up-take by plants through phytoremediation. Int J Chem Eng 2011:1–36

Taylor GJ (1987) Exclusion of metals from the symplasm: a possible mechanism of metal tolerance in higher plants. J Plant Nutr 10:1213–1222

Terry N, Carlson C, Raab TK, Zayed A (1992) Rates of selenium volatilization among crop species. J Environ Qual 21:341–344

Tiedemann AR, Klemmedson JO (2004) Responses of desert grassland vegetation to mesquite removal and regrowth. J Range Manag 57:455–465

Tordoff GM, Baker AJM, Willis AJ (2000) Current approaches to the revegetation and reclamation of metalliferous mine wastes. Chemosphere 41:219–228

United States Environmental Protection Agency (2001) Brownfields technology primer: selecting and using phytoremediation for site cleanup. USEPA, Washington, DC, p 46

United States Environmental Protection Agency (USEPA). 2000. Introduction to phytoremediation. EPA 600/R-99/107. U.S. Environmental Protection Agency, Office of Research and Development, Cincinnati

Vamerali T, Bandiera M, Mosca G (2010) Field crops for phytoremediation of metal-contaminated land: a review. Environ Chem Lett 8:1–17

Verma P, George KV, Singh HV, Singh SK, Juwarkar A, Singh RN (2006) Modeling rhizofiltration: heavy-metal uptake by plant roots. Environ Model Assess 11:387–394

Wei S, Zhou Q, Wang X (2005) Identification of weed plants excluding the uptake of heavy metals. Environ Int 31:829–834

Wong MH, Wenzel W, Bunkowski M, Puschenreiter M, Horak O (2003) Rhizosphere characteristics of indigenously growing nickel hyperaccumulator and excluder plants on serpentine soil. Environ Pollut 123:131–138

Wuana RA, Okieimen FE (2011) Heavy metals in contaminated soils: a review of sources, chemistry, risks and best available strategies for remediation. Commun Soil Sci Plant Anal 42:111–122

Xia HP, Ao HX, Liu SZ, He DQ (1999) Application of the vetiver eco-engineering for the prevention of highway slippage in south China. Proceedings of the 1st Asia-Pacific conference on ground and water bioengineering for erosion control and slope stabilization, Manila, 19–21 April 1999, pp 522–527

Yang X, Feng Y, He Z, Stoffella PJ (2005) Molecular mechanisms of heavy metal hyperaccumulation and phytoremediation. J Trace Elem Med Biol 18:339–353

Ye ZH, Wong JWC, Wong MH (2000) Vegetation response to lime and manure compost amendments on acid lead/zinc mine tailings: a greenhouse study. Restor Ecol 8:289–295

Zayed A, Pilon Smits E, DeSouza M, Lin ZQ, Terry N (2000) Remediation of selenium polluted soils. In: Bañuelos G (ed) Phytoremediation of contaminated soil and water. Lewis, Boca Raton, pp 61–83

Zhu Y, Pilon Smits EAH, Tarun A, Weber SU, Juanin L, Terry N (1999a) Cadmium tolerance and accumulation in Indian mustard is enhanced by overexpressing glutamyl cysteine synthetase. Plant Physiol 121:1169–1177

Zhu YL, Zayed AM, Quian JH, de Souza M, Trry N (1999b) Phytoaccumulation of trace elements by wetland plants: II. Water Hyacinth J Environ Qual 28:339–344

Chapter 11
Credibility of In Situ Phytoremediation for Restoration of Disturbed Environments

Cynthia Wong-Arguelles, Angel Josabad Alonso-Castro, Cesar Arturo Ilizaliturri-Hernandez, and Candy Carranza-Alvarez

11.1 Introduction

Soil and water pollution are a major environmental issue in the world.

Urbanization and industrialization contribute to the increase of contaminants (hydrocarbons, potentially toxic metals, pesticides, etc.) into the environment (Bhat et al. 2018a). These pollutants cannot be degraded, and therefore they are accumulated in living organisms, as well as in water, air, and soil. Sometimes, these pollutants can be degraded by some microorganisms. Industrial wastes, without previous treatment, are often disposed into water bodies. These contaminants might be incorporated into the food chain and cause a risk for human health (Hazrat and Ezzat 2013).

Mining and smelting are important economic activities in Mexico (INEGI 2010). The spreading of mining by-products contaminates surrounding soils, water streams, and air. Specifically, mining activity releases metals and metalloids to the environment (Machado et al. 2013). Approximately 100 million tons of mine wastes are

C. Wong-Arguelles
Programas Multidisciplinarios de Posgrado en Ciencias Ambientales, Universidad Autónoma de San Luis Potosí, San Luis Potosí, S.L.P., Mexico

A. J. Alonso-Castro
Departamento de Farmacia, División de Ciencias Naturales y Exactas, Universidad de Guanajuato, Guanajuato, Mexico

C. A. Ilizaliturri-Hernandez
Centro de Investigación Aplicada en Ambiente y Salud (CIACyT-CIAAS), Universidad Autónoma de San Luis Potosí, San Luis Potosí, S.L.P., Mexico

C. Carranza-Alvarez (✉)
Unidad Académica Multidisciplinaria Zona Huasteca, Universidad Autónoma de San Luis Potosí, Cd. Valles, S.L.P., Mexico
e-mail: candy.carranza@uaslp.mx

K. R. Hakeem et al. (eds.), *Bioremediation and Biotechnology*,
https://doi.org/10.1007/978-3-030-35691-0_11

generated in Mexico, each year, which cause pollution of soil, water, and air (SEMARNAT 2010). Mining industries that extract Ag, Pb, and Zn, among other trace metals, pour their residues into lakes and rivers, and part of this water is used for crop farming (Armienta and Rodríguez 1996; Ramos and Siebe 2006). In addition, the presence of these potentially toxic metals could reduce land productivity (Prieto et al. 2005). Mine tailings in Mexico represent also an important ecological problem due the dispersion of pollutants (Cortés et al. 2013). The most hazardous metals in Mexico are Hg, As, Pb, and Cr. The central states of Mexico, such as Zacatecas, Querétaro, Hidalgo, and San Luis Potosi, have been affected by the pollution of soils and water with potentially toxic metals (Razo et al. 2004; González et al. 2012; Hernández et al. 2012; Levresse et al. 2012; Martínez et al. 2013; Rivera et al. 2013; Covarrubias and Peña 2017). In addition, chronic diseases in humans caused by Hg contamination have been reported (Hernández et al. 2012; Martínez et al. 2013). Nevertheless, investigation into the health implications of contamination by potentially toxic metals remains to be carried out in many areas of Mexico.

This chapter summarizes the information obtained, by scientific sources, about in situ phytoremediation studies carried out in Mexico. The information here described will be useful for planning the remediation of contaminated sites by potentially toxic metals in Mexico.

11.2 In Situ Phytoremediation

Some plants have developed mechanisms to adapt and grow in contaminated sites. In situ phytoremediation involves the use of plants for the removal of organic and inorganic contaminants from soil and water and is a long-term process, nondisruptive, environmentally friendly and cost-effective (Baker et al. 1994). This process considers the level of contamination in the study site and the output of contaminants obtained after the study and includes the immobilization of toxic agents and its subsequent accumulation in roots (Pilon 2005).

Phytoremediation reduces the spread of pollutants in air and water since plants may be harvested and removed from the contaminated site for disposal and recovery of contaminants. A plant used for phytoremediation must have the following characteristics: high accumulation of contaminants, high biomass, good adaptation to prevailing environmental and climatic conditions, fast growth, and high translocation from roots to shoots (Bonanno 2013; Hazrat and Ezzat 2013).

In situ phytoremediation is based on the extraction of pollutants from the environment under natural conditions. Therefore, this complex process is influenced by physicochemical and biological parameters. The physicochemical parameters include metal availability, pH, dissolved oxygen, sediment type, pollutant loading, temperature, salinity, organic matter, weather, redox status, cation exchange capacity, and mobilization of these contaminants in soil/water, among others (Karami et al. 2011; Leblebici et al. 2011; Salt et al. 1995). The biological parameters include high growth rate, plant biomass efficiency of xylem loading, microbial biota, and deep root system, among others (Giordani et al. 2005).

The efficiency of phytoremediation systems, both for water and soil, mostly depends on the type of plants to be used. Plants must have a great capacity to tolerate and/or accumulate contaminants or transform pollutants to a lesser toxic form (Flores et al. 2003).

11.3 Native Plants Used in In Situ Phytoremediation

The use of native plants with high tolerance and capacity to accumulate or stabilize the metal to be removed is a very convenient approach, which reduces the physical dispersion of contaminants (Cortés et al. 2013; González and González-Chávez 2006; Salas et al. 2009; Sánchez et al. 2015). Native plant species have demonstrated their higher capacity for survival under environmental stresses (Antosiewicz 1992; Haque et al. 2009; Jamil et al. 2009). Thus, indigenous plants could be used as biosensors and/or bioremediators of potentially toxic metal-polluted zones (Aldrich et al. 2003; Clemens 2006; Gardea-Torresdey et al. 2005; Ma et al. 2001; Wang et al. 2002; Zhang et al. 2007; Peralta-Videa et al. 2009). Metal distribution in plant tissues is influenced by different translocation processes among plant species (Barman et al. 2000; Cai and Ma 2003; Gulz et al. 2005; Tian et al. 2009; Xiong 1998; Bhat et al. 2018b). The identification of native plant species adapted to grow on polluted areas could help to remediate sites contaminated by potentially toxic metals (Cortés et al. 2013).

11.4 Accumulation of Trace Metals by Plants Under Field Conditions in Mexico

In Mexico there are many sites contaminated with potentially toxic metals. In many studies, the levels of trace metals in water/soil surpass the levels considered as toxic for humans and plants (González et al. 2012; Mireles et al. 2004; Puga et al. 2006). Most of the reports correspond to sites polluted with trace metals by mining activities. Huge mine spoils can be found throughout the country, but the most important mining regions are in North and Central Mexico. The studies analyzed in this review correspond to plants growing in arid and semiarid regions. Most of these plants correspond to crop plants, small shrubs, or small trees. We did not find studies carried out in Southern Mexico, where the geographical conditions correspond to areas of rainforest.

The content of trace metals found in plants is correlated with the concentration of trace metals in soil/water (Carranza et al. 2008; Levresse et al. 2012; Mireles et al. 2004; Santos et al. 2012). The bioavailability of trace elements is another factor that influences their incorporation in plant tissues. Only some reports of in situ phytoremediation in Mexico included the physicochemical characteristics and the content of trace elements in water and/or soil. It is important to point out that further studies should consider carrying out chemical analysis in soil and water to provide complementary information about the accumulation of trace elements in plants and the mobilization process of these elements in soil-plant or water-plant systems.

The metal availability and accumulation of trace metals in plants from soil/water is highly dependent on the metal speciation in soil/water, pH, age of the plants, its capacity to be transported in water, and more particularly the plant metabolism efficiency or ecotypes. The trace metal accumulation depends on the predominant dissolved chemical species and the trace metal mechanisms of mobilization at the sediment-water interface (Carranza et al. 2008; González and González-Chávez 2006; Levresse et al. 2012).

In most of the reports of in situ phytoremediation in Mexico, the accumulation of trace metals is informed in the roots (Carranza et al. 2008; Carrión et al. 2012; Hernández et al. 2012; Martínez et al. 2013; Mauricio et al. 2010; Mireles et al. 2004; Santos et al. 2012; Zarazúa et al. 2013). This might be explained because this is the tissue most exposed to the existing trace metals in water/sediment (Carranza et al. 2008; Franco et al. 2010; González and González-Chávez 2006). In addition, the translocation process to aerial parts could last during months. On the contrary, Puga et al. (2006) obtained that most of the trace metals studied (As and Zn) were found in aerial parts.

Differences in accumulation were observed in different plant species collected from the same sites (Carmona et al. 2016; Cortés et al. 2013; Franco et al. 2010; González and González-Chávez 2006; Levresse et al. 2012; Salas et al. 2009). This might be explained to the tolerance mechanisms developed for each plant species to accumulate/exclude potentially toxic metals.

11.5 Hyperaccumulator Plants

Plants capable to hyperaccumulate trace metals are defined as those that meet the following criteria: bioaccumulation factor (BF) higher than 1000 μg metal g dry weight^{-1} and translocation factor (TF) higher than 1 (Baker 1981; Mireles et al. 2004). Some of the plant species cited in this review, including *Hydrocotyle ranunculoides*, *Parietaria pensylvanica*, and *Commelina diffusa*, could be considered as hyperaccumulators for Zn (Carmona et al. 2016; Zarazúa et al. 2013) and *Rorippa nasturtium-aquaticum* (synonym *Nasturtium officinale* W.T. Aiton) for Cu. From these plant species, two of them are worldwide distributed (*Hydrocotyle ranunculoides* and *Rorippa nasturtium-aquaticum*), whereas two of them are native from the American continent (*Parietaria pensylvanica* and *Commelina diffusa*). In our knowledge, no studies about the mechanism of metal accumulation have been carried out with these plant species. Therefore, it will be interesting to provide information that helps to understand, in a molecular level, how the hyperaccumulation of trace element by these plant species is performed. In addition, we could not find reports about the mechanism(s) of trace metal accumulation in other plant species cited in this work. This clearly indicates the need to perform studies for understanding how the accumulation of trace elements in plants is carried out.

11.6 Methods, Perspectives, and Future Needs

Most of phytoremediation studies are carried out under laboratory conditions, and only few studies evaluate the ability of phytoextraction under field conditions. This review provides information of extraction of trace metals and potentially toxic metals by plants from polluted sites under field conditions in Table 11.1. The information was searched by consulting the following electronic sources: ScienceDirect, Scopus, Web of Science, SpringerLink, SciELO, PubMed, and Google scholar. Scientific reports were searched from the databases using the following keywords: plant, phytoremediation, phytoextraction, and Mexico. Articles written in Spanish were also considered in this work. The publications considered in this review dated from 1981 to 2018.

The study of organic pollutants, including chlorinated solvents, linear halogenated hydrocarbons, and volatile organic carbons, among others, remains to be carried out in Mexico. In situ phytoremediation studies that analyze the accumulation and/or transformation of organic toxic substances should be considered. It is well-known that contamination by pesticides in Mexico is a great threat for human health. Plants might be considered as a possible solution for the removal of these persistent contaminants. The information on the mechanisms by which plants cited in this review transform such compounds is scarce (López et al. 2005).

In addition, the probable mechanism for metal extraction in plants is also lacking. Some plants exposed to high levels of contaminants have developed physiological mechanisms for their adaption and growth under stressful conditions. Some of these mechanisms include the exudation of low molecular weight organic acids (LMWOAs), which contribute in the detoxification of some trace metals (Li et al. 2013; Tu et al. 2004). Nevertheless, the mechanism(s) of trace metal accumulation of plants cited in this study should be studied.

The plant-microbe interaction is also a topic to be evaluated. Some microorganisms such as *Rhizophagus irregularis* and *Funneliformis mosseae* (Hassan et al. 2013) can enhance phytoremediation in different manners: (1) expediting growth of plant biomass, (2) increasing or (3) decreasing metal availability in soil, (4) facilitating metal translocation from soil to root, and (5) inducing the translocation from root to shoots (Ma et al. 2011; Rajkumar et al. 2012).

The chelating agents, such as ethylene bis[oxyethylenetrinitrilo] tetraacetic acid (EGTA), ethylenediamine-*N*,*N*′bis(*o*-hydroxyphenyl)acetic acid (EDDHA), ethylenediaminetriacetic acid (EDTA), N-(2-hydroxyethyl)-ethylenediaminetriacetic acid (HEDTA), diethylenetetraminepentaacetatic acid (DTPA), and diethylenetetraminepentaacetatic acid-calcium chloride dihydrate-triethanolamine

Table 11.1 In situ phytoremediation in Mexico

Plant species	Site of the study	Physicochemical characteristics in water/soil	Trace metal accumulation (mg/kg)	Reference
Typha latifolia, Scirpus americanus	Artificial lagoon highly polluted by municipal and industrial wastewater in San Luis Potosi	pH of sediment: 7.6–9.0 pH of the water 8.0–8.5 pE 0.28–5.0 <u>Sediments</u> Pb 20.3–55.2 mg/kg Cd 5.3–13.5 mg/kg Cr 16.1–116.7 mg/kg Fe 2247—11,953 mg/kg	*Typha latifolia* (whole plant) Pb 18 ± 0.8 mg/kg Cr 110 ± 13.1 mg/kg Cd 4.6 ± 0.08 mg/kg Mn 1651.9 ± 106 mg/kg Fe 669.2 ± 7.1 mg/kg Pb 25.3 ± 1.5 mg/kg (root)	Carranza et al. (2008)
Asclepias linaria, Euphorbia sp., *Haplopappus venetus, Salvia microphylla, Stevia salicifolia, Gnaphalium arizonicum, Brickellia veronicifolia, Bouvardia terniforlia, Jatropha dioica, Teloxys graveolens, Polygonium aviculare, Tagetes lunulata*	Plants and soils of four sites at Zacatecas, México: 1. El Bote 2. San Martín 3. Fresnillo 4. Noria de Angeles	Soil Cd 9–93 mg/kg Zn 16–176 mg/kg	<u>Aerial parts</u> *Teloxys graveolens* Cd 42.9 mg/kg Ni 37.0 mg/kg Pb 125 mg/kg *Jatropha dioica* Cd 30.5 C mg/kg Ni 28 mg/kg Zn 6250 mg/kg *Polygonium aviculare* Pb 124 mg/kg Zn 9230 mg/kg *Brickellia veronicifolia* Zn 6000 mg/kg *Solanum elaeagnifolium* Pb 148 mg/kg	González and González-Chávez (2006)

Green tubercle fruits of *Medicago sativa*	Plants and soils of Tláhuac, DF, and Mixquiahuala, Hidalgo, México	Approximation [mean in soils] Cr 60–90 mg/kg Mn 120–135 mg/kg Co 55–65 mg/kg Ni 25–40 mg/kg Cu 35–100 mg/kg Zn 55–90 mg/kg Pb 40–65 mg/kg	*Medicago sativa* (roots) Mn 240–247 mg/kg Fe 493–691 mg/kg Ni 18–105 mg/kg Cu 224–229 mg/kg Zn 198–585 mg/kg Pb 33–54 mg/kg	Mireles et al. (2004)
22 plant species (*Buddleja*, *Dasylirion*, *Gymnosperma*, *Pinus*, *Asphodelus*, *Nicotiana*, *Yucca*, *Schinus*, *Juniperus*, *Eucalyptus*, *Echinocactus*, *Opuntia*, *Berberis*, *Larrea*)	Sediments and waste tailings, plants and water from the Wadley Sb District (real de Catorce, SLP)	Waste tailings Sb 6563–13,743 mg/kg Fe 996–3799 mg/kg Zn 32–84 ppm mg/kg As 13–57 ppm mg/kg Hg 15–53 ppm mg/kg Groundwater wells Fe 0.30–0.90 mg/kg Zn 0.0089–0.0369 mg/kg	<u>Seasonal plants</u> Fe 116–618 mg/kg Sb 1–447 mg/kg Zn 15.5–350 mg/kg Mn 8–218 mg/kg Cu 1.5–56 mg/kg Ni 0.5–43.5 mg/kg <u>Perennial plants</u> Fe 86–589 mg/kg Mn 4–54 mg/kg Zn 4–113 mg/kg Sb 1–20 mg/kg	Levresse et al. (2012)
Ricinus communis, *Tithonia diversifolia*	Silver and gold mining sites, Guanajuato, México	Mining soil Pb 362 ± 8 mg/kg Cu 221 ± 1.0 mg/kg Ag 42.20 ± 0.65 mg/kg Cd 0.373 ± 0.010 mg/kg pH 7.7–8.0	*R. communis*(roots) Ag 0.141 ± 0.009 mg/kg Cd 0.123 ± 0.008 mg/kg Cu 2.6 ± 0.07 mg/kg Pb 2.74 ± 0.06 mg/kg *T. diversifolia* (roots) Ag 0.035 ± 0.004 mg/kg Cd 0.015 ± 0.002 mg/kg Cu 0.592 ± 0.02 mg/kg Pb 0.805 ± 0.018 mg/kg	Figueroa et al. (2008)

(continued)

Table 11.1 (continued)

Plant species	Site of the study	Physicochemical characteristics in water/soil	Trace metal accumulation (mg/kg)	Reference
Thymophylla setifolia, Karwinskia humboldtiana, Trixis angustifolia, Brickellia veronicifolia, Asphodelus fistulosus, Ambrosia artemisiifolia, Amaranthus hybridus, Simsia amplexicaulis, Viguiera dentata, Parthenium bipinnatifidum, Flaveria angustifolia, Flaveria trinervia, Sporobolus indicus	Contaminated mine tailings in Villa de La Paz, San Luis Potosí, México	pH 7.2–8.1 Organic C 11–93 g/kg Total N 0.1–4.3 g/km As 36–8420 mg/kg Pb 16–754 mg/kg Cu 2–1154 mg/kg Zn 81–1766 mg/kg	Shoots *Flaveria angustifolia* As198.5 mg/kg *Sporobolus indicus* Cu 77.7 mg/kg *Ambrosia artemisiifolia* Zn 405.7 mg/kg Roots *Flaveria angustifolia* As 189.8 mg/kg Pb 17.8 mg/kg Zn 138.4 mg/kg *Sporobolus indicus* Cu 97.5 mg/kg	Franco et al. (2010)
Thymus vulgaris, Origanum majorana, Petroselinum crispum, Melissa officinalis, Chenopodium ambrosioides, Coriandrum sativum, Brassica oleracea, Dysphania ambrosioides, Cucurbita ficifolia, Lactuca sativa, Beta vulgaris subsp. *vulgaris, Sechium edule, Chenopodium nuttalliae, Physalis philadelphica, Allium cepa, Solanum lycopersicum, Capsicum* sp., *Raphanus raphanistrum, Opuntia ficus-indica,* *Citrus X sinensis, Prunus pérsica, Punica granatum, Musa × paradisiaca, Psidium guajava, Persea americana, Citrus × Limon*	Agricultural and water zone in Zimapán, Hidalgo	Water As 0.04–0.48 mg/L Soil As 0.717–20.878 mg/kg	*Dysphania ambrosioides* (leaves) As 12.39 ± 6.15 mg/kg *Cucurbita ficifolia* (fruit) As 10.74 ± 1.0 mg/kg *Melissa officinalis* (leaves) As 20.13 ± 0.55 mg/kg	Prieto et al. (2005)

Nicotiana glauca, Flaveria pubescens, Schinus molle, Casuarina sp., *Tecoma stans, Prosopis* sp., *Cenchrus ciliaris, Maurandya antirrhiniflora, Ricinus communis, Opuntia lasiacantha*	"La Negra" mine, Maconi, Cadereyta de Montes, Querétaro	Tailings As 183–14,660 mg/kg Cd 45–308 mg/kg Pb 327–1754 mg/kg Cu 149–459 mg/kg Zn 448–505 mg/kg Soil As 88–5990 mg/kg Cd 5–129 mg/kg Pb 169–3638 mg/kg Cu 159–1254 mg/kg Zn 1431–13,488 mg/kg	Whole plants *Nicotiana glauca* As 91.94 mg/kg Cd 106.07 mg/kg Cu 95.17 mg/kg Zn 1984.48 mg/kg *Flaveria pubescens* As 9.21 mg/kg Cd 25.64 mg/kg Pb 222.89 mg/kg Cu 102.46 mg/kg) Zn 755.82 mg/kg *Tecoma stans* As 9.22 mg/kg Cd 31.68 mg/kg Zn 942.80 mg/kg *Opuntia lasiacantha* As 12.87 mg/kg *M. antirrhiniflora* Pb 203.27 mg/kg *Prosopis* sp. Cu 63.64 mg/kg	Santos et al. (2012)
Acacia farnesiana, Juniperus deppeana, Baccharis glutinosa, Prosopis juliflora, Cynodon dactylon	Tailings in San Francisco del Oro, Chihuahua	Soil As 552.66 y 4222.78 mg/kg Zn 1086.90 y 3361.50 mg/kg	*Baccharis glutinosa* *As 68.85 mg/kg* (leaves) *As 12.52 mg/kg* (root) *Cynodon dactylon* As *27.55 mg/kg* (root) Zn 302.18 mg/kg (leaves) Zn 225.81 mg/kg (root) *Acacia farnesiana* As 43.85 mg/kg (root)	Puga et al. (2006)

(continued)

Table 11.1 (continued)

Plant species	Site of the study	Physicochemical characteristics in water/soil	Trace metal accumulation (mg/kg)	Reference
25 species in 7 families (Asteraceae, Poaceae)	Slagheaps in "Dos Carlos," Pachuca, Hidalgo	Slag Zn 45.640 mg/kg Pb 14.710 mg/kg Cd 0.981 mg/kg Ni 0.168 mg/kg	<u>Whole plants</u> *Solanum corymbosum* Cu 6 mg/kg *Brickellia veronicifolia* Pb 5 mg/kg Zn 20 mg/kg *Atriplex suberecta* Cd 1 mg/kg *Cynodon dactylon* Mn 69 mg/kg *Bouteloua curtipendula* Ni 4 mg/kg	Hernández et al. (2009)
22 different pioneering plant species from 11 families were found growing on the slag Heap: *Dalea* sp., *Melampodium divaricatum*, *Galinsoga parviflora*, *Trifolium goniocarpum*, *Eragrostis intermedia*, *Phytolacca icosandra*, *Lopezia racemosa*, *Crotalaria longirostrata*, *Phytolacca icosandra*, *Bidens odorata*, *Tagetes micrantha*, *Rhynchelytrum repens*, *Dalea obreniformis*, *Aeschynomene villosa*, *Eleusine indica*, *Euphorphia ocymoidea*, *Crusea longiflora*, *Anoda cristata*, *Sida rhombifolia*, *Anagallis arvensis*, *Jaegeria hirta*, *Cuphea rocumbens*, *Ipomoea pedicellaris*, *Crotalaria rotundifolia*	Slag heaps of Mina "La Guitarra," Temascaltepec, Morelos	pH 4.3–7.5 OM 1–119.4 g/kg Available (DTPA) Cd <0.12–0.78 mg/kg	<u>Leaves tissues</u> *Aeschynomene villosa* Cd 0.52 mg/kg *Crusea longiflora* Cd 0.56 mg/kg *Anagallis arvensis* Cd 0.62 mg/kg *Lopezia racemosa* Cd 0.70 mg/kg	González et al. (2009)

Zea mays	Agricultural areas from the municipality of Guadalupe, Zacatecas	pH 7.602–8.304 OM 1.294–4.062% TN 0.062–0.208% EC 0.538–3.570 dS/m Agricultural soils Pb 5660.25 mg/kg As 289.9 mg/kg Hg 505.9 mg/kg Zn 10086.5 mg/kg Cu 1323.82 mg/kg Fe 55330.6 mg/kg Mn 1792.39 mg/kg	*Zea mays (roots)* Pb 293.24 mg/kg As 98.15 mg/kg Zn 849.74 mg/kg Cu 213.63 mg/kg Mn 629.71 mg/kg	González et al. (2012)
Zea mays	Mine zone of San Joaquín, Querétaro, México (soil, plant, air)	Hg Sediments 67 ± 148 mg/kg Forest soil 8.5 ± 15.5 mg/kg Agricultural soil 53 ± 92 mg/kg Tailing + soil 488 ± 554 mg/kg Skarn tailing 13 ± 8.2 mg/kg Hydrothermal tailing 721 ± 1162 mg/kg Rain water Hg 2.87 a 60.74 mg/kg	Zea mays (hg) 0.2–8.7 mg/kg (shoot) 0.2–8.2 mg/kg (leaves) 0.06–1.0 mg/kg (steams) 0.04–0.24 mg/kg (grains)	Hernández et al. (2012)
Zea mays	Mine zone of San Joaquín, Querétaro, México (air, soil, plant, and water)	Hg Forests soils 0.2–69 mg/kg Agricultural soils 0.5–314 mg/kg Mine tailing 2.4–4164 mg/kg Sediments 0.6–687 mg/kg Drinking water 0.01–0.17 mg/kg Air 22–153 ng/m^3	Zea mays (hg) 0.09–6.2 mg/kg (roots) 0.08–2.0 mg/kg (stems) 0.10–8.2 mg/kg (leaves) 0.04–0.90 mg/kg (grains)	Martínez et al. (2013)

(continued)

Table 11.1 (continued)

Plant species	Site of the study	Physicochemical characteristics in water/soil	Trace metal accumulation (mg/kg)	Reference
Hydrocotyle ranunculoides	Lerma River	Water pH 5.9–7.1 T° 11–21 °C CE 562–820 μS/cm OD 0.30–0.75 mg/L Cr 0.014 mg/kg Mn 1.41 mg/kg Fe 4.5 mg/kg Cu 0.015 mg/kg Zn 0.097 mg/kg Pb 0.070 mg/kg	*H. ranunculoides* (aerial parts) Cr 1.18 mg/kg Mn 541 mg/kg Fe 816 mg/kg Cu 7.36 mg/kg Zn 103 mg/kg Pb <1.1 mg/kg *H. ranunculoides* (submerged parts) Cr 12.8 mg/kg Mn 1752 mg/kg Fe 8929 mg/kg Cu 21.53 mg/kg Zn 126 mg/kg Pb 5.27 mg/kg	Zarazúa et al. (2013)
Prosopis laevigata, *Acacia farnesiana*	Zimapán mining area	Tailings As 4000–32,000 mg/kg Water (soluble) As 1.07–19 mg/kg	*Prosopis laevigata* (As) 20.1–78.2 mg/kg (leaves-tailings) 17.9–64 mg/kg (leaves-slags) 3.6–20.8 mg/kg (twigs-background) 1400 mg/kg (roots-tailings pile) *Acacia farnesiana* (As) 22.6–225 mg/kg (twigs-tailings) 8.5–15.8 mg/kg (twigs-background)	Armienta et al. (2008)

Eichhornia crassipes	Waterways in Xochimilco (ANP), San Gregorio Atlapulco	Water Urban site V 0.061 mg/kg Cu 0.0048 mg/kg Pb 0.011 mg/kg Agricultural site Mn 0.0654 mg/kg Cr 0.002 mg/kg Cd 0.0002 mg/kg Tourist site Zn 0.9581 mg/kg Sr 0.2433 mg/kg As 0.0076 mg/kg Ni 0.0060 mg/kg	Urban site (submerged parts) Al 2292 ± 23 mg/kg Fe 1660.4 ± 18 mg/kg Mn 587.3 ± 10 mg/kg Ti 248.8 ± 230 mg/kg Zn 135.5 ± 17 mg/kg Pb 7.7 ± 6 mg/kg Sn 2.4 ± 6 mg/kg As 2.2 ± 16 mg/kg Co 1.5 ± 12 mg/kg Cd 0.7 ± 9 mg/kg Agricultural site (aerial parts) Sr 116 ± 12 mg/kg Tourist site (submerged parts) V 154.2 ± 11 mg/kg Cr 58.1 ± 7 mg/kg Cu 27.3 ± 12 mg/kg Ni 33.2 ± 9 mg/kg	Carrión et al. (2012)
Brickellia sp., *Gnaphalium chartaceum*, *Senecio salignus*, *Wigandia urens*, *Guardiola tulocarpus*, *Jacaranda mimosifolia*, *Juniperus flaccida*, *Ficus goldmanii*, *Cheilanthes* sp., *Russelia aff. villosa*, and *Alvaradoa amorphoides*	"La Concha" tailings Heaps in Mexcala, Tlaxco	Tailings pH 7.7 ± 0.1 OM 0.3 ± 5% Cu 254 ± 26 mg/kg Mn 4520 ± 225 mg/kg Zn 3735 ± 143 mg/kg Pb 3291 ± 358 mg/kg	*G. chartaceum* Cu 121 mg/kg (whole plant) Cu 74 mg/kg (shoots) Cu 21 mg/kg (leaves) Mn 744 mg/kg (flowers) Pb 2901 mg/kg (flowers) Pb 946 mg/kg (shoots) Zn 4906 mg/kg (flowers) *Wigandia urens* Zn ≈ 2300 mg/kg (flowers) Pb ≈ 800 mg/kg (leaves) *Senecio salignus* Zn ≈ 1000 mg/kg (steams)	Cortés et al. (2013)

(continued)

Table 11.1 (continued)

Plant species	Site of the study	Physicochemical characteristics in water/soil	Trace metal accumulation (mg/kg)	Reference
Scirpus americanus	"San Germán," Léon, Guanajuato (receptor of industrial wastewater)	Sediments Cr 409 mg/kg As 8 mg/kg Se 3 mg/kg Cd 3 mg/kg	*Scirpus americanus* (whole plant) Cr 971 mg/kg As 58 mg/kg Cd 65 mg/kg Se 90 mg/kg *Scirpus americanus* (aerial tissues) Cr 1 mg/kg As 9 mg/kg Cd 23 mg/kg Se 5 mg/kg *Scirpus americanus* (roots) Cr 970 mg/kg As 49 mg/kg Cd 41 mg/kg Se 85 mg/kg	Mauricio et al. (2010)

17 families *Yucca* sp., *Amaranthus hybridus*, *Bidens odorata*, *Heterosperma pinnatum*, *Pseudognaphalium inornatum*, *Tithonia tubiformis*, *Brassica campestris*, *Buddleja scordioides*, *Buddleja tomentella*, *Cerdia congestiflora*, *Chenopodium graveolens*, *Salsola tragus*, *Ipomoea longifolia*, *Cucurbita foetidissima*, *Sicyos laciniatus*, *Lupinus campestris*, *Sphaeralcea angustifolia*, *Acacia schaffneri*, *Prosopis laevigata*, *Mimosa aculeaticarpa*, *Mirabilis jalapa*, *Proboscidea louisianica*, *Chloris virgata*, *Eragrostis cilianensis*, *Eragrostis mexicana*, *Pennisetum villosum*, *Sporobolus airoides*, *Salix bonplandiana*, *Physalis hederifolia*, *Solanum elaeagnifolium*	"Francisco I. Madero," Zacatecas (mining tailings)	Soil (Pb) *Amaranthus hybridus* 2898 ± 195 mg/kg *Brassica campestris* 2457 ± 237 mg/kg *Buddleja scordioides* 1048 ± 104 mg/kg *Cerdia congestiflora* 1119 ± 114 mg/kg *Cucurbita foetidissima* 1075 ± 108 mg/kg *Lupinus campestris* 699 ± 95 mg/kg *Mimosa aculeaticarpa* 1528 ± 144 mg/kg *Acacia schaffneri* 1052 ± 87 mg/kg	Shoots (Pb) *Amaranthus hybridus* 2208 ± 136 mg/kg *Brassica campestris* 1095 ± 84 mg/kg *Buddleja scordioides* 1378 ± 153 mg/kg *Cerdia congestiflora* 1175 ± 126 mg/kg *Cucurbita foetidissima* 357 ± 69 mg/kg *Lupinus campestris* 615 ± 40 mg/kg *Mimosa aculeaticarpa* 759 ± 69 mg/kg *Acacia schaffneri* 953 ± 73 mg/kg *Eragrostis mexicana* 63 mg/kg	Salas et al. (2009)

(continued)

Table 11.1 (continued)

Plant species	Site of the study	Physicochemical characteristics in water/soil	Trace metal accumulation (mg/kg)	Reference
Eleocharis sp., *Baccharis salicifolia, Machaeranthera gypsitherma, Flaveria chlorifolia, Brickellia veronicifolia, Baccharis neglecta, Samolus ebracteatus, Nicotiana glauca, Anemopsis califórnica*	Chihuahua (Parral and Naica mine tailings)	Soil (As) 63 ± 0.08–2100 ± 0.19 mg/kg. Water (As) 110–181 ug/L	*Eleocharis* sp. (shoot) As 301 ± 0.72 mg/kg *Brickellia veronicifolia* (shoot) As 59 ± 0.74 mg/kg	Flores et al. (2003)
Ricinus communis	Zimapán, Hidalgo (San Francisco, Los Gómez, and Santa María mine tailings)	EC 488–26, 576 μmhos/cm LOI 2.8–16% pH 3.0–7.4	*Ricinus communis* (shoots) Ni ≈ 18 mg/kg Cu ≈ 26 mg/kg Zn ≈ 180 mg/kg Pb ≈ 68 mg/kg Cd ≈ 2.7 mg/kg Mn ≈ 180 mg/kg *Ricinus communis (roots)* Ni ≈ 25 mg/kg Cu ≈ 48 mg/kg Zn ≈ 590 mg/kg Pb ≈ 170 mg/kg Cd ≈ 8 mg/kg Mn ≈ 100 mg/kg	Ruiz Olivares et al. (2013)

Acmella repens, Apium graveolens, Argemone grandiflora, Aster subulatus, Bacopa monnieri, Bidens pilosa, Calyptocarpus vialis, Chenopodium ambrosioides (49 species)	Zimapán, Hidalgo (Santiago stream)	Water pH 7.1–7.8 Soil pH 7.7–8.5 Zn 638–29,190 mg/kg (rhizosphere) Cd 16–71 mg/kg Pb 201–3991 mg/kg	*R. nasturtium-aquaticum* Pb 70,31 mg/kg (roots) Pb 31 mg/kg) (stem) Pb 46 mg/kg (leaves) Cu 350 mg/kg (stem) *Acmella repens* Cu 140 mg/kg (roots) *P. major* Cu 140 mg/kg (aerial parts) *C. diffusa* Cu 200 mg/kg (roots) *Zn 5086 mg/kg (leaves)* *V. litoralis* *Cu 200 mg/kg (roots)* *P. pensylvanica* *Zn 5257 mg/kg (roots)* *Zn 7630 mg/kg (stem)* *F. trinervia* *Zn 1839 mg/kg (stem)*	Carmona et al. (2016)
Platanus mexicana, Cnidoscolus multilobus, Asclepias curassavica, Solanum diversifolium, Thelypteris kunthii, Equisetum hyemale, Xanthosoma robustum, Pluchea symphytifolia	Molango, Hidalgo (Mn mine)	pH 7.65–8.19 Mn substrate 11637.50 ± 263.39–106104.17 ±3755.36 mg/kg	*Cnidoscolus multilobus* Mn 1055.80 ± 22.27 mg/kg (root) *Asclepius curassavica* Mn 1507.69 ± 9.78 mg/kg (leaf) Mn 1130.05 ± 235 mg/kg (flower) *Thelypteris kunthii* Mn 6842.36 ± 355.43 mg/kg (root) *Equisetum hyemale* Mn 5266.30 ± 102.30 mg/kg (root)	Juárez et al. (2010)

(continued)

Table 11.1 (continued)

Plant species	Site of the study	Physicochemical characteristics in water/soil	Trace metal accumulation (mg/kg)	Reference
Agave lechuguilla, Buchloe dactyloides, Dyssodia setifolia, Euphorbia prostrata, Parthenium incanum, Zinnia acerosa	Villa de la Paz, San Luis Potosí	Rhizospheric soil 20–2400 mg/kg (As) 20–188 mg/kg Pb 13–23 mg/kg cu 7–33 mg/kg Zn	*Euphorbia prostrata* (whole plant) As ≈165 mg/kg Pb ≈ 66 mg/kg Cu ≈ 88 mg/kg Zn ≈ 300 mg/kg *Parthenium incanum* (whole plant) As ≈130 mg/kg Pb ≈ 34 mg/kg Cu ≈ 76 mg/kg Zn ≈ 350 mg/kg *Zinnia acerosa* (whole plant) As ≈93 mg/kg Pb ≈ 49 mg/kg Cu ≈ 70 mg/kg Zn ≈ 340 mg/kg *Dyssodia setifolia* (whole plant) Cu ≈ 70 mg/kg	Machado et al. (2013)
Pteridium sp., *Juniperus* sp., *Cuphea lanceolata, Dichondra argéntea, Brickellia veronicifolia, Ruta graveolens, Dalea bicolor* *Viguiera dentata, Aster gymnocephalus, Gnaphalium* sp., *Crotalaria pumila, Flaveria trinervia*	San Francisco and Santa María mine tailings, Zimapán, Hidalgo	Mine tailings Cu 800 mg/kg (both) Zn and Pb 4000 mg/kg (both) Cd 157 mg/kg (San Francisco) 120 mg/kg (Santa María)	*Viguiera dentata* Zn 2231 ± 29 mg/kg (unwashed shoot) Cd 21 ± 3 mg/kg (unwashed shoot) *Gnaphalium* sp. Mn 338 ± 36 mg/kg (unwashed shoot) *Flaveria trinervia* Pb 419 ± 25 mg/kg (unwashed shoot)	Sánchez et al. (2015)

Fabronia ciliaris	Metropolitan Zone of Toluca Valley	Epiphytic mosses Mn 62.54 ± 2.20 mg/kg Zn 34.88 ± 0.76 mg/kg Pb 4.76 ± 0.04 mg/kg Soil Mn 852 mg/kg Zn 86 mg/kg Pb 15 mg/kg	*Fabronia ciliaris* Mn 292 ± 10 mg/kg Zn 406 ± 20 mg/kg Pb 112 ± 5 mg/kg	Zarazúa et al. (2013)
Thalassia testudinum	Yum Balam Reserve, Yucatán	Groundwater Fe > 400 mg/kg Cd >4 mg/kg Cr ≈ 1 mg/g	*Thalassia testudinum* (leaves) Fe 119–445.7 mg/kg Cd 0.2–5 mg/kg Cr 0.4–0.7 mg/kg *Thalassia testudinum* (root-rhizome) Fe 141.4–504.3 mg/kg Cd 0.2–1.2 mg/kg Cr 0.5–1.1 mg/kg	Avelar et al. (2013)
Z. mays, Ambrosia psilostachya, Chenopodium ambrosioides, Cynodon dactylon, Polygonum hydropiperoides, Wigandia urens	Molango, Hidalgo (Mn mine)	ND	*Ambrosia psilostachya* Mn 89.8 ± 27.07 mg/kg (roots)	Rivera et al. (2013)

(DTPA-$CaCl_2$-TEA), among others, have been used to increase the phytoextraction process (Cortés et al. 2013; Nowack et al. 2006). Ruiz Olivares et al. (2013) suggested that the use of DTPA could increase the metal stabilization shown by *Ricinus communis*, and Cortés et al. (2013) found that *Gnaphalium chartaceum*, *Senecio salignus*, and *Wigandia urens* had Pb-BFs (in relation to DTPA-extraction) in the range between 6.4 and 18.5. However, the use of these agents on in situ studies might cause groundwater pollution due to metal mobilization (Wenzel et al. 2003). The use of these chelator agents could improve trace metal accumulation in mine tailings.

The taxonomical identification of plants species should be encouraged. In many cases, the botanical species is not reported. Misidentification of plant species could cause misinterpretation of results. Botanical personal should be incorporated in phytoremediation studies carried out under controlled and field conditions.

Finally, the speciation of metals determines the bioavailability of contaminants in soil and water, before carrying out the phytoremediation study. Chemical forms of trace metals are needed to be investigated for evaluating their possible mobility, bioavailability, toxicity in the environment, and the possible interactions with soil particles. The ability of metals and metalloids to form complexes with compounds present in water and soil plays an important role in increasing their bioavailability and uptake (Babula et al. 2009).

11.7 Conclusion

In situ phytoremediation studies are scarce. Special attention should be paid on native plants to propose strategies for the remediation of soil and water contaminated with potentially toxic metals. In addition, complementary studies should include the evaluation of the accumulation of pesticides and other organic toxic substances in plants. In addition, the molecular mechanisms of accumulation have not been described.

References

Aldrich MV, Gardea-Torresdey JL, Peralta-Videa JR, Parsons JG (2003) Uptake and reduction of Cr (VI) to Cr (III) by mesquite (*Prosopis* spp.): chromate–plant interaction in hydroponics and solid media studied using XAS. Environ Sci Technol 37:1859–1864

Antosiewicz DM (1992) Adaptation of plants to an environment polluted with heavy metals. Acta Soc Bot Pol 61:281–299

Armienta MA, Rodríguez R (1996) Arsénico en el Valle de Zimapán, México: Problemática Ambiental. Rev MAPFRE Seguridad 63:33–43

Armienta MA, Ongley LK, Rodríguez R, Cruz O, Mango H, Villaseñor G (2008) Arsenic distribution in mesquite (*Prosopis laevigata*) and huizache (*Acacia farnesiana*) in the Zimapán mining area, México. Geochem Exp Environ Anal 8:191–197

Avelar M, Bonilla B, Merino M, Herrera JA, Ramirez J, Rosas H, Martínez A (2013) Iron, cadmium, and chromium in seagrass (*Thalassia testudinum*) from a coastal nature reserve in karstic Yucatán. Environ Monit Assess 185:7591–7603

Babula P, Adam V, Opatrilova R, Zehnalek J, Havel L, Kizek R (2009) Uncommon heavy metals, metalloids and their plant toxicity: a review. In: Lichtfouse E (ed) Organic farming, pest control and remediation of soil pollutants, Sustainable agriculture reviews. Springer, New York, pp 275–309

Baker AJM (1981) Accumulators and excluders-strategies in the response of plants to heavy metals. J Plant Nutr 3:643–654

Baker AJM, Reeves RD, Hajar ASM (1994) Heavy metal accumulation and tolerance in British populations of the metallophyte *Thlaspi caerulescens* J. & C. Presl (Brassicaceae). New Phytol 127:61–68

Barman SC, Sahu RK, Bhargava SK, Chaterjee C (2000) Distribution of heavy metals in wheat, mustard, and weed grown in field irrigated with industrial effluents. Bull Environ Contam Toxicol 64:489–496

Bhat RA, Beigh BA, Mir SA, Dar SA, Dervash MA, Rashid A, Lone R (2018a) Biopesticide techniques to remediate pesticides in polluted ecosystems. In: Wani KA, Mamta (eds) Handbook of research on the adverse effects of pesticide pollution in aquatic ecosystems. IGI Global, Hershey, pp 387–407

Bhat RA, Dervash MA, Qadri H, Mushtaq N, Dar GH (2018b) Macrophytes, the natural cleaners of toxic heavy metal (THM) pollution from aquatic ecosystems. In: Environmental contamination and remediation. Cambridge Scholars, Cambridge, pp 189–209

Bonanno G (2013) Comparative performance of trace element bioaccumulation and biomonitoring in the plant species *Typha domingensis*, *Phragmites australis* and *Arundo donax*. Ecotox Environ Safe 97:124–130

Cai Y, Ma LQ (2003) Metal tolerance, accumulation, and detoxification in plants with emphasis on arsenic in terrestrial plants. In: Cai Y, Braids O (eds) Biogeochemistry of environmentally important trace elements, ACS symposium series, vol 835. University Press, London, pp 95–114

Carmona E, Carrillo R, González Mdel CA, Vibrans H, Yáñez L, Delgado A (2016) Riparian plants on mine runoff in Zimapan, Hidalgo, Mexico: useful for phytoremediation? Int J Phytoremediation 18:861–868

Carranza C, Alonso AJ, Alfaro MC, García RF (2008) Accumulation and distribution of heavy metals in *Scirpus americanus* and *Typha latifolia* from an artificial lagoon in San Luis Potosí, México. Water Air Soil Poll 188:297–309

Carrión C, Ponce C, Cram S, Sommer I, Hernández M, Vanegas C (2012) Aprovechamiento potencial del lirio acuático (*Eichhornia crassipes*) en Xochimilco para fitorremediación de metales. Agrociencia 46:609–620

Clemens S (2006) Toxic metal accumulation, responses to exposure and mechanisms of tolerance in plants. Biochimie 88:1707–1719

Cortés JEV, Mugica AV, González CMCA, Carrillo GR, Gordillo MM, Mier MV (2013) Natural revegetation of alkaline tailing heaps at Taxco, Guerrero, México. Int J Phytoremediation 15:127–141

Covarrubias SA, Peña JJ (2017) Contaminación Ambiental por metales pesados en México: Problemática y estrategias de fitorremediación. Rev Int Contam Ambient 33:7–21

Figueroa JAL, Wrobel K, Afton S, Caruso JA, Gutierrez J, Wrobel K (2008) Effect of some heavy metals and soil humic substances on the phytochelatin production in wild plants from silver mine areas of Guanajuato, Mexico. Chemosphere 70:2084–2091

Flores E, Alarcón MT, González S, Olguín EJ (2003) Arsenic tolerating plants from mine sites and hot springs in the semi-arid region of Chihuahua, Mexico. Acta Biotechnol 23:113–119

Franco MO, Vásquez MS, Patiño A, Dendooven L (2010) Heavy metals concentration in plants growing on mine tailings in Central Mexico. Bioresour Technol 101:3864–3869

Gardea-Torresdey JL, Peralta-Videa JR, De La Rosa G, Parsons JG (2005) Phytoremediation of heavy metals and study of the metal coordination by X-ray absorption spectroscopy. Coord Chem Rev 249:1797–1810

Giordani C, Cecchi S, Zanchi C (2005) Phytoremediation of soil polluted by nickel using agricultural crops. Environ Manag 36:675–681

González RC, González-Chávez MCA (2006) Metal accumulation in wild plants surrounding mining wastes. Environ Pollut 144:84–92

González MC, Carrillo R, Gutiérrez MC (2009) Natural attenuation in a slag heap contaminated with cadmium: the role of plants and arbuscular mycorrhizal fungi. J Hazard Mater 161:1288–1298

González O, Gómez JM, Ruíz EA (2012) Plants and soil contamination with heavy metals in agricultural areas of Guadalupe, Zacatecas, Mexico. In: Srivastava JK (ed) Environmental contamination. Intech Open, London, pp 37–50

Gulz PA, Gupta SK, Schulin R (2005) Arsenic accumulation of common plants from contaminated soils. Plant Soil 272:337–347

Haque N, Peralta-Videa JR, Duarte-Gardea M, Gardea-Torresdey JL (2009) Differential effect of metals/metalloids on the growth and element uptake of mesquite plants obtained from plants grown at a copper mine tailing and commercial seeds. Bioresour Technol 100:6177–6182

Hassan SE, Hijri M, St-Arnaud M (2013) Effect of arbuscular mycorrhizal fungi on trace metal uptake by sunflower plants grown on cadmium contaminated soil. New Biotechnol 30:780–787

Hazrat AK, Ezzat S (2013) Phytoremediation of heavy metals–concepts and applications. Chemosphere 91:869–881

Hernández E, Mondragón E, Cristóbal D, Rubiños JE, Robledo E (2009) Vegetación, residuos de mina y elementos potencialmente tóxicos de un jal de Pachuca, Hidalgo, México. Rev Chapingo Ser Cienc For Ambient 15:109–114

Hernández G, García R, Solís S, Martínez S, Mercado I, Ramírez M, Solorio G (2012) Presencia del Hg total en una relación suelo-planta-atmósfera al sur de la Sierra Gorda de Querétaro, México. Rev Esp Cienc Quim Biol 15:5–15

INEGI (2010) National Institute of Statistics Geography and Informatics. Mining in Mexico. Sect Stat Series 24:156

Jamil S, Abhilash PC, Singh N, Sharma PN (2009) *Jatropha curcas*: a potential crop for phytoremediation of coal fly ash. J Hazard Mater 172:269–275

Juárez LF, Lucho CA, Vázquez GA, Cerón NM, Beltrán RI (2010) Manganese accumulation in plants of the mining zone of Hidalgo, Mexico. Bioresour Technol 101:5836–5841

Karami N, Clemente R, Moreno E, Lepp NW, Beesley L (2011) Efficiency of green waste compost and biochar soil amendments for reducing lead and copper mobility and uptake to ryegrass. J Hazard Mater 191:41–48

Leblebici Z, Aksoy A, Duman F (2011) Influence of salinity on the growth and heavy metal accumulation capacity of *Spirodela polyrrhiza* (Lemnaceae). Turk J Biol 35:215–220

Levresse G, Lopez G, Tritlla J, López EC, Chavez AC, Salvador EM, Corona R (2012) Phytoavailability of antimony and heavy metals in arid regions: the case of the Wadley Sb district (San Luis, Potosí, Mexico). Sci Total Environ 427–428:115–125

Li T, Tao Q, Liang C, Shohag MJ, Yang X, Sparks DL (2013) Complexation with dissolved organic matter and mobility control of heavy metals in the rhizosphere of hyperaccumulator *Sedum alfredii*. Environ Pollut 182:248–255

López S, Gallegos ME, Flores LJP, Rojas MG (2005) Mecanismos de fitorremediación de suelos contaminados con moléculas orgánicas xenobióticas. Rev Int Contam Ambient 21:91–100

Ma LQ, Komar KM, Tu C, Zhang WH, Cai Y, Kennelley ED (2001) A fern that hyperaccumulates arsenic. A hardy, versatile, fast-growing plant helps to remove arsenic from contaminated soils. Nature 409:579

Ma Y, Prasad MNV, Rajkumar M, Freitas H (2011) Plant growth promoting rhizobacteria and endophytes accelerate phytoremediation of metalliferous soils. Biotechnol Adv 29: 248–258

Machado B, Calderón J, Moreno R, Rodríguez JS (2013) Accumulation of arsenic, lead, copper, and zinc, and synthesis of phytochelatins by indigenous plants of a mining impacted area. Environ Sci Pollut Res 20:3946–3955

Martínez S, Hernández G, Ramírez ME, Martínez J, Solorio G, Solís S, García R (2013) Total mercury in terrestrial systems (air-soil-plant-water) at the mining region of San Joaquín, Queretaro, Mexico. Geofis Int 52:43–58
Mauricio A, Peña JJ, Maldonado M (2010) Isolation and characterization of hexavalent chromium-reducing rhizospheric bacteria from a wetland. Int J Phytoremediation 12:317–334
Mireles A, Solıs C, Andrade E, Lagunas-Solar M, Pina C, Flocchini RG (2004) Heavy metal accumulation in plants and soil irrigated with wastewater from Mexico City. Nucl Instrum Methods Phys Res B 219:187–190
Nowack B, Schulin R, Robinson B (2006) Critical assessment of chelant-enhanced metal phytoextraction. Environ Sci Technol 40:5225–5232
Peralta-Videa JR, Lopez ML, Narayan M, Saupe G, Gardea-Torresdey J (2009) The biochemistry of environmental heavy metal uptake by plants: implications for the food chain. Int J Biochem Cell Biol 41:1665–1677
Pilon SE (2005) Phytoremediation. Annu Rev Plant Biol 56:15–39
Prieto F, Judith G, Hernández C, Ángeles MDL, Gaytán J, Enrique I, Lechuga MDLA (2005) Acumulación en tejidos vegetales de arsénico proveniente de Aguas y Suelos de Zimapán, Estado de Hidalgo, México. Bioagro 17:129–135
Puga S, Sosa M, De la Mora A, Pinedo C, Jiménez J (2006) Concentraciones de As y Zn en vegetación nativa cercana a una presa de jales. Rev Int Contam Ambie 22:75–82
Rajkumar M, Sandhya S, Prasad MNV, Freitas H (2012) Perspectives of plant-associated microbes in heavy metal phytoremediation. Biotechnol Adv 30:1562–1574
Ramos YR, Siebe CD (2006) Estrategia para identificar jales con potencial de riesgo ambiental en un distrito minero: estudio de caso en el Distrito de Guanajuato, México. Rev Mex Cienc Geol 23:54–74
Razo I, Carrizales L, Castro J, Diaz-Barriga F, Monroy M (2004) Arsenic and heavy metal pollution of soil, water and sediments in a semi-arid climate mining area in Mexico. Water Air Soil Poll 152:129–152
Rivera F, Juárez LV, Hernández SC, Acevedo OA, Vela G, Cruz E, De León F (2013) Impacts of manganese mining activity on the environment: interactions among soil, plants, and arbuscular mycorrhiza. Arch Environ Contam Toxicol 64:219–227
Ruiz Olivares A, Carrillo-González R, González-Chávez Mdel C, Soto Hernández RM (2013) Potential of castor bean (*Ricinus communis* L.) for phytoremediation of mine tailings and oil production. J Environ Manag 114:316–323
Salas MA, Manzanares E, Letechipia C, Vega HR (2009) Tolerant and hyperaccumulators autochthonous plant species from mine tailing disposal sites. Asian J Exp Sci 23:27–32
Salt DE, Blaylock M, Kumar NPAB, Dushenkov V, Ensley BD (1995) Phytoremediation: a novel strategy for the removal of toxic metals from the environment using plants. Biotechnology 13:468–474
Sánchez AS, Carrillo R, González MDCA, Rosas GH, Vangronsveld J (2015) Phytobarriers: plants capture particles containing potentially toxic elements originating from mine tailings in semi-arid regions. Environ Pollut 205:33–42
Santos J, Castro A, Huezo J, Torres L (2012) Arsenic and heavy metals in native plants at tailings impoundments in Queretaro, Mexico. Phys Chem Earth 37–39:10–17
SEMARNAT (2010) Environmental and Natural Resources Secretariat. National Program for Prevention and Integral Management of Residues 117
Tian D, Zhu F, Yan W, Fang X, Xiang W, Deng X, Wang G, Peng C (2009) Heavy metal accumulation by Panicled Goldenrain tree (*Koelreuteriapaniculata*) and common Elaeocarpus (*Elaeocarpus decipens*) in abandoned mine soils in southern China. J Environ Sci 21:340–345
Tu S, Ma L, Luongo T (2004) Root exudates and arsenic accumulation in arsenic hyperaccumulating *Pteris vittata* and non-hyperaccumulating *Nephrolepis exaltata*. Plant Soil 258:9–19
Wang J, Zhao FJ, Meharg AA, Raab A, Feldmann J, McGrath SP (2002) Mechanisms of arsenic hyperaccumulation in *Pteris vittata*. Uptake kinetics, interactions with phosphate, and arsenic speciation. Plant Physiol 130:1552–1561

Wenzel WWR, Unterbrunner P, Sommer SP (2003) Chelate-assisted phytoextraction using canola (*Brassica napus* L.) in outdoors pot and lysimeter experiments. Plant Soil 249:83–96

Xiong ZT (1998) Lead uptake and effects on seed germination and plant growth in a Pb hyperaccumulator *Brassica pekinensis* Rupr. Bull Environ Contam Toxicol 60:285–291

Zarazúa G, Poblano J, Tejeda S, Ávila P, Zepeda C, Ortiz H, Macedo G (2013) Assessment of spatial variability of heavy metals in metropolitan zone of Toluca valley, Mexico, using the biomonitoring technique in mosses and TXRF analysis. Sci World J:426492

Zhang XH, Liu J, Huang HT, Chen J, Zhu YN, Wang DQ (2007) Chromium accumulation by the hyperaccumulator plant *Leersia hexandra* Swartz. Chemosphere 67:1138–1143

Chapter 12
Role of White Willow (*Salix alba* L.) for Cleaning Up the Toxic Metal Pollution

Junaid Ahmad Malik, Aadil Abdullah Wani, Khursheed Ahmad Wani, and Muzaffer Ahmad Bhat

12.1 Introduction

Plants and plant communities are very important to humans and their environment. The quality of the air can be greatly influenced by plants. Plants can stop the movement of dust and pollutants. Trees have been proposed as a minimal effort, manageable, and environmentally stable answer for phytoremediation of trace metal debased land. The utilization of willow and poplar species in phytoremediation is promising (Dickinson 2000). Also, willows are easy to propagate and proliferate as well as quickly developing. They are likewise metal tolerant, perpetual, and with a broad root framework with high rate of evapotranspiration that can soothe pollutants (Hammer et al. 2003; Wilkinson 1999).

There are about 450 species of *Salix* genus, and all the species are easily propagated, fast-growing, and tolerant to varied soil conditions. The ability of *Salix* to resprout after harvesting of aboveground biomass, along with significant transpiration rates and potential production of energy biomass, makes it an effective group of plants for phytoremediation purposes. *Salix* species are apropos as a heavy metal phytoextractors due to their high element accumulation, high heavy metal

J. A. Malik (✉) · K. A. Wani
Department of Zoology, Government Degree College, Bijbehara, Anantnag, Jammu and Kashmir, India

A. A. Wani
Directorate of Sericulture Development Department, Srinagar, Jammu and Kashmir, India

M. A. Bhat
Department of Zoology and Applied Aquaculture, Barkatullah University, Bhopal, Madhya Pradesh, India

K. R. Hakeem et al. (eds.), *Bioremediation and Biotechnology*,
https://doi.org/10.1007/978-3-030-35691-0_12

transport to the shoots, and high biomass production. A number of *Salix* spp. have varied genetic variability within the genus. Some species are known to colonize contaminated soil (Greger and Landberg 1999). *Salix* is a diverse genus and has the capacity to absorb and resist heavy metal ions. Compared to other plants, *Salix* are quite effective in phytoextraction of heavy metal pollutants (Laureysens et al. 2004; Kocik et al. 2007). Environmental studies reveal that phytoremediation by some *Salix* clones can be a useful tool in technical replenishment approaches in soil remediation (Vande et al. 2007). *Salix* not being a hyperaccumulator plant, a proportion of clones can develop quick in intensely contaminated regions (Schaff et al. 2003; Vervaeke et al. 2003). *Salix* growth in contaminated soil can be supported or restrained by the soil conditions and the presence of other plants or weather conditions. Efficiency of phytoextraction with *Salix* use in sullied and uncontaminated areas among the additional gears depends on species or even diversity, soil situations, and plant oldness (Newman and Reynolds 2004; Schaff et al. 2003; Berndes et al. 2004).

Modern procedure, rural preparations, mining, and other human exercises have brought about extensive defilement of soils with substantial metals (Goyal et al. 2008; Atafar et al. 2010). The primary anthropogenic wellsprings of overwhelming metals happen because of human exercises, for example, mining, broad utilization of composts, and sewage creation. In the most recent years, the capability of some tree species in expelling metal particles from soil has gotten increasingly more consideration. Soils contaminated with metals may compromise biological systems and human well-being. In common habitats, overwhelming and heavy metals are available at low fixations without making huge harmful impacts to living beings. Notwithstanding, their expanded gathering in soil and water may have genuine ramifications for plants, creatures, and human well-being. Some substantial metal-tainted soils might be tidied up by developing plants which gather the toxicity, at that point reaping the plants and discarding them in a "sheltered zone." This sort of innovation, known as "phytoremediation," speaks to an innocuous and minimal effort method, lacking of particular reactions (Cunningham and Owen 1996; Ledin 1998). Salicaceae are well-thought-out as good contenders for a phytoremediation approach (Marmiroli et al. 2011) due to their easy propagation, great lenience against metal pollutants, and high annual biomass production. *Salix alba* plant species, which have high biomass production, can be proposed for usage in phytoremediation technology. The ecophysiological reaction of common and viable willow replicas evaluates the effects of the Zn treatment on the photosynthetic process by the analysis of CO_2 integration and chlorophyll (Chl.*a*) fluorescence. Among the overwhelming metals, cadmium (Cd) is of specific worry because of its generally high portability in soils and potential danger to biota at low fixations. In spite of the fact that Cd is normally present in soil at very low sums, elevated amounts of cadmium have been accounted in some conditions (Das et al. 1997).

12.2 Reaction of *Salix alba* L. to Heavy and Toxic Metals

Of the absolute vegetation accessible, trees offer ease, reasonable, and naturally stable answer for phytoremediation of trace metal tainted land. In addition the utilization of yields, for example, willow species, has likewise been observed to guarantee in phytoremediation as they gather Cd and Zn in their over-the-ground biomass. So willows could be utilized for collecting the supplements as well as the overwhelming metals as they are easy to propagate and spread with a broad root framework with high evapotranspiration rates that can settle toxins. Willows can develop on tainted soils and have high biomass generation by which they can amass great amounts of PTE in the roots and so on. Understanding their significance in overwhelming metal gathering, Belarus planted willow trees for the adjustment and phytoremediation of polluted soils, in post-mining scenes and on sterile landfills. Yet, In India ponders on phytoremediation potential under field conditions stay restricted (Hazrat et al. 2013).

Oxidative stress is seen in plants when exposed to heavy metals that subsequently lead to their cell harm and unsettling influence in cell ionic homeostasis. Applied pollutants impact negatively to the plant and cause confined chlorotic as well as necrotic variations of shoot leaves. These progressions are obvious generally on more youthful greeneries in little fixes, on plants treated with higher concentrations of contamination blend, and on higher connected diesel fuel treatment. There is significant reduction of plant growth in response to all the applied pollutants (Table 12.1). Pollutant level concentration is an important aspect that determines the mark of growth discount. The blend of heavy metals connected at lower fixations (m) determines the smallest negative impact on the growth. The most grounded decrease of photosynthetic rate (PR) and transpiration rate (TR) is brought about by the treatment of heavy metals in higher concentrations. Correlation between photosynthetic rate and growth parameters has proven to be positive, but without statistical significance (Table 12.2).

Table 12.1 Leaf developmental parameters of *S. alba* in connection with toxin treatments

Treatment	Leaf mass (g)	No. of leaves per plant
Control	5.5 ± 0.4	43.3
Cd	2.0 ± 0.2	29.2
Ni	1.9 ± 0.2	33.2
Pb	1.6 ± 0.1	29.2
M	1.6 ± 0.2	27.5

Table 12.2 Correlations (*r*) between photosynthetic rate and growth restrictions

Growth parameter	Photosynthetic rate
Plant height	0.69
Mass of the shoot	0.46
Mass of the leaf	0.46
Number of leaves	0.50

Cadmium, nickel, and lead (Pb) highly accumulate in the roots of the plant. Higher accumulation and more uptakes of cadmium (Cd) and nickel (Ni) have been seen in the plant when the pollutant is treated in higher concentrations. Cadmium (Cd) and nickel (Ni) are progressively aggregated when blended substantial metal treatment is applied in comparison with individual overwhelming metal treatments. The singular treatment of Lead (Pb-EDTA) has categorically the most significant accumulation level of the pollutant in all the parts of the plant while as the presence of Cd and Ni in treated soil reduces the Pb accumulation which indicates the antagonistic relation between the heavy metals. Gerhardt et al. (2009) propose that individual lethal impact of overwhelming metals has shown strong additive effect. Toxicity is showed through leaf chlorosis, putrefaction, decreased biomass of all plant organs, and unsettling influences in photosynthetic CO_2 absorption. This was also verified later by Atagana (2011) on *Salix alba* and other plant species. The success of plant adjustment to toxin stress depends both on stable photosynthetic action and water use productivity. According to a previous study result, *S. Alba* genotype could have adequate practicality and practicability at the locales with lower connected groupings of poisons (pollutants) like cadmium <2.15 mg/kg, nickel <70.8 mg/kg, and Pb-EDTA <116.1 mg/kg.

Mühlbachová (2009) established that the use of EDTA in substantial metal debased soil, at beginning times, can produce poisonous consequences for the soil microorganisms which lessens their general biomass. In any case, further research is required to clarify the job of EDTA in co-polluted condition.

12.3 Heavy Metal Accumulation

Accumulation of heavy metals in the environment is a potential risk to living system due to their uptake by plants and subsequent introduction into the food chain. The accumulation of heavy metals in soils and subsequently to the plants proves an increasing concern because of the potential human health risks. The accumulation of heavy metals in plants depends upon plant species, soil properties, and the efficiency of different plants in absorbing metals. It can be evaluated by either plant uptake or soil-to-plant transfer factors of the metals (Rattan et al. 2005). The extent of accumulation of heavy metals is a direct proportionate of the concentration of the same in the soil. While taking the medium, cadmium, concentration in the soil under consideration, *Salix alba* is characterized by high extent of metal concentration. Accumulation of cadmium is very intensive and proves to be toxic. Copper (Cu) is also accumulated in the medium level by the *Salix alba*, while mercury (Hg) is highly accumulated by the plant. *Salix alba* has an intensive accumulation capability toward lead and zinc as well. The factors which influence the metalloid mobility in the soil are pH, concentration, and composition of organic compounds. Metals can be easily absorbed by the *Salix alba* when they are easily bioavailable and can be disseminated along the subsequent food chain which in turn causes mutagenic effects (Hazrat et al. 2013).

The higher zinc and cadmium concentration is found in leaves than in the stem. Pertinently, the zinc and cadmium fixations in the over-the-ground portions of the plant are highly variable which indicates that these two elements are not transferred to the aboveground parts of willows in a reasonable concentration (Vangronsveld et al. 2009). However the dynamics of the gathering, transport, and resilience is explicit for each metal. Another study by Aktaruzzaman et al. (2013) reveals the soil-to-plant transfer factor (TF), and it can be concluded that Pb and Cd are high accumulators. The study puts the tendency of TF for heavy metals in the order of Pb > Cd > Zn > Cu > Cr. The inoculation of microorganisms such as *Streptomyces* sp., *Agromyces* sp., and *C. finlandica* is known to increase the accumulation of cadmium and zinc to shoots by increasing the bioavailability of Cd, Zn, and potassium (K) in polluted soil (Maria et al. 2011). The treatments of mycorrhizae are also known to increase the accumulation of Cu and the shoot biomass (Cloutier et al. 2014). The *Salix alba* L. in coordination with microorganisms and fungus prove to be the sources of enhancement for the process of phytoremediation. Purdy and Smart (2008) envisaged that the hydroponic experiments along with the phosphates have decreased the toxicity and increased the accumulation of metals in the shoot tissues of *Salix alba.*

12.4 Phytoextraction

The translocation and transport of metals to the aboveground parts of a plant is an effective biochemical process to remediate the polluted sites. The damaging effect caused by the metal accumulations on root physiology is reduced more by the efficient utilization from the root to the aboveground parts. This in turn increases the efficiency of the plant toward metal uptake allowing the metal removal from the site very efficiently.

Since the 1980s, the techniques for the process of bioremediation have been used. Initially microorganisms were used for the purpose of metal remediation via biosorption and by reducing metals to their lower redoxes (Lovley and Coates 1997). *Salix* phytoremediation capabilities have been researched from the past two decades along with their associated fungus and bacteria. The coordination of microorganisms with the woody hyperaccumulators enhances the phytoremediation and can also be considered booming. Metals are often made bioavailable by the microorganisms by acidifying the soil near roots. The enhanced extraction of Cd and Zn by *Salix alba* L. (and other species) through inoculation of rhizobacteria is an example for the improved metal accumulation using microorganisms (Maria et al. 2011). The contaminated soil is acidified by the bacteria in the rhizosphere which increases the bioavailability of the metals. The metals are then easily removed through the bacterial sulfate reduction (White and Gadd 1997).

Phytoextraction is the process of plants by which they expel perilous components or mixes from the soil or water, most generally substantial metals, metals that have a high compactness and might be harmful. These extracted metals prove to be toxic for the plants and the animals as well. Heavy metals react with a number of chemicals which are essential for the plant and may cause the adverse effects on the cellular metabolism. They can likewise break different molecules into much increasingly responsive species which in turn disrupt biological processes. There are some fundamentals through which the plant (and the metal) undergoes to extract the heavy metals in the soil or water:

1. The metal must dissolve.
2. The plant roots must retain the substantial metal.
3. The plant must chelate (e.g., metal-EDTA chelate) the metal to secure itself and make the metal progressively transportable.
4. Movement of the chelated metal to a site where it is safely stored.
5. The plant must acclimatize to any costs by the carriage or storage of the metal.

There is much interest on the plant species which are capable of accumulating the high concentrations of toxic elements from the soil. Hyperaccumulators are the plants which have capabilities of accumulating metal concentrations in excess of multiple times greater than the normal species (Shen et al. 1997). The cropping on the metal contaminated land with hyperaccumulators results in a potentially hazardous biomass. Meanwhile, one of the major drawbacks of hyperaccumulators is slow growth rate of the plant. Application of chelating agents is known to cause disadvantageous effects on treated plants. A study by Cooper et al. (1999) envisages that the copper and zinc concentrations of several herbaceous plants increase with the application of chelates (nitrilotriacetate, NTA, and EDTA), but at the same time the report points that the increases were perplexed by the decrease in the dry weight of the plant.

Ion concentrations strongly affect the metal uptake by *Salix alba*. For instance, phosphorus restricts the commitment and translocation of zinc and lead ions. Hence the nutritious aspect of the soil has a substantial consequence on the bioavailability of the noxious metals to *Salix alba*. *Salix alba* is known to solubilize metals by radiating protons from the roots to acidify the rhizosphere.

The withholding and drive of heavy metals in the soils are administered by various processes, e.g., cation exchange and specific adsorption are the major machineries regulating Ni, Cd, and Zn movements. However the organic complexation of Cr (chromium), Cu, and Pb is extensive. The processes which may affect heavy metal retention, uptake, translocation, and mobility are seen to control the bioavailability of the metals. The heavy metals while on uptake may have damaging effects as they block the functional groups of the polynucleotides and displace the essential metal ions from biomolecules. The absorbed heavy metals also denature the enzymes and disrupt the cell organelles along with the rupture of their membrane (Ross and Kaye 1994).

12.5 How *Salix alba* L. Is Eligible for Phytoremediation

Willows are the peculiar phytoextractors as they accumulate and tolerate metals. Moreover they are known to form the dominant vegetation in the upper watersheds and crunches at higher reaches. Many characteristics of the *Salix alba* make it the paramount phytoremediation agent. A successful phytoremediator requires a high translocation rate from the root to the shoot (Greger and Landberg 1999). High biomass, metal translocation ability, and rapid growth and development of the *Salix alba* L. make it the useful phytoextractor for the polluted soils (Pulford and Watson 2003). The *Salix* spp. have broad genetic variability and quick growth promptness. The metals which are accumulated in the upper biomass are ideal for harvesting and the metals can be removed permanently. *Salix alba* has an additional cash benefit of producing woody biomass that can be used as fuel and hence the plant is a preferred one for the cultivation.

Phytoextraction is a machinery of the plants by which the contaminants taken up by the plant are used into a harvestable portion. It has been viewed as a less expensive and helpful choice for low defiled destinations in comparison with other methods such as excavation and soil washing (Jadia and Fulekar 2009; Pilon-Smits 2005). Metal uptake and phytoextraction by the plants is dependent upon rates of the uptake and the bioavailability of the metal to the plant (Pilon-Smits 2005). The elimination of the metals from the soil accomplished through the harvest usually consists of aboveground parts. The use of hyperaccumulators has been suggested in the process of phytoextraction (Baker et al. 1994) as they are known to take up maximum quantities of an explicit metal. Hyperaccumulators amass >0.01% of cadmium, >0.1% of copper, or >1.0% of zinc in dry mass of leaves (Baker et al. 1994). Willows accumulate elevated amounts of Cadmium and Zinc that is why the ash from willow contains 10 times higher cadmium concentrations than from the other forest trees (Brieger et al. 1992).

Phytoextraction of heavy metals by *Salix* has been investigated in hydroponic studies (Watson et al. 2003), in pot trials (Meers et al. 2007*),* in contaminated soils (Bissonnette et al. 2010), and in biosolid-amended soils (Maxted et al. 2007). Table 12.3 is a research data analysis done by Landberg and Greger in 1994 and 1996, which shows the uptake and transport of Cd, Zn, and Cu by *Salix alba.*

According to the data reported by Greger and Landberg (1999), *Salix alba* is reasonable as a phytoextractor on low to tolerably metal sullied soil because of the way that the plant has high accumulation, high transport to the shoot of heavy or substantial metals, and high take-up, at least for Cd, Zn, and Cu. *Salix alba* likewise has high biomass generation and is as of now in economic use (for bioenergy). In addition there is also a method for removing cadmium from the ash of the plant. The plant can also be bred to the clones which are not attractive to the animals.

Harvesting near the end of the season is known to expel the greatest amount of the metals. The concentration of the metals in the wood and the bark seems to remain steady over the season or is increasing at the end of a season. The biomass

Table 12.3 Differences in concentration in shoot, root, and transport of cadmium, copper, and zinc in different replicas of *Salix alba* L.

Metal	Metal µg (g dry weight)$^{-1}$		Transport %
	Shoot	Root	
Cadmium	0.1–7.9	4.3–302	1–68
Copper	0.4–8.6	16–353	1–22
Zinc	14–1776	65–1950	11–70

is also increasing at the termination of the period. The annual uptake ratio by the *Salix alba* subsequently increases by the co-harvesting of leaves as well. It is known that Cd availability in the soil can decrease with time using *Salix* cultures, but the rate of the decline is poorly known. Pulford et al. (2002) have portrayed that the concentration of EDTA-extractable Cd, Cu, Ni, and Zn in sewage sludge-amended soil is higher at the site of *Salix alba* soil than in any other cultivation or in the unplanted areas. With the leaf fall, a considerable amount of metals are recycled in the soil-stand surface of the *Salix alba* in comparison with the stem harvesting. Hence a promising preference for the management is to harvest the leaves notwithstanding the wood which would decrease the danger of the natural pecking order collection that is food chain.

12.6 Phytoremediation Potential

Miroslaw et al. (2010) tested two *Salix* species, *Salix viminalis* and *Salix alba*, which were already cloned for the study of phytoremediation and accumulation of the heavy metals. The dormant cuttings were planted straight in the unprepared soil in rows at a distance of 0.5 m and were assessed during the two consecutive seasons. The *Salix* clones were as follows:

1. *S. alba.*
2. *S. viminalis* "v1" (two samples).
3. *S. viminalis* "v2".
4. *S. viminalis* "v3".
5. *S. viminalis* "v4".
6. *S. viminalis* "v5".
7. *S. viminalis* "v6".
8. *S. viminalis* "v7" (two samples).
9. *S. viminalis* "v8".

Soil tests were gathered toward the start of the test around the specimens at first from the whole study region so as to decide the level of soil homogeneity. The bioaccumulation factors (BAFs) were determined as the proportion of substantial metal fixation in *Salix* shoots to focus on this metal in soil. The investigation of substantial metal substance in plant material and soil was directed by electrothermal atomic absorption spectrometry (ETAAS) utilizing a Varian SpectraAA 200 spectrometer. Overall concentration of chosen substantial metals in individual *Salix* shoots was essentially differing. Results in mean qualities are exhibited in Table 12.4. So as to decide accumulation proficiency, bioaccumulation factors (BAFs) for every taxon (Table 12.4) were determined, and the proportion of accumulated metals in the plant, what's more in the soil, was determined. The positioning was readied contemplating contrasts in concentrated substantial metals in the particular plants.

Table 12.4 Concentration of heavy metals [mg kg^{-1}] in examined soil and *Salix* sprouts

Salix clone	Heavy metal Cd	Cu	Pb	Zn	Clone position	Accuml. abilities (simultaneous)
S. viminalis "*v1a*"	1.9834 (I)	5.1842 (M)	2.8392 (M)	52.8259 (I)	1	*S. alba*
S. viminalis "*v1b*"	2.4739 (I)	5.9482 (M)	2.3566 (M)	46.5762 (I)	2	*S. viminalis* "v8"
S. viminalis "*v4*"	1.9472 (I)	6.7877 (M)	3.3764 (M)	53.7994 (I)	3	*S. viminalis* "v4"
S. viminalis "*v6*"	3.4822 (I)	4.0445 (M)	4.8492 (M)	56.2389 (I)	4	*S. viminalis* "v3"
S. viminalis "*v5*"	2.7445 (I)	5.4829 (M)	3.3816 (M)	50.7029 (I)	5	*S. viminalis* "v2"
S. viminalis "*v7a*"	2.4925 (I)	6.4239 (M)	4.9942 (M)	57.1304 (I)	6	*S. viminalis* "v7b"
S. viminalis "*v7b*"	2.8237 (I)	6.0821 (M)	5.2814 (M)	63.4672 (I)	7	*S. viminalis* "v1b"
S. viminalis "*v3*"	3.6488 (I)	7.6725 (M)	3.4873 (M)	48.4779 (I)	8	*S. viminalis* "v7a"
S. viminalis "*v2*"	2.0036 (I)	6.8247 (M)	2.8342 (M)	59.3849 (I)	9	*S. viminalis* "v6"
S. viminalis "*v8*"	2.0462 (I)	6.2894 (M)	4.0508 (M)	56.9821 (I)	10	*S. viminalis* "v5"
S. alba	2.4837 (I)	5.8745 (M)	6.8372 (M)	59.9242 (I)	11	*S. viminalis* "v1a"
LSD	0.0262	0.0296	0.0334	0.0338		

The rank of clones in accumulation of all metals concurrently (Miroslaw et al. 2010)
I intensive accumulation (BAFs >1), *M* medium accumulation (1 > BAFs >0.1)

Fig. 12.1 A view of the *Salix alba* at a wetland near Nilandrusu, Bijbehara

Hence it is evident from the studies of Miroslaw et al. (2010) that *Salix alba* is having a high translocation rate from the root to the shoot and same with the metal accumulation and the extent of the phytoextraction as well, which makes it the useful phytoextractor for the polluted soils (Fig. 12.1).

12.7 Conclusion

In view of the present scenario of urban environmental contamination, there is a need for planting the *Salix alba* and other plant species. Plants with the capability of mitigating heavy metal contaminations would be a reasonable option for cleanup of all metal substances to acceptable levels. *Salix* species are promising for the practice of phytoremediation of filthy land. Willows are laid to proliferate and propagate fast-growing, metal-tolerant, persistent crops, with a widespread root system and high evaporation and transpiration rates that are known to decontaminate the land and subsequently control to stop metal transference to other cubicles of the ecosystem. Phytoremediation by *Salix alba* (having a huge economic value) offers eco-friendly machinery for conventional remediation of the heavy metals from the soil as the plant is having a deep-root system as well as high biomass yields. The first step for developing an effective phytoremediation method with the hybridization of *Salix alba* is the complete understanding of the growth dynamics and the physiological machines which administer the relations among a unwavering plant genotype and pollutants. The genetically identical stem cuttings of *Salix alba* are a potential advantageous apparatus for air quality monitoring.

References

Aktaruzzaman M, Fakhruddin ANM, Chowdhury MAZ, Fardous Z, Alam MK (2013) Accumulation of heavy metals in soil and their transfer to leafy vegetables in the region of Dhaka Aricha highway, Savar, Bangladesh. Pak J Biol Sci 16(7):332–338

Atafar Z, Mesdaghinia A, Nouri J, Homaee M, Yunesian M (2010) Effect of fertilizer application on soil heavy metal concentration. Environ Monit Assess 160(1–4):83–89

Atagana HI (2011) Bioremediation of co-contamination of crude oil and heavy metals in soil by phytoremediation using *Chromolaena odorata* (L.)King & HE Robinson. Water Air Soil Pollut 215:261–271

Baker AJM, McGrath SP, Sidoli CMD, Reeves RD (1994) The possibility of in situ heavy metal decontamination of polluted soils using crops of metal-accumulating plants. Resour Conserv Recyl 11:41–49

Berndes G, Fredrikson F, Borjesson P (2004) Cadmium accumulation and Salix-based phytoextraction on arable land in Sweden. Agric Ecosyst Environ 103:20–23

Bissonnette L, St-Arnaud M, Labrecque M (2010) Phytoextraction of heavy metals by two Salicaceae clones in symbiosis with arbuscular mycorrhizal fungi during the second year of a field trial. Plant Soil 332:55–67. https://doi.org/10.1007/s11104-009-0273-x

Brieger G, Wells JR, Hunter RD (1992) Content in fly ash ecosystem. Water Air Soil Pollut 63:87–103

Cloutier HB, Turmel MC, Mercier C, Courchesne F (2014) The sequestration of trace elements by willow (*Salix purpurea*)—which soil properties favor uptake and accumulation. Environ Sci Pollut Res 21(6):4759–4771

Cooper EM, Sims JT, Cunningham JW, Berti WR (1999) Chelate-assisted phytoextraction of lead from contaminated soils. J Environ Qual 28:1709–1719

Cunningham SD, Owen DW (1996) Promises and prospects of phytoremediation. Plant Physiol 110(3):715–719

Das P, Samantaray S, Rout GR (1997) Studies on cadmium toxicity in plants: a review. Environ Pollut 98(1):29–36

Dickinson NM (2000) Strategies for sustainable woodland on contaminated soils. Chemosphere 41:259–263

Gerhardt KE, Huang XD, Glick BR, Greenberg BM (2009) Phytoremediation and rhizoremediation of organic soil contaminants: potential and challenges. Plant Sci 176:20–30

Goyal P, Sharma P, Srivastava S, Srivastava MM (2008) *Saraca indica* leaf powder for decontamination of Pb: removal, recovery, adsorbent characterization and equilibrium modelling. Int J Environ SciTech 5(1):27–34

Greger M, Landberg T (1999) Use of willow in phytoextraction. Int J Phytoremediation 1:115–123

Hammer D, Kayser A, Keller C (2003) Phytoextraction of Cd and Zn with *Salix viminalis* in field trials. Soil Use Manag 19:187–192

Hazrat A, Khan E, Anwar SM (2013) Phytoremediation of heavy metal concepts and applications. Chemosphere 91-7:869–881

Jadia CD, Fulekar MH (2009) Phytoremediation of heavy metals: recent techniques. Afr J Biotechnol 8:921–928

Kocik A, Truchan M, Rozen A (2007) Application of willows (*Salix viminalis*) and earthworms (*Eisenia fetida*) in sewage sludge treatment. Eur J Soil Biol 43:327–331

Laureysens I, Blust R, De Temmerman L, Lemmens C, Ceulemans R (2004) Clonal variation in heavy metal accumulation and biomass production in a poplar coppice culture: I. seasonal variation in leaf, wood and bark concentrations. Environ Pollut 131:485–494

Ledin S (1998) Environmental consequences when growing short rotation forest in Sweden. Biomass Bioenergy 15(1):49–55

Lovley DR, Coates JD (1997) Bioremediation of metal contamination. Environ Biotechnol 8:285–289

Maria D, SusannaRAR KM, Sessitsch A, Wenzel WW, Gorfer M, Strauss J, Puschenreiter M (2011) Interactions between accumulation of trace elements and macronutrients in *Salix caprea* after inoculation with Rhizosphere microorganisms. Chemosphere 84(9):1256–1261

Marmiroli M, Pietrini F, Maestri E et al (2011) Growth, physiological and molecular traits in Salicaceae trees investigated for phytoremediation of heavy metals and organics. Tree Physiol 31:1319–1334

Maxted AP, Black CR, West HM, NMJ C, SP MG, Young SD (2007) Phytoextraction of cadmium and zinc by *Salix* from soil historically amended with sewage sludge. Plant Soil 290:157–172. https://doi.org/10.1007/s11104-006-9149-5

Meers E, Vandecasteele B, Ruttens A, Vangronsveld J, Tack FMG (2007) Potential of five willow species (*Salix* spp.) for phytoextraction of heavy metals. Environ Exp Bot 60:57–68. https://doi.org/10.1016/j.envexpbot.2006.06.008

Miroslaw M, Pawel R, Iwona R, Zygmunt K, Piotr G, Kinga S, Katarzyna S, Agnieszka S (2010) Biomass productivity and phytoremediation potential of *Salix alba* and *Salix viminalis*. Biomass Bioenergy 34:1410–1418

Mühlbachová G (2009) Microbial biomass dynamics after addition of EDTA into heavy metal contaminated soils. Plant Soil Environ 55(12):544–550

Newman LA, Reynolds CM (2004) Phytodegradation of organic compounds. Curr Opin Biotechnol 15:225–230

Pilon-Smits E (2005) Phytoremediation. Annu Rev Plant Biol 56:15–39. https://doi.org/10.1146/annurev.arplant.56.032604.144214

Pulford ID, Riddel-Black D, Stewart C (2002) Heavy metal uptake by willow clones from sewage sludge-treated soil: the potential for phytoremediation. Int J Phytoremediation 4:59–72

Pulford ID, Watson C (2003) Phytoremediation of heavy metal-contaminated land by trees-a review. Environ Int 29(4):529–540

Purdy JJ, Smart LB (2008) Hydroponic screening of shrub willow (*Salix* spp.) for arsenic tolerance and uptake. Int J Phytoremediation 10(6):515–528

Rattan RK, Datta SP, Chhonkar PK, Suribabu K, Singh AK (2005) Long-term impact of irrigation with sewage effluents on heavy metal content in soils, crops and groundwater: a case study. Agric Ecosyst Environ 109:310–322
Ross SM, Kaye KJ (1994) The meaning of metal toxicity in soil-plant systems. In: Ross SM (ed) Toxic metals in soil-plant systems. Wiley, New York, pp 27–61
Schaff SD, Pezeshki SR, Shields FD (2003) Effects of soil conditions on survival and growth of black willow cuttings. Environ Manag 31:748–763
Shen ZG, Zhao FJ, McGrath SP (1997) Uptake and transport of zinc in the hyperaccumulator *Thlaspi caerulescens* and the non-hyperaccumulator *Thlaspi ochroleucum*. Plant Cell Environ 20:898–906
Vande WI, Van CN, Van-de CL, Verheyen K, Lemeur R (2007) Short-rotation forestry of birch, maple, poplar and willow in Flanders (Belgium) II. Energy production and CO_2 emission reduction potential. Biomass Bioenergy 31:276–283
Vangronsveld J, Herzig R, Weyens N, Boulet J, Adriaensen K, Ruttens A (2009) Phytoremediation of contaminated soils and groundwater: lessons from the field. Environ Sci Pollut Res 16(7):76–94
Vervaeke P, Luyssaert S, Mertens J, Meers E, Tack FMG, Lust N (2003) Phytoremediation prospects of willow stands on contaminated sediment: a field trial. Environ Pollut 126:275–282
Watson C, Pulford ID, Riddell-Black D (2003) Development of a hydroponic screening technique to assess heavy metal resistance in willow (*Salix*). Int J Phytoremediation 5:333–349
White C, Gadd GM (1997) An internal sedimentation bioreactor for laboratory scale removal of toxic metals from soil leachates using biogenic sulphide precipitation. J Ind Microbiol Biotechnol 18(6):414–421
Wilkinson G (1999) Poplars and willows for soil erosion control in New Zealand. Biomass Bioenergy 16:263–274

Chapter 13
Mycoremediation: A Sustainable Tool for Abating Environmental Pollution

Sajad Ahmad Raina, Nesrine Ben Yahmed, Rouf Ahmad Bhat, and Moonisa Aslam Dervash

13.1 Introduction

"Environmental contamination due to heavy metals (by anthropogenic and industrial activities)" has posed substantial and irreversible damage to "aquatic environs." The sources include "mining" and "fusion operations of minerals," discharges from storage batteries and car exhausts, manufacture and bulk use of fertilizers and pesticides, (Bhat et al. 2017b). Commonly found metals and metalloids "(lead, chromium, mercury, uranium, selenium, zinc, arsenic, cadmium, silver, gold, and nickel)" pollute the water and which are of concern because of their relatively high toxicity. In addition to being dangerous to human health, they are detrimental to "fauna and flora" and are "recalcitrant in nature." Therefore, there is a need to develop new approaches or strategies to minimize or even eliminate metals from the environment.

Various physical, chemical, and "biological processes are commonly used to remove heavy metals from industrial wastewaters before they are discharged into the environment" (Fomina and Gadd 2014; Mehmood et al. 2019). Traditional physicochemical methods "(such as electrochemical treatment, ion exchange, precipitation, osmosis, evaporation and absorption) are not much profitable with the notion that some of them are not even environment friendly" (Mulligan et al. 2001; Kadirvelu et al. 2002). On the other hand, bioremediation processes show promising results for the elimination of metals, even when they are present in very low

S. A. Raina (✉)
Department of Environmental Science, Bhagwant University, Ajmer, Ajmer, Rajasthan, India

N. B. Yahmed
Laboratotoir LIP-MB INSAT, LR11ES26, Universite de Carthage, INSAT-BP 676, Centre Urbain Nord, Carthage Cedex, Tunisie

R. A. Bhat · M. A. Dervash
Department of Environmental Science, Sri Pratap College, Cluster University Srinagar, Srinagar, Jammu and Kashmir, India

K. R. Hakeem et al. (eds.), *Bioremediation and Biotechnology*,
https://doi.org/10.1007/978-3-030-35691-0_13

concentrations where physical-chemical elimination procedures do not work. Besides, bioremediation is an eco-friendly and economically viable option that is based on "the high binding capacity of metals of biological agents, which can remove heavy metals from contaminated sites with high efficiency." In this fashion, microorganisms can be counted among "biological tools for the elimination of metals by concentrating, removing and recovering heavy metals from contaminated aquatic environments" (Riggle and Kumamoto 2000). Scientific temper using microorganisms as an alternative strategy to conventional treatments for the absorption of heavy metals in contaminated water possesses a broader scope (Tsezos and Volesky 1981). Microbial bioremediation is incredibly helpful in an array of conditions ranging from the highly dilute solutions to the extreme conditions. Although "the mechanism associated with the biosorption of metals by microorganisms are not yet well understood, studies show that they play an imperative function in the absorption of metals and that this action implies 'accumulation or resistance.'"

Fungi (mushrooms) are very economical, effectual, and eco-friendly way to help eliminate a wide range of toxins from damaged wastewater or module environments. The toxins are an array of entities like "heavy metals, persistent organic pollutants, textile dyes, chemicals and the tanning wastewater industry, petroleum derived fuels, polyaromatic hydrocarbons, pharmaceutical and personal care products, pesticides and herbicides," in the earth, freshwater, and marine environments. The bioremediation by-products can be useful materials themselves, such as enzymes (Strong and Burgess 2007) and edible or medicinal mushrooms (Kulshreshtha et al. 2014), which makes the process of repair even profitable.

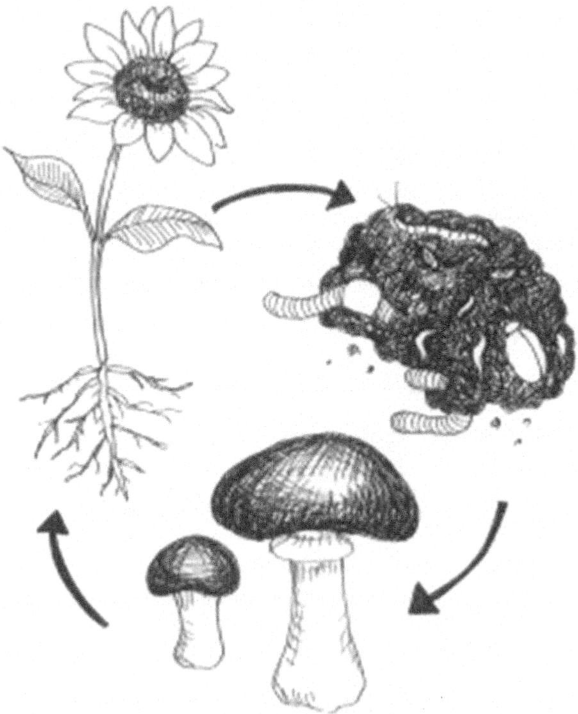

13.2 Mycoremediation Treatment Techniques

Mushrooms employ different strategies to detoxify contaminated sites and eventually stimulate the environment. These methods include the following.

13.2.1 Biodegradation

It is the utmost "degradation and recycling of complex molecules" to their "mineral components." It is the progression that leads to the absolute mineralization of the complex compounds to simpler substances (such as CO_2, H_2O, NO_3, and other inorganic compounds) by biotic organisms, in particular microbes.

13.2.2 Biosorption

The procedure of abolition of metals/xenobiotics from the environment (by fungi) is called biosorption which is considered as "an alternative to the reclamation of industrial effluents, as well as to the recovery of metals present in the effluent." It is a procedure based on the absorption of "metal ions/contaminants/xenobiotics" from the "effluent by live or dry biomass" that often shows a discernible tolerance to metals and other adverse conditions.

13.2.3 Bioconversion

"The conversion of industrial/agro-industrial mud into other beneficial forms is called bioconversion."

13.3 Contaminants/Pollutants

Mycoremediation is commonly used worldwide by scientific fraternity in waste treatment to remove contaminants and pollutants such as the following.

13.3.1 *Heavy Metal Pollution, Effects, and Their Remediation with Fungi*

Heavy metals are detrimental to humans, for example, Hg, Pb, Cd, Cu, Ni, and Co (Pierzynski et al. 2000), depending on quantum and period of exposure. Heavy metal pollution causes many harmful effects encompassing all life forms, viz., fauna, flora, and essential microbes, and also poses negative impact on human health (Ilyina et al. 2003; Chakraborty et al. 2013). Heavy metal ions possess strong bonds "(electrostatic attraction and high binding affinities)" with the same sites to which metal ions normally unite in various cellular structures, causing "the destabilization of structures and biomolecules (cell wall enzymes, DNA and RNA), which stimulates replication defects and consequential mutagenesis, hereditary genetic disorders and cancers" (Perpetuo et al. 2011). "Heavy metals are noteworthy pollutants because they are toxic, non-biodegradable in the environment and easily accumulate in living organisms" (Zhuang and Gao 2013).

Water contamination with xenioboitics (heavy metals) occurs through natural and anthropogenic activities, mainly associated with industrialization. The "natural and anthropogenic sources of some of the most widely studied heavy metals as environmental pollutants, along with a brief list of their adverse health effects (Table 13.1)

Table 13.1 Contamination sources, uses, and adverse health effects of some heavy metals

	Source			
Metal	Natural	Anthropogenic	Uses	Adverse health effects
Cd	Zn and Pb minerals, phosphate rocks	Mining waste, electroplating, battery plants	Automobile exhaust	Respiratory, cardiovascular, renal effects
Cr	Chromite mineral	Electroplating, metal alloys, industrial sewage, anticorrosive products	Pesticides, detergents	Mental disturbance, cancer, ulcer, hypokeratosis
Cu	Sulfides, oxides carbonates	Electroplating, metal alloys, domestic and industrial waste, mining waste, pesticides	Most uses are based on electrical conductor properties	Anemia and other toxicity effects induced indirectly through interaction with other nutrients
Pb	Galena mineral	Battery plants, pipelines, coal, gasoline, pigments	Batteries, alloys	Neurotoxic
Ni	Soils	Metal alloys, battery plants, industrial waste, production of vegetable oils	Batteries, electronics, catalysts	Skin allergies, lung fibrosis, diseases of the cardiovascular system
Zn	Minerals (sulfides, oxides, and silicates)	Metal alloys, pigments, electroplating, industrial waste, pipelines	Fertilizers, plastics, pigments	Abdominal pain, nausea, vomiting and diarrhea, gastric irritation, headache, irritability, lethargy, anemia

and their applications" (Rajendran et al. 2003). Although the bioremediation studies generally consider the type of metal complex in the environment, but, the levels of toxicity depends on the chemical form of respective metal which may include species and cationic/anionic complexes (hydroxylated or complexed with Cl), and their oxidation states vary according to the pH and the composition of the soil.

Metal contamination is very common because of wide functional range like galvanic, textile, leather and paint (Bhatia et al. 2017). Wastewater in these areas is often used for agricultural purposes, so in addition to the immediate damage to the ecosystem in which it is reflected, metals find their route into the creatures and humans through the trophic linkages (food chain). Mycoremediation is one of the most economical, most effective, and friendly solutions to deal with various environmental problems (Joshi et al. 2011). Many mushrooms are hyper-accumulators, which mean that they are able to concentrate toxins in their fruiting bodies for later disposal. This is usually true for people who have been exposed to pollutants for a long time and have developed a high tolerance, and it occurs through bio-absorption on the surface of the cells, which urges that the metals enter the mycelium passively via intracellular diffusion (Gazem and Nazareth 2013). A variety of fungi, "such as *Pleurotus*, *Aspergillus* and *Trichoderma* has been shown to be effective in the removal of lead" (Gazem and Nazareth 2013; Joshi et al. 2011), cadmium (Joshi et al. 2011), nickel (Cecchi et al. 2017), chrome (Joshi et al. 2011), mercury (Kurniati et al. 2014), arsenic (Singh et al. 2015), copper (Gazem and Nazareth 2013; Zotti et al. 2014), boron (Taştan et al. 2016), iron, and zinc (Vaseem et al. 2017) in the marine environment, wastewater, and earth. The ability of some fungi to extract metals from the soil can also be useful as bioindicators and can be a problem when the fungus is edible. For example, "the shaggy ink cap (*Coprinus comatus*), a common edible mushroom from the northern hemisphere, can be an excellent bio-indicator of mercury and accumulate it in various trophic levels with certain toxicity" (Falandysz 2016).

13.3.2 Considerations on the Metal's Uptake Capacity of Microorganisms

The path through which metal orients to a particular polluted region has foremost importance in connection to the effectiveness of a "bioremediation process." For instance, the inoculation of "sediments by microorganisms" will measure the core route of metal contamination. Although the free metal ions present in sediment waters are generally considered to be the most bioavailable form of metals. Therefore, "the accumulation of metal is influenced by the feeding behavior of microorganisms" (Fukunaga and Anderson 2011). After the "ingestion of heavy metals," a procedure of "excretion and/or detoxification of metals begins to avoid possible toxic effects." However, "the microorganisms will not suffer the toxic effects of the presence of metals if they are stored in detoxified forms" (Fukunaga and Anderson 2011). Furthermore, "the metal-biomass interaction depends on the type of metal that can bind to ligands containing O_2, N and S while this can be a simple description of the mechanisms involved, it can act as a starting point to propose new approaches to efficiency of absorption of metal by microorganisms".

Otherwise, "microorganisms can synthesize metal proteins, such as binding MT or PC, and proteins are strongly linked to the capacity of adsorption, accumulation and metal resistance".

The metal-binding proteins (on the outside of the cell membrane) aid in attachment of metal ions. These ions help in transportation within the cytosol, where metallochaperones (specialized protein chelators) transfer metal to the suitable recipient protein. It is a well-established fact that "the binding sites of the metal binding proteins have been improved for other proteins such as heterologous metalloproteins using genetic techniques". Some researchers have developed metalloproteins known as "heterologous metalloproteins with higher affinity and metal binding capacity and / or specificity and selectivity, which is expressed in bacteria to improve its metal absorption capacity". Furthermore, "the technique modifies the proteins on the cell surface in a heterogeneous using recombinant DNA, emerged as a new approach to improve adsorption capacity" (Saleem et al. 2008). Both bacteria and yeasts have been studied for this purpose, "glutathione (GSH) ligands, phytochelatin (PC) related GSH rich metallothionein cysteine and synthetic fitoquelaminas (ECN) possess attributes to improve heavy metal bioaccumulation" (Saleem et al. 2008). For example, "recombinant bacterial cloned mercury strain of the operon that encodes the regulatory gene (MERR) and other genes involved in transport was constructed. The strain showed high resistance to mercury detoxification of mercury ions within the cell" (Saleem et al. 2008).

It is a well-established fact that the expression of metal binding proteins or peptides in microorganisms to improve the accumulation and/or tolerance of heavy metals has great potential (Joshi et al. 2011). Various peptides and proteins have been explored with different resistance mechanisms (Mejare and Bülow 2001), for example, "the production of peptides from the family of metal-binding proteins, such as MT or phytochelatin (PC); the regulation of the intracellular concentration of metals, with the expression of protein transporters of metal-ligand complexes from the cytoplasm towards the inside of the vacuoles; and the flow of metal ions through the ion channels present in the cell wall" (Perpetuo et al. 2011). Consequently, genes with remarked tolerance to metal toxicity are often encoded in transposons or plasmids, which facilitates their dispersion from one cell to another (Perpetuo et al. 2011). Eventually, "tolerance is attributed to the activity that produces bacterial resistance to metals, either by pumping the active flow of toxic metals out of the cell, or by enzymatic detoxification (usually through the redox chemistry), where a lethal ion becomes a less toxic metal ion or less available" (Perpetuo et al. 2011).

It has been also explored, several metal-binding peptides have been studied for the purpose of increasing Cd resistance or accumulation by *E. coli* cells" (Mejare and Bülow 2001). Natural Cd-binding proteins and peptides (such as MT and PC) are very rich in cysteine residues. Furthermore, it is known that histidines have a high affinity for transition metal ions such as Zn^{2+}, Co^{2+}, Ni^{2+} and Cu^{2+} (Mejare and Bülow 2001). Therefore, to bind the Cd; several peptides comprising different cysteine or histidine sequences can be employed and, therefore, Cd tolerance and accumulation could be improved in *E. coli* cells and would be interesting to evaluate peptides and Cd-binding proteins designed to form bacteria more resistant to the environment, such as *Pseudomonas*, for possible use in bioremediation (Mejare and Bülow 2001).

It has been also explored, "hexavalent chromium is mobile, highly toxic and is considered a priority environmental pollutant, whereas, chromatin reductase present in chromium-resistant bacteria has the potential to be used in the bioremediation process because it is known to catalyze the reduction of Cr (VI) to Cr (III)" (Thatoi et al. 2014). Furthermore, "the enzymatic reduction of Cr (VI) in Cr (III) involves the transfer of electrons from electron donors, such as NADPH to Cr (VI) with the simultaneous generation of reactive oxygen species (ROS)" (Thatoi et al. 2014). The microbial consortiums that possess the ability to reduce Cr (VI) are known as chromium-reducing bacteria (CRB) (Thatoi et al. 2014) with the attribute that "Gram-positive CRB shows a significant tolerance to Cr (VI) toxicity even at high concentrations, while Gram-negative bacteria are much more sensitive to Cr (VI)" (Thatoi et al. 2014). Some genes responsible for Cr (VI) resistance have been determined in bacteria, for example, "the chrR gene located on the *P. aeruginosa* chromosome confers chromate resistance. *Ochrobactrum tritici* contains several genes associated with chromic resistance, ie chrB, chrA, chrC, chrF and ruvB" (Thatoi et al. 2014). The presence of enzymes that play a role in reducing Cr (VI) has been reported for several microorganisms, and enzymes (such as quinone reductase, nitroreductases, and NADPH-dependent enzymes) vary in their ability to transform the chromate and involve different pathways (Thatoi et al. 2014). Several bacteria also possess unique enzymes that reduces the chromium Cr (VI) through membrane-bound reductases, such as flavin reductase, cytochromes and hydrogenases (Thatoi et al. 2014); "these enzymes can be part of the electron transport system and use chromate as a terminal electron acceptor" (Thatoi et al. 2014).

An extensive array of microbes has so far been considered for the development of a proficient technology for the removal of heavy metal ions from contaminated effluents (Shafi et al. 2018), and the capability of distinctive microorganisms (algae, bacteria, fungi, and yeasts) to remove heavy metals from certain environments (Shafi et al. 2018) is depicted in Table 13.2.

13.4 Remediation of Organic Contaminants

Mushrooms are among the primary saprotrophs in an ecosystem and are capable in the degradation of matter. Trudge putrefaction fungi, mainly white caries, exude "extracellular enzymes" and acids that putrefy "lignin and cellulose." These are "long-chain organic compounds" (carbon-based) and similar to various "organic pollutants." Fungi are very effective in case of "polycyclic aromatic hydrocarbons" (IPA), complex "organic compounds" with fused and highly "stable polycyclic aromatic ring" (Batista-García et al. 2017), and also in "marine environments"

Table 13.2 Sorption potential of certain microorganisms to remove heavy metals

Microorganism	Type	Metal	Reference
Algae	*Ascophyllum nodosum*	Pb, NiPb, Cu, Cd, Zn	Holan and Volesky (1994); Romera et al. (2007)
	Chlorella pyrenoidosa	U	Singhal et al. (2004)
	Cladophora fascicularis	Pb	Deng et al. (2007)
	Fucus vesiculosus	CrPb, Cd	Holan and Volesky (1994); Singhal et al. (2004); Murphy et al. (2008)
	Hydrodictyon, *Oedogonium*, and *Rhizoclonium*	V, As	Saunders et al. (2012)
	Spirogyra and *Cladophora*	Pb, Cu	Lee and Chang (2011)
	Spirogyra and *Spirulina*	Cr, Cu, Fe, Mn, Zn	Mane and Bhosle (2012)
Bacteria	*Bacillus cereus*	Cr	Kanmani et al. (2012)
	Burkholderia	Cd, Pb	Jiang et al. (2008)
	Kocuria flava	Cu	Achal et al. (2011)
	Pseudomonas veronii	Cd, Zn, Cu	Murphy et al. (2008)
	Sporosarcina ginsengisoli	As	Singhal et al. (2004)
	Stenotrophomonas	Au	Song et al. (2008)
Fungi	*Agaricus bisporus*	Cd, Zn	Nagy et al. (2014)
	Aspergillus fumigatus	Pb	Ramasamy et al. (2011)
	Aspergillus versicolor	Cr, Ni, Cu	Tastan et al. (2010)
	Aspergillus, *Mucor*, *Penicillium*, and *Rhizopus*	Cd, Cu, Fe	Fulekar et al. (2012)
	Aspergillus niger, *Aspergillus foetidus*, and *Penicillium simplicissimum*	Ni, Co, Mo, V, Mn, Fe, W, Zn	Anahid et al. (2011)
	Ganoderma lucidum and *Penicillium*	Ar	Loukidou et al. (2003)
	Penicillium canescens	Cr	Say et al. (2003)
Yeasts	*Candida tropicalis*	Cd, Cr, Cu, Ni, Zn	Mattuschka et al. (1993)
	Candida utilis	Cd	Kujan et al. (2006)
	Pichia guilliermondii	Cu	Mattuschka et al. (1993)
	Saccharomyces cerevisiae	Cr, Ni, Cu, Zn	Machado et al. (2010)
	Streptomyces longwoodensis	Pb	Friss and Myers-Keith (1986)

(Passarini et al. 2011). The enzymes taking part in ligninolytic degradation include "lignin peroxidase," "versatile peroxidase," "peroxidase," "manganese lipase," "laccase," and sometimes "intracellular enzymes," especially "cytochrome P450" (Deshmukh et al. 2016; Pozdnyakova 2012).

Other toxins that fungi can degrade into harmless compounds include petroleum-derived fuels (Young et al. 2015), "phenols in wastewater" (Batista-García et al. 2017), and "polychlorinated biphenyls" (PCB) in soils "contaminated with *Pleurotus ostreatus*" (Stella et al. 2017).

13.4.1 Degradation of Pesticides

"Pesticide contamination can be long term and have a significant impact on the decomposition processes and, therefore, in the nutrient cycle" (Magan et al. 2010; Mushtaq et al. 2018) and its "degradation can be expensive and difficult (Bhat et al. 2018a). The most widely used to facilitate the "degradation of pesticides" are white caries fungi and ligninolytic extracellular enzymes (laccase and manganese peroxidase). Examples include the "insecticide endosulfan" (Rivero et al. 2012), "imazalil," "thiophanate-methyl," "ortho-phenylphenol," "diphenylamine," "chlorpyrifos" (Karas et al. 2011), and "atrazine" wastewater (Chan-Cupul et al. 2016).

13.4.2 Degradation of Toxic Chemical Dyes

Dyes are used in several "industrial sectors," such as "paper printing" and "fabrics." Most of them are recalcitrant to breakdown and, in some cases, like some "carcinogenic or toxic azo dyes" (Bhattacharya et al. 2011). The mechanism by which "fungi degrade" these "dyes" is their "lignolytic," particularly "laccase enzymes," so that white fungi are more commonly used.

Mycoremediation has demonstrated economical and effective remedy dye technology such as malachite green, nigrosin, and basic fuchsine with "*Aspergillus niger* and *Phanerochaete chrysosporium*" (Rani et al. 2014); "Congo red," a recalcitrant carcinogenic dye for biodegradation processes (Bhattacharya et al. 2011); and direct blue 14 (using *Pleurotus*) (Singh et al. 2013).

13.5 Advantages of Mycoremediation

Fungi, with the help of their "non-specific enzymes," can degrade various kinds of "harmful pollutants." They are used as follows:

- "Pharmaceuticals and fragrances" that are normally "recalcitrant" to the "degradation" of bacteria, such as "paracetamol," whose "decomposition products are toxic."
- Traditional "water treatment," using "*Mucor hiemalis*" and "phenols and pigments" from the "wastewater" of the "wine distillery."

Besides, "mycoremediation" is a cost-effective technique for "removal of hazardous substances" form any "pollutant affected" environs. Because of this potential, it can be a viable tool for "small-scale industries" and will act as "microfiltration of domestic wastewater."

13.5.1 Synergy with Phytoremediation

Phytoremediation is a "plant-based technology" to decontaminate polluted environs (Bhat et al. 2018b). Most plants can form a "symbiosis association with fungi," which is beneficial to the organisms involved in this kind of interaction, and this relationship is called "mycorrhiza." "Mycorrhizal fungi," in particular "arbuscular mycorrhizal fungi" (AMF), can appreciably improve the "power of remediation" in some plants (Khanday et al. 2016; Bhat et al. 2017a). This is mainly because the "stress experienced by plants" due to contaminants is significantly reduced in the presence of AMF, so that they can grow and produce more biomass (Rabie 2005). Fungi "interactions also provide more nutrition (particularly phosphorus) and promotes the general health of the rapidly expanding mycelium plant can also greatly extend the area of influence of the rhizosphere (hifósfera), which provides access to more nutrients and pollutants" (Sofi et al. 2017) and also help in exploring bacteria community for degradation of harmful pollutants (Rabie 2005; Bhatti et al. 2017). This kind of relationship has been proved to be useful for remediation of lead pollutants. For instance, *Rhizophagus intraradices* and *Robinia pseudoacacia are involved in decontamination of lead from the soils contaminated with lead toxicity.* (Yang et al. 2016). Furthermore *Rhizophagus intraradices* with Glomus have versiform incoulado in vetiver for lead removal (Bahraminia et al. 2016), AMF and *Calendula officinalis* in soil "contaminated with cadmium and lead" (Tabrizi et al. 2015; Singh et al. 2018), and it was generally effective to increase the capacity of bioremediation plants for metals (Yang et al. 2015; Li et al. 2013), fuel oil (Xun et al. 2014; Hernández-Ortega et al. 2012), and IPA (Rabie 2005). In wetlands, "AMF greatly promotes biodegradation of organic pollutants such as benzene, methyl tert-butyl ether and underground ammonia when inoculated into *Phragmites australis*" (Fester 2013).

13.6 Mechanisms of Bioremediation

Bioremediation can be divided into two categories: "biosorption and bioaccumulation" as depicted in Fig. 13.1. "Biosorption is a rapid and reversible passive adsorption mechanism" (Gadd and White 1993; Ahalya et al. 2003). Metals are retained by the "physical-chemical interaction (ion exchange, adsorption, complexation, precipitation and crystallization) between the metal and the functional groups present on the cell surface" (Gadd and White 1993; Ahalya et al. 2003; Volesky 2004; Gadd 2009; Fosso-Kankeu and Mulaba-Bafubiandi 2014). Numerous factors can manipulate "the biosorption of metals, such as pH, ionic strength, biomass concentration, temperature, particle size and the presence of other ions in the solution" (Volesky 2004). "Live and dead biomass can be produced for biosorption because it is independent of cellular metabolism" (Gadd and White 1993; Ahalya et al. 2003). On the other hand, "bioaccumulation includes intra and extracellular processes in

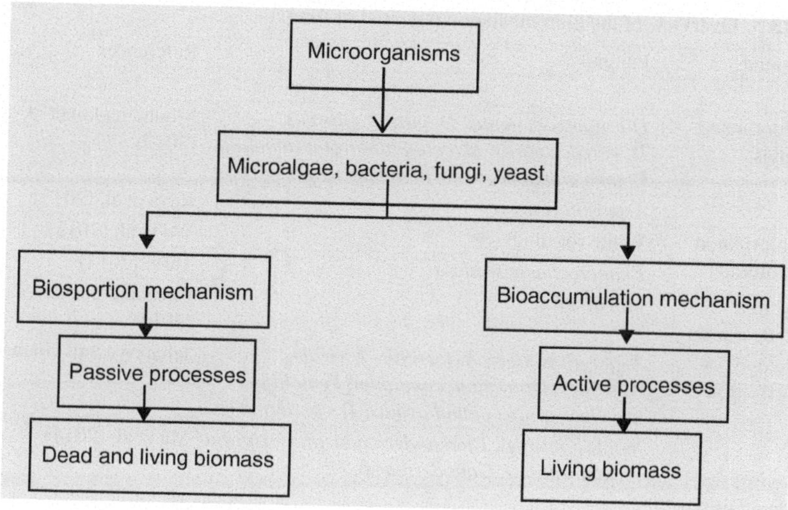

Fig. 13.1 "Microorganisms employed in the bioremediation and processes/mechanisms involved in the case of dead and living biomass" (Gadd and White 1993)

which passive absorption plays a limited and not very well defined role" (Gadd and White 1993). Therefore, an ample mass of living organism are produced by way of bioaccumulation.

13.7 Bioremediation Potential of Fungi

Fungi have been shown to play an important role in the "bioremediation of a variety of contaminants such as POPs, textile dyes, petroleum hydrocarbons, effluent from pulp and paper for leather tanning, PAHs, pesticides, PPCP" (Table 13.3). "Filamentous fungi such as *Aspergillus*, *Curvularia*, *Acrimonium* and *Pythium* have been studied for their ability to tolerate metal" (Akhtar et al. 2013). It has been reported as "members of the *Basidiomycota*, such as *T. versicolor* and black rot *Pleurotus ostreatus*, to degrade PAH in solid-state fermentation (SSF) models during growth in agri-food waste, such as orange peel" (Rosales et al. 2013). "Bioremediation / bleaching of colored effluents from the sugar sector, textile dye, bleached kraft paste plant, effluent from tanning has been reported in the case of fungi belonging to various groups, including *Aspergillus*, *Penicillium* and alkaline rot fungi, which indicates a different preference of the substrate of these mushrooms" (Jebapriya and Gnanadoss 2013; Huang et al. 2014; Reya et al. 2013); "decaffeinated coffee pulp may contain mushrooms in controlled conditions with additional nutrients for applications in the preparation of feed or for the production of bioethanol, as discussed in the case of fungi such as *Aspergillus restrictus*,

Table 13.3 Overview of the bioremediation potential of fungi

Compound	Fungi	References
POPs		
Polychlorinated biphenyls	*Doratomyces nanus*, *D. purpureofuscus D. verrucisporus*, *Myceliophthora thermophila*, *Phoma eupyrena*, and *Thermoascus crustaceus*	Mouhamadou et al. (2013)
	Aspergillus niger	Reya et al. (2013)
Polychlorinated dibenzofurans	White rot fungi	Wu et al. (2013)
	Phanerochaete sordida	Turlo (2014)
Phenylurea herbicide diuron	*Mortierella*	Ellegaard-Jensen et al. (2013)
Textile dye decolorization	*Aspergillus niger, A. foetidus, T. viride, A. sojae, Geotrichum candidum, Penicillium* sp., *Pycnoporus cinnabarinus, Trichoderma* sp.	Jebapriya and Gnanadoss (2013)
	White-rot fungi, *Bjerkandera adusta, Ceriporia metamorphosa, Ganoderma* sp.	Ma et al. (2014)
Petroleum products		
Crude oil	*A. niger, Rhizopus* sp., *Candida* sp., *Penicillium* sp., *Mucor* sp.	Damisa et al. (2013)
Gasoline	*Exophiala xenobiotica*	Isola et al. (2013)
Bleached Kraft pulp mill effluent	*Rhizopus oryzae* and *Pleurotus sajor-caju*	Duarte et al. (2013)
Effluent from leather tanning	*Aspergillus flavus, Aspergillus* sp., *A. niger,* and *Aspergillus jegita*	Bennett et al. (2013); Reya et al. (2013)
PAH		
Diphenyl ether	White-rot fungi, *Pleurotus ostreatus, Trametes versicolor*	Rosales et al. (2013); Wu et al. (2013)
Anthracene	*Armillaria* sp.	Hadibarata et al. (2013)
Naphthalene	White-rot fungi, *Pleurotus eryngii*	Hadibarata et al. (2013)
PPCP		
Caffeine	*Chrysosporium keratinophilum, Gliocladium roseum, Fusarium solani, A. restrictus, Penicillium,* and *Stemphylium*	Nayak et al. (2013)
Citalopram, fluoxetine, sulfamethoxazole	*Bjerkandera* sp. R1, *Bjerkandera adusta,* and *Phanerochaete chrysosporium*	Rodarte-Morales et al. (2011)
Fungicide		
Metalaxyl and Folpet	*Gongronella* sp. and *R. stolonifer*	Martins et al. (2013)
Pesticide		
Chlorinated hydrocarbons; heptachlor	*P. ostreatus*	Purnomo et al. (2013)
Chlorpyrifos	*Aspergillus terreus*	Silambarasan and Abraham (2013)
Heavy metals	*Aspergillus, Curvularia, Acrimonium, Pythyme,* and *Aspergillus flavus*	Akhtar et al. (2013) Kurniati et al. (2014)

Chrysosporium keratinophilum, *Fusarium solani*, *Gliocladium roseum*, *Penicillium* and *Stemphylium*" (Nayak et al. 2013). "Bioremediation in the presence of fungi *A. niger* and *P. chrysosporium* exhibited substantial removal of oil hydrocarbons from contaminated soil in short time gasoline and diesel short incubation, as indicated by improved removal of total organic carbon" (Maruthi et al. 2013). The "elimination of chlorpyrifos and its metabolite 3, 5, 6-trichloro-2-pyridinol (TCP) for the fungal soil strain *A. niger* JAS1 contaminated even in the absence additional nutrient with the complete elimination of both metabolites" (Silambarasan and Abraham 2013). "TCP degradation from the degrading chlorpyrifos strain was a significant finding considering the antimicrobial nature and the catabolite repression property exhibited by TCP."

13.8 Fungal Enzymes in Bioremediation

Fungi have healthy morphology and metabolic capacity that can provide sustainable alternative for the remediation of contaminated sites. Fungi are also known to grow in harsh condition, thus can play significant role in the degradation of recalcitrant pollutants. Different enzymes from fungi like amylases, cellulases, xylanases, lipases, proteases, peroxidases, laccases, and catalases have a great scientific, industrial importance and also have significant capacity of depollution and bioremediation. Fungus is the only organism that breaks down wood by developing white threadlike structure called mycelium that release enzymes necessary for the decomposition of lignin and cellulose. Organic wastes from food industries as well as residential areas are rich in polymeric substances such as cellulose, lipid, fats, and protein and are hydrolyzed by different enzymes. Organic waste (vegetables, food, kitchen waste, algae) can be used for the production of industrially valuable products like biogas, biofuel, and fatty acids (Khardenavis et al. 2013). Enzymes from fungus can also be utilized for efficient treatment of municipal waste (Marco et al. 2013). Kitchen waste after pretreatment can serve efficient feedstock for solid-state fermentation of cellulase enzymes produced by *Aspergillus niger* (Bansal et al. 2012). Besides, locally isolated strain of *Aspergillus fumigatus* produces cellulase enzyme that can degrade macroalgae and its valorization for bioethanol production (Ben Yahmed et al. 2016, 2018). Pretreatment of waste is beneficial in many ways such as biological pretreatment of green algae *Ulva* sp. Using SSF of *Aspergillus fumigatus* was also exploited for the bioremediation of green tides and the biogas production (Ben Yahmed et al. 2017). Moreover, their effect on the enhancement of saccharification efficiency of biomasses was studied by applying consortium of fungi (*Armillaria gemina* and *Pholiota adiposa)* for the hydrolysis of rice straw and willow biomasses (Dhiman et al. 2015). Various important enzymes are produced by microorganisms for the degradation of the waste, but among them ligninolytic enzymes released by white-rot fungi degrade not only lignocellulosic substrate but also compounds including dyes (Novotný et al. 2004).

Extracellular enzymes only include peroxidases and laccases (Janusz et al. 2013). Persistent organic pollutants cannot be removed easily, but extracellular enzymes (laccases and some fungal class II peroxidases) produced by white-rot basidiomycetes can degrade these toxic pollutants. Toxic pollutants are released into the environment at rapid rate without any efficient treatment, but enzymes produced from the fungi have shown great potential (efficiency and selectivity) in degrading toxic pollutant in economical and eco-friendly manner (Viswanath et al. 2014). White-rot fungi are responsible for transformation of persistent pollutants like pesticide by releasing ligninolytic enzyme and promote microbial activity by providing nutrients. Table 13.4 presented various applications of fungal enzymes in bioremediation field. Currently, a lot of researches are interested in developing modified enzymes through recombinant expression of genes from white-rot fungi and protein engineering techniques for environment-friendly degradation of different toxic pollutants (Deshmukh et al. 2016; Janusz et al. 2013).

Table 13.4 Fungal enzymes and its various applications in bioremediation

Enzymes	"Bioremediation applications"	"Fungi-producing enzymes"	References
Laccases	*Decolorization of dyes* Decolorization of several synthetic dyes such as Azure B and Brilliant Blue R in low nitrogen medium Partial decolorization of two azo dyes and complete decolorization of two triphenylmethane dyes (bromophenol blue and malachite green) Degradation of the dye Navy blue HER Degradation of triarylmethane, indigoid, azo, and athraquinonic dyes used in dyeing textiles Degradation of 92% in the Azo Black Reactive 5 dye Decolorization of effluent containing textile indigo dye as well as 23 industrial dyes *Degradation of xenobiotics* Oxidation of alkenes Immobilization of soil pollutants by coupling to soil humic substances *Pulp and paper industry*	*Flavodon flavus* *Pycnoporus sanguineus* *Trichosporon beigelii NCIM-3326* *Trametes hirsuta* *P. chrysosporium* *P.chrysosporium* and *Curvularia lunata* *Trametes hirsuta* ND	Soares et al. (2001) Pointing and Vrijmoed (2000) Saratale et al. (2009) Kaushik (2015) Enayatizamir et al. (2011) Rita de Cássia et al. (2013)

(continued)

Table 13.4 (continued)

Enzymes	"Bioremediation applications"	"Fungi-producing enzymes"	References
	Decolorization of flexographic inks in presence of synthetic and artificial mediators in recycled paper industry Decolorization of the alkaline effluents of the pulp and paper industry Dechlorination and removal of chorophenols and chlorolignins from bleach effluents Reduction of the kappa number of pulp and improvement of the papermaking properties of pulp *Effluent treatment* Degradation of bisphenol A Detoxification of agricultural byproducts including olive mill wastes or coffee pulp Decolorization of the effluent from a Kraft paper mill bleach plant Degradation of xenoestrogen nonylphenol	Ascomycete (*Myceliophthora thermophila*), Basdiomycetes (*Trametes villosa, Coriolopsis rigida, Pycnoporus coccineus*) *Coriolopsis gallica* ND[a] ND *Fusarium incarnatum* *Lentinula edodes* *Trametes versicolor B7* *Clavariopsis aquatica*	Niku-Paavola (2000) Ahn et al. (2002) Fillat et al. (2012) Calvo et al. (1998) Milstein et al. (1988) Bajpai (1999); Wong et al. (2000) Chhaya and Gupte (2013) D'Annibale et al. (2000) Bajpai et al. (1993) Junghanns et al. (2005)
Peroxidases	Degradation of toxic compounds Decolorization of azo and phthalocyanine dye Oxidation of recalcitrant dyes (e.g., Azure B), phenolic substrate 2,6-dimethoxyphenol and non-phenolic aromatic compounds (Reactive Black B) Delignification	White-rot and Basidiomycetes fungi *Bjerkandera adusta* *Mycetinis scorodonius, Auricularia auricula-judae, Exidia glandulosa, Mycena epipterygia* Basidiomycetes fungi (*Pleurotus ostreatus* sensu *Cooke, Coriolus versicolor (L.) Quel., Tyromyces albidus (Schaeff.) Donk,* and *Trametes gallica*	Deshmukh et al. (2016) Baratto et al. (2015) Liers et al. (2010) Hong et al. (2011)
Catalases	Bioremediation of metal contaminated sites Bioremediation of oil contaminated soil	*Aspergillus foetidus* ND	Chakraborty et al. (2013) Lin et al. (2009)

[a]**Not determined**

13.9 Conclusions

A huge quantity of effluents are produced with the urbanization, industrial development, and population explosion mainly in developing countries. Effluent produced is rich in different types of pollutant which are persistent and in small concentration can pose serious risk to the environment. Conventional techniques for the purification of contaminated water and waste management are not efficient in removing different toxic pollutants; thus focus is shifting toward eco-friendly techniques like mycoremediation. Microorganism needs to be exploited widely for the remediation of toxic pollutants from the contaminated water and degradation of the municipal, agriculture, and industrial waste. Microorganism has great potential to remove pollutant from effluent by bioaccumulation, biosorption, or degradation by the release of different enzymes. Microorganism like fungi is efficient in degrading the organic waste by releasing various enzymes like cellulases, oxidases, phosphatases, chitinases, and proteases that convert the waste into valuable product. Thus, enzymes released from fungi should be widely utilized for degradation of pollutants in a cost-effective and environment-friendly manner.

References

Achal V, Pan X, Zhang D (2011) Remediation of copper-contaminated soil by *Kocuria flava* CR1, based on microbially induced calcite precipitation. Ecol Eng 37(10):1601–1605

Ahalya N, Ramachandra TV, Kanamadi RD (2003) Biosorption of heavy metals. Res J Chem Environ 7(4):4544–4552

Ahn M-Y, Dec J, Kim J-E, Bollag J-M (2002) Treatment of 2, 4-dichlorophenol polluted soil with free and immobilized laccase. J Environ Qual 31:1509–1515

Akhtar S, Mahmood-ul-Hassan M, Ahmad R, Suthor V, Yasin M (2013) Metal tolerance potential of filamentous fungi isolated from soils irrigated with untreated municipal effluent. Soil Environ 32:55–62

Anahid S, Yaghmaei S, Ghobadinejad Z (2011) Heavy metal tolerance of fungi. Scientia Iranica 18(3):502–508

Bahraminia M, Zarei M, Ronaghi A, Ghasemi-Fasaei R (2016) Effectiveness of arbuscular mycorrhizal fungi in phytoremediation of lead- contaminated soil by vetiver grass. Int J Phytoremediation 18(7):730–737

Bajpai P (1999) Application of enzymes in the pulp and paper industry. Biotechnol Prog 15:147–157. https://doi.org/10.1021/bp990013k

Bajpai P, Mehna A, Bajpai PK (1993) Decolorization of Kraft bleach plant effluent with the white rot fungus Trametes versicolor. Process Biochem 28:377–384. https://doi.org/10.1016/0032-9592(93)80024-B

Bansal N, Tewari R, Soni R, Soni SK (2012) Production of cellulases from *Aspergillus niger* NS-2 in solid state fermentation on agricultural and kitchen waste residues. Waste Manag 32:1341–1346. https://doi.org/10.1016/j.wasman.2012.03.006

Baratto MC, Juarez-Moreno K, Pogni R, Basosi R, Vazquez-Duhalt R (2015) EPR and LC-MS studies on the mechanism of industrial dye decolorization by versatile peroxidase from *Bjerkandera adusta*. Environ Sci Pollut Res 22:8683–8692

Batista-García RA, Kumar VV, Ariste A, Tovar-Herrera OE, Savary O, Peidro-Guzmán H, González-Abradelo D, Jackson SA, Dobson ADW, Sánchez-Carbente MDR, Folch-Mallol JL,

Leduc R, Cabana H (2017) Simple screening protocol for identification of potential mycoremediation tools for the elimination of polycyclic aromatic hydrocarbons and phenols from hyperalkalophile industrial effluents. J Environ Manag 198(Pt 2):1–11

Ben Yahmed N, Jmel MA, Ben Alaya M, Bouallagui H, Marzouki MN, Smaali I (2016) A biorefinery concept using the green macroalgae Chaetomorpha linum for the coproduction of bioethanol and biogas. Energy Convers Manag 119:257–265. https://doi.org/10.1016/j.enconman.2016.04.046

Ben Yahmed N, Carrere H, Marzouki MN, Smaali I (2017) Enhancement of biogas production from Ulva sp. by using solid-state fermentation as biological pretreatment. Algal Res 27:206–214. https://doi.org/10.1016/j.algal.2017.09.005

Ben Yahmed N, Berrejeb N, Jmel MA, Jazzar S, Marzouki MN, Smaali I (2018) Efficient biocatalytic conversion of stranded green macroalgal biomass using a specific cellulases-based cocktail. Waste Biomass Valoriz. doi:https://doi.org/10.1007/s12649-018-0397-4

Bennett RM, Cordero PRF, Bautista GS, Dedeles GR (2013) Reduction of hexavalent chromium using fungi and bacteria isolated from contaminated soil and water samples. Chem Ecol 29:320–328

Bhat RA, Dervash MA, Mehmood MA, Bhat MS, Rashid A, Bhat JIA, Singh DV, Lone R (2017a) Mycorrhizae: a sustainable industry for plant and soil environment. In: Varma A et al (eds) Mycorrhiza-nutrient uptake, biocontrol, ecorestoration. Springer International, Berlin, pp 473–502

Bhat RA, Shafiq-ur-Rehman MMA, Dervash MA, Mushtaq N, Bhat JIA, Dar GH (2017b) Current status of nutrient load in dal Lake of Kashmir Himalaya. J Pharm Phytochem 6(6):165–169

Bhat RA, Beigh BA, Mir SA, Dar SA, Dervash MA, Rashid A, Lone R (2018a) Biopesticide techniques to remediate pesticides in polluted ecosystems. In: Wani KA, Mamta (eds) Handbook of research on the adverse effects of pesticide pollution in aquatic ecosystems. IGI Global, pp 387–407

Bhat RA, Dervash MA, Qadri H, Mushtaq N, Dar GH (2018b) Macrophytes, the natural cleaners of toxic heavy metal (THM) pollution from aquatic ecosystems. In: Environmental contamination and remediation. Cambridge Scholars, Cambridge, pp 189–209

Bhatia D, Sharma NR, Singh J, Kanwar RS (2017) Biological methods for textile dye removal from wastewater: a review. Crit Rev Environ Sci Technol 47(19)

Bhattacharya S, Das A, Mangai G, Vignesh K, Sangeetha J (2011) Mycoremediation of congo red dye by filamentous fungi. Braz J Microbiol 42(4):1526–1536

Bhatti AA, Haq S, Bhat RA (2017) Actinomycetes benefaction role in soil and plant health. Microb Pathog 111:458–467

Calvo AM, Copa-Patino JL, Alonso O, Gonzalez AE (1998) Studies of the production and characterization of laccase activity in the basidiomycete Coriolopsis gallica, an efficient decolorizer of alkaline effluents. Arch Microbiol 171:31–36

Cecchi G, Roccotiello E, Piazza D, Simone, Riggi A, Mariotti MG, Zotti M (2017) Assessment of Ni accumulation capability by fungi for a possible approach to remove metals from soils and waters. J Environ Sci Health 52(3):166–170

Chakraborty S, Mukherjee A, Das TK (2013) Biochemical characterization of a lead-tolerant strain of *Aspergillus foetidus*: an implication of bioremediation of lead from liquid media. Int Biodeterior Biodegrad 84:134–142. https://doi.org/10.1016/j.ibiod.2012.05.031

Chan-Cupul W, Heredia-Abarca G, Rodríguez-Vázquez R (2016) Atrazine degradation by fungal co-culture enzyme extracts under different soil conditions. J Environ Sci Health 51(5):298–308

Chhaya U, Gupte A (2013) Possible role of laccase from Fusarium incarnatum UC-14 in bioremediation of Bisphenol A using reverse micelles system. J Hazard Mater 254:149–156

D'Annibale A, Stazi SR, Vinciguerra V, Giovannozzi Sermanni G (2000) Oxirane-immobilized Lentinula edodes laccase: stability and phenolics removal efficiency in olive mill wastewater. J Biotechnol 77:265–273. https://doi.org/10.1016/S0168-1656(99)00224-2

Damisa D, Oyegoke TS, Ijah UJJ, Adabara NU, Bala JD, Abdulsalam R (2013) Biodegradation of petroleum by fungi isolated from unpolluted tropical soil. Int J Appl Biol Pharm Technol 4:136–140

Deng L, Su Y, Su H, Wang X, Zhu X (2007) Sorption and desorption of lead (II) from wastewater by green algae Cladophora fascicularis. J Hazard Mater 143(1–2):220–225

Deshmukh R, Khardenavis AA, Purohit HJ (2016) Diverse metabolic capacities of fungi for bioremediation. Indian J Microbiol 56:247–264. https://doi.org/10.1007/s12088-016-0584-6

Dhiman SS, Haw JR, Kalyani D, Kalia VC, Kang YC, Lee JK (2015) Simultaneous pretreatment and saccharification: green technology for enhanced sugar yields from biomass using a fungal consortium. Bioresour Technol 179:50–57. https://doi.org/10.1016/j.biortech.2014.11.059

Duarte K, Justino CI, Pereira R, Panteleitchouk TS, Freitas AC, Rocha-Santos TA, Duarte AC (2013) Removal of the organic content from a bleached Kraft pulp mill effluent by a treatment with silica–alginate–fungi biocomposites. J Environ Sci Heal A Tox Hazard Subst Environ Eng 48:166–172

Ellegaard-Jensen L, Aamand J, Kragelund BB, Johnsen AH, Rosendahl S (2013) Strains of the soil fungus *Mortierella* show different degradation potentials for the phenylurea herbicide diuron. Biodegradation 24:765–774

Enayatizamir N, Tabandeh F, Rodriguez-Couto S, Yakhchali B, Alikhani HA, Mohammadi L (2011) Biodegradation pathway and detoxification of the diazo dye reactive black 5 by Phanerochaete chrysosporium. Bioresour Technol 102:10359–10362. https://doi.org/10.1016/j.biortech.2011.08.130

Falandysz J (2016) Mercury bio-extraction by fungus *Coprinus comatus*: a possible bioindicator and mycoremediator of polluted soils? Environ Sci Pollut Res Int 23(8):7444–7451

Fester T (2013) Arbuscular mycorrhizal fungi in a wetland constructed for benzene-, methyl tert-butyl ether- and ammonia-contaminated groundwater bioremediation. Microb Biotechnol 6(1):80–84

Fillat U, Prieto A, Camarero S, Martínez ÁT, Martínez MJ (2012) Biodeinking of flexographic inks by fungal laccases using synthetic and natural mediators. Biochem Eng J 67:97–103. https://doi.org/10.1016/j.bej.2012.05.010

Fomina M, Gadd GM (2014) Biosorption: current perspectives on concept, definition and application. Bioresour Technol 160:3–14

Fosso-Kankeu E, Mulaba-Bafubiandi AF (2014) Implication of plants and microbial metalloproteins in the bioremediation of polluted waters: a review. Phys Chem Earth 67–69:242–252

Friss N, Myers-Keith P (1986) Biosorption of uranium and lead by *Streptomyces longwoodensis*. Biotechnol Bioeng 28:21–28

Fukunaga A, Anderson MJ (2011) Bioaccumulation of copper, lead and zinc by the bivalves *Macomona liliana* and *Austrovenus stutchburyi*. J Exp Mar Biol Ecol 396(2):244–252

Fulekar MH, Sharma J, Tendulkar A (2012) Bioremediation of heavy metals using biostimulation in laboratory bioreactor. Environ Monit Assess 184(12):7299–7307

Gadd GM (2009) Biosorption: critical review of scientific rationale, environmental importance and significance for pollution treatment. J Chem Technol Biotechnol: 13–28

Gadd GM, White C (1993) Microbial treatment of metal pollution—a working biotechnology? Trends Biotechnol 11:353–359

Gazem MAH, Nazareth S (2013) Sorption of lead and copper from an aqueous phase system by marine-derived *Aspergillus* species. Ann Microbiol 63(2):503–511

Hadibarata T, Teh ZC, Zubir MM, Khudhair AB, Yusoff AR, Salim MR, Hidayat T (2013) Identification of naphthalene metabolism by white-rot fungus *Pleurotus eryngii*. Bioprocess Biosyst Eng 24:728–732

Hernández-Ortega HA, Alarcón A, Ferrera-Cerrato R, Zavaleta-Mancera HA, López-Delgado HA, Mendoza-López MR (2012) Arbuscular mycorrhizal fungi on growth, nutrient status, and total antioxidant activity of Melilotus albus during phytoremediation of a diesel-contaminated substrate. J Environ Manag 95(Suppl):S319–S324

Holan ZR, Volesky B (1994) Biosorption of lead and nickel by biomass of marine algae. Biotechnol Bioeng 43(11):1001–1009

Hong Y, Dashtban M, Chen S, Song R, Qin W (2011) Enzyme production and lignin degradation by four Basidiomycetous Fungi in submerged fermentation of peat containing medium. Int J Biol 4. https://doi.org/10.5539/ijb.v4n1p172

Huang J, Fu Y, Liu Y (2014) Comparison of alkali-tolerant fungus *Myrothecium* sp. IMER1 and white-rot fungi for decolorization of textile dyes and dye effluents. J Bioremed Biodegr 5:1–5

Ilyina A, Castillo Sanchez MI, Villarreal Sanchez JA, Ramirez EG, Candelas RJ (2003) Isolation of soil bacteria for bioremediation of hydrocarbon contamination. Bull Moscow Univ 44(1):88–91

Isola D, Selbmann L, de Hoog GS, Fenice M, Onofri S, Prenafeta-Boldú FX, Zucconi L (2013) Isolation and screening of black fungi as degraders of volatile aromatic hydrocarbons. Mycopathologia 175:369–379

Janusz G, Kucharzyk KH, Pawlik A, Staszczak M, Paszczynski AJ (2013) Fungal laccase, manganese peroxidase and lignin peroxidase: gene expression and regulation. Enzyme Microb Technol 52:1–12. https://doi.org/10.1016/j.enzmictec.2012.10.003

Jebapriya GR, Gnanadoss JJ (2013) Bioremediation of textile dye using white-rot fungi: a review. Int J Curr Res Rev 5:1–13

Jiang CY, Sheng XF, Qian M, Wang QY (2008) Isolation and characterization of heavy metal resistant Burkholderia species from heavy metal contaminated paddy field soil and its potential in promoting plant growth and heavy metal accumulation in metal polluted soil. Chemosphere 72:157–164

Joshi PK, Swarup A, Maheshwari S, Kumar R, Singh N (2011) Bioremediation of Heavy Metals in Liquid Media Through Fungi Isolated from Contaminated Sources. Indian J Microbiol 51(4):482–487

Junghanns C, Moeder M, Krauss G, Martin C, Schlosser D (2005) Degradation of the xenoestrogen nonylphenol by aquatic fungi and their laccases. Microbiology 151:45–57. https://doi.org/10.1099/mic.0.27431-0

Kadirvelu K, Senthilkumar P, Thamaraiselvi K, Subburam V (2002) Activated carbon prepared from biomass as adsorbent: elimination of Ni (II) from aqueous solution. Bioresour Technol 81:87–90

Kanmani P, Aravind J, Preston D (2012) Remediation of chromium contaminants using bacteria. Int J Environ Sci Technol 9:183–193

Karas PA, Perruchon C, Exarhou K, Ehaliotis C, Karpouzas DG (2011) Potential for bioremediation of agro-industrial effluents with high loads of pesticides by selected fungi. Biodegradation 22(1):215–228

Kaushik G (2015) Bioremediation of industrial effluents: distillery effluent. In: Applied environmental biotechnology: present scenario and future trends. Springer, Berlin, pp 19–32

Khanday M, Bhat RA, Haq S, Dervash MA, Bhatti AA, Nissa M, Mir MR (2016) Arbuscular mycorrhizal fungi boon for plant nutrition and soil health. In: Hakeem KR, Akhtar J, Sabir M (eds) Soil science: agricultural and environmental prospectives. Springer International, Berlin, pp 317–332

Khardenavis AA, Wang JY, Ng WJ, Purohit HJ (2013) Management of various organic fractions of municipal solid waste via recourse to VFA and biogas generation. Environ Technol 34:2085–2097. https://doi.org/10.1080/09593330.2013.817446

Kujan P, Prell A, Safár H, Sobotka M, Rezanka T, Holler P (2006) Use of the industrial yeast Candida utilis for cadmium sorption. Folia Microbiol 51(4):257–260

Kulshreshtha S, Mathur N, Bhatnagar P (2014) Mushroom as a product and their role in mycoremediation. AMB Express 4:29

Kurniati E, Arfarita N, Imai T, Higuchi T, Kanno A, Yamamoto K, Sekine M (2014) Potential bioremediation of mercury-contaminated substrate using filamentous fungi isolated from forest soil. J Environ Sci 26:1223–1231

Lee YC, Chang SP (2011) The biosorption of heavy metals from aqueous solution by Spirogyra and *Cladophora filamentous* macroalgae. Bioresour Technol 102(9):5297–5304

Li S-P, Bi Y-L, Kong W-P, Wang J, Yu H-Y (2013) Effects of the arbuscular mycorrhizal fungi on environmental phytoremediation in coal mine areas. Huan Jing Ke Xue Huanjing Kexue 34(11):4455–4459

Liers C, Bobeth C, Pecyna M, Ullrich R, Hofrichter M (2010) DyP-like peroxidases of the jelly fungus Auricularia auricula-judae oxidize nonphenolic lignin model compounds and high-redox potential dyes. Appl Microbiol Biotechnol 85:1869–1879. https://doi.org/10.1007/s00253-009-2173-7

Lin X, Li X, Sun T, Li P, Zhou Q, Sun L, Hu X (2009) Changes in microbial populations and enzyme activities during the bioremediation of oil-contaminated soil. Bull Environ Contam Toxicol 83:542–547
Loukidou MX, Matis KA, Zouboulis AI, Liakopoulou-Kyriakidou M (2003) Removal of As(V) from wastewaters by chemically modified fungal biomass. Water Res 37(18):4544–4552
Ma L, Zhuo R, Liu H, Yu D, Jiang M, Zhang X, Yang Y (2014) Efficient decolorization and detoxification of the sulfonated azo dye Reactive Orange 16 and simulated textile wastewater containing Reactive Orange 16 by the white-rot fungus *Ganoderma* sp. En3 isolated from the forest of Tzu-chin Mountain in China. Biochem Eng J 82:1–9
Machado MD, Soares EV, Soares HM (2010) Removal of heavy metals using a brewer's yeast strain of *Saccharomyces cerevisiae*: chemical speciation as a tool in the prediction and improving of treatment efficiency of real electroplating effluents. J Hazard Mater 180(1–3):347–353
Magan N, Fragoeiro S, Bastos C (2010) Environmental factors and bioremediation of xenobiotics using white rot fungi. Mycobiology 38(4):238–248
Mane PC, Bhosle AB (2012) Bioremoval of some metals by living algae spirogyra sp. and Spirulina sp. from aqueous solution. Int J Environ Res 6(2):571–576
Marco E, Font X, Sánchez A, Gea T, Gabarrell X, Caminal G (2013) Co-composting as a management strategy to reuse the white-rot fungus Trametes versicolor after its use in a biotechnological process. Int J Environ Waste Manag 11:100. https://doi.org/10.1504/ijewm.2013.050637
Martins MR, Pereira P, Lima N, Cruz-Morais J (2013) Degradation of Metalaxyl and Folpet by filamentous fungi isolated from Portuguese (Alentejo) vineyard soils. Arch Environ Contam Toxicol 65:67–77
Maruthi YA, Hossain K, Thakre S (2013) *Aspergillus flavus*: a potential bioremediator for oil contaminated soils. Eur J Sustain Dev 2:57–66
Mattuschka B, Junghaus K, Straube G (1993) Biosorption of metals by waste biomass. In: Torma AE, Apel ML, Brierley CL (eds) Biohydrometallurgical technologies, vol 2. The Minerals, Metals & Materials Society, Warrendale
Mehmood MA, Qadri H, Bhat RA, Rashid A, Ganie SA, Dar GH, Shafiq-ur-Rehman (2019) Heavy metal contamination in two commercial fish species of a trans-Himalayan freshwater ecosystem. Environ Monit Assess 191:104. https://doi.org/10.1007/s10661-019-7245-2
Mejare M, Bülow L (2001) Metal-binding proteins and peptides in bioremediation and phytoremediation of heavy metals. Trends Biotechnol 19(2):67–73
Milstein O, Haars A, Majcherczyk A, Trojanowski J, Tautz D, Zanker H, Hüttermann A (1988) Removal of chlorophenols and chlorolignins from bleaching effluent by combined chemical and biological treatment. Water Sci Technol 20:161–170. https://doi.org/10.2166/wst.1988.0019
Mouhamadou B, Faure M, Sage L, Marçais J, Souard F, Geremia RA (2013) Potential of autochthonous fungal strains isolated from contaminated soils for degradation of polychlorinated biphenyls. Fungal Biol 117:268–274
Mulligan CN, Yong R, Gibbs BF (2001) Remediation technologies for metal contaminated soils and groundwater: an evaluation. Eng Geol 60(1–4):193–207
Murphy V, Hughes H, McLoughlin P (2008) Comparative study of chromium biosorption by red, green and brown seaweed biomass. Chemosphere 70(6):1128–1134
Mushtaq N, Bhat RA, Dervash MA, Qadri H, Dar GH (2018) Biopesticides: the key component to remediate pesticide contamination in an ecosystem. In: Environmental contamination and remediation. Cambridge Scholars, Cambridge, pp 152–178
Nagy B, Mânzatu C, Maicaneanu A, Indolean C, Lucian BT, Majdik C (2014) Linear and nonlinear regression analysis for heavy metals removal using Agaricus bisporus macrofungus. Arab J Chem 10:S3569–S3579
Nayak V, Pai PV, Pai A, Pai S, Sushma YD, Rao CV (2013) A comparative study of caffeine degradation by four different fungi. Biorem J 17:79–85
Niku-Paavola M-L, Viikari L (2000) Enzymatic oxidation of alkenes. J Mol Catal B Enzym 10:435–444

Novotný C, Svobodová K, Erbanová P, Cajthaml T, Kasinath A, Lang E, Šašek V (2004) Ligninolytic fungi in bioremediation: extracellular enzyme production and degradation rate. Soil Biol Chem 36:1545–1551. https://doi.org/10.1016/j.soilbio.2004.07.019

Passarini MRZ, Rodrigues MVN, da Silva M, Sette LD (2011) Marine-derived filamentous fungi and their potential application for polycyclic aromatic hydrocarbon bioremediation. Mar Pollut Bull 62(2):364–370

Perpetuo EA, Souza CB, Nascimento CAO (2011) Engineering bacteria for bioremediation. In: Carpi A (ed) Progress in molecular and environmental bioengineering—from analysis and modeling to technology applications. In Tech, Rijeka, pp 605–632

Pierzynski GM, Sims JT, Vance GF (2000) Soil and environmental quality. CRC, Boca Raton

Pointing SB, Vrijmoed L (2000) Decolorization of azo and triphenylmethane dyes by Pycnoporus sanguineus producing laccase as the sole phenoloxidase. World J Microbiol Biotechnol 16:317–318

Pozdnyakova NN (2012) Involvement of the ligninolytic system of white-rot and litter-decomposing fungi in the degradation of polycyclic aromatic hydrocarbons. Biotechnol Res Int 2012:243217

Purnomo AS, Mori T, Putra SR, Kondo R (2013) Biotransformation of heptachlor and heptachlor epoxide by white-rot fungus *Pleurotus ostreatus*. Int Biodeterior Biodegrad 82:40–44

Rabie GH (2005) Role of arbuscular mycorrhizal fungi in phytoremediation of soil rhizosphere spiked with poly aromatic hydrocarbons. Mycobiology 33(1):41–50

Rajendran P, Muthukrishnan J, Gunasekaran P (2003) Microbes in heavy metal remediation. Indian J Exp Biol 41:935–944

Ramasamy RK, Congeevaram S, Thamaraiselvi K (2011) Evaluation of isolated fungal strain from e-waste recycling facility for effective sorption of toxic heavy metal Pb (II) ions and fungal protein molecular characterization-a Mycoremediation approach. Asian J Exp Biol Sci 2(2):342–347

Rani B, Kumar V, Singh J, Bisht S, Teotia P, Sharma S, Kela R (2014) Bioremediation of dyes by fungi isolated from contaminated dye effluent sites for bio-usability. Braz J Microbiol 45(3):1055–1063

Reya I, Lakshmi Prabha M, Renitta E (2013) Equilibrium and kinetic studies on biosorption of Cr (VI) using novel *Aspergillus jegita* isolated from tannery effluent. Res J Chem Environ 17:72–78

Riggle PJ, Kumamoto CA (2000) Role of a Candida albicans P1-type ATPase in resistance to copper and silver ion toxicity. J Bacteriol 182:4899–4905

Rita de Cássia M, de Barros Gomes E, Pereira N Jr, Marin-Morales MA, KMG M, de Gusmão NB (2013) Biotreatment of textile effluent in static bioreactor by Curvularia lunata URM 6179 and *Phanerochaete chrysosporium* URM 6181. Bioresour Technol 142:361–367

Rivero A, Niell S, Cesio V, Cerdeiras MP, Heinzen H (2012) Analytical methodology for the study of endosulfan bioremediation under controlled conditions with white rot fungi. J Chromatogr B 907:168–172

Rodarte-Morales AI, Feijoo G, Moreira MT, Lema JM (2011) Degradation of selected pharmaceutical and personal care products (PPCPs) by white-rot fungi. World J Microbiol Biotechnol 27:1839–1846

Romera E, González F, Ballester A, Blázques MI, Munoz JA (2007) Comparative study of biosorption of heavy metals using different types of algae. Bioresour Technol 98(17):3344–3353

Rosales E, Pazos M, Ángeles SM (2013) Feasibility of solid-state fermentation using spent fungi-substrate in the biodegradation of PAHs. CLEAN Soil Air Water 41:610–615

Saleem M, Brim H, Hussain S, Arshad M, Leigh MB (2008) Perspectives on microbial cell surface display in bioremediation. Biotechnol Adv 26(2):151–161

Saratale R, Saratale G, Chang J-S, Govindwar S (2009) Decolorization and biodegradation of textile dye navy blue HER by *Trichosporon beigelii* NCIM-3326. J Hazard Mater 166:1421–1428

Saunders RJ, Paul NA, Hu Y, de Nys R (2012) Sustainable sources of biomass for bioremediation of heavy metals in wastewater derived from coal-fired power generation. PLoS One 7(5):e36470
Say R, Yimaz N, Denizli A (2003) Removal of heavy metal ions using the fungus *Penicillium canescens*. Adsorpt Sci Technol 21(7):643–650
Shafi S, Bhat RA, Bandh SA, Shameem N, Nisa H (2018) Microbes: key agents in the sustainable environment and cycling of nutrients. In: Environmental contamination and remediation. Cambridge Scholars, Cambridge. 152-179-188
Silambarasan S, Abraham J (2013) Ecofriendly method for bioremediation of chlorpyrifos from agricultural soil by novel fungus *Aspergillus terreus* JAS1. Water Air Soil Pollut 224:1369
Singh MP, Vishwakarma SK, Srivastava AK (2013) Bioremediation of Direct Blue 14 and Extracellular Ligninolytic Enzyme Production by White Rot Fungi: *Pleurotus* Spp. Biomed Res Int 2013:1801–1856
Singh M, Srivastava PK, Verma PC, Kharwar RN, Singh N, Tripathi RD (2015) Soil fungi for mycoremediation of arsenic pollution in agriculture soils. J Appl Microbiol 119(5):1278–1290
Singh DV, Bhat JIA, Bhat RA, Dervash MA, Ganei SA (2018) Vehicular stress a cause for heavy metal accumulation and change in physico-chemical characteristics of road side soils in Pahalgam. Environ Monit Assess 190:353. https://doi.org/10.1007/s10661-018-6731-2
Singhal RK, Joshi S, Tirumalesh K, Gurg RP (2004) Reduction of uranium concentration in well water by *Chlorella* (*Chlorella pyrenoidosa*) a fresh water algae immobilized in calcium alginate. J Radio Analyt Nucl Chem 261:73–78
Soares GMB, Costa-Ferreira M, Pessoa de Amorim MT (2001) Decolorization of an anthraquinone-type dye using a laccase formulation. Bioresour Technol 79:171–177. https://doi.org/10.1016/S0960-8524(01)00043-8
Sofi NA, Bhat RA, Rashid A, Mir NA, Mir SA, Lone R (2017) Rhizosphere mycorrhizae communities an input for organic agriculture. In: Varma A et al (eds) Mycorrhiza-nutrient uptake, biocontrol, ecorestoration. Springer International, Berlin, pp 387–413
Song HP, Li XG, Sun JS, Xu SM, Han X (2008) Application of a magnetotatic bacterium, *Stenotrophomonas* sp. to the removal of Au(III) from contaminated wastewater with a magnetic separator. Chemosphere 72:616–621
Stella T, Covino S, Čvančarová M, Filipová A, Petruccioli M, D'Annibale A, Cajthaml T (2017) Bioremediation of long-term PCB-contaminated soil by white-rot fungi. J Hazard Mater 324(Pt B):701–710
Strong PJ, Burgess JE (2007) Bioremediation of a wine distillery wastewater using white rot fungi and the subsequent production of laccase. Water Sci Technol 56(2):179–186
Tabrizi L, Mohammadi S, Delshad M, Zadeh M, Babak (2015) Effect of arbuscular mycorrhizal fungi on yield and phytoremediation performance of pot Marigold (*Calendula officinalis* L.) under heavy metals stress. Int J Phytoremediation 17(12):1244–1252
Tastan BE, Ertugrul S, Donmez G (2010) Effective bioremoval of reactive dye and heavy metals by *Aspergillus versicolor*. Bioresour Technol 101(3):870–876
Taştan BE, Çakir DN, Dönmez G (2016) A new and effective approach to boron removal by using novel boron-specific fungi isolated from boron mining wastewater. Water Sci Technol 73(3):543–549
Thatoi H, Das S, Mishra J, Rath BP, Das N (2014) Bacterial chromate reductase, a potential enzyme for bioremediation of hexavalent chromium: a review. J Environ Manag 146:383–399
Tsezos M, Volesky B (1981) Biosorption of uranium and thorium. Biotechnol Bioeng 23:583–604
Turlo J (2014) The biotechnology of higher fungi-current state and perspectives. Folia Biol Oecol 10:49–65
Vaseem H, Singh VK, Singh MP (2017) Heavy metal pollution due to coal washery effluent and its decontamination using a macrofungus, *Pleurotus ostreatus*. Ecotoxicol Environ Saf 145:42–49
Viswanath B, Rajesh B, Janardhan A, Kumar AP, Narasimha G (2014) Fungal laccases and their applications in bioremediation. Enzyme Res 2014:163242. https://doi.org/10.1155/2014/163242
Volesky B (2004) Sorption and biosorption. BV-Sorbex, Montreal

Wong KKY, Richardson JD, Mansfield SD (2000) Enzymatic treatment of mechanical pulp fibers for improving papermaking properties. Biotechnol Prog 16:1025–1029. https://doi.org/10.1021/bp000064d

Wu J, Zhao Y, Liu L, Fan B, Li M (2013) Remediation of soil contaminated with decabrominated diphenyl ether using white-rot fungi. J Environ Eng Landsc Manag 21:171–179

Xun F, Xie B, Liu S, Guo C (2014) Effect of plant growth-promoting bacteria (PGPR) and arbuscular mycorrhizal fungi (AMF) inoculation on oats in saline-alkali soil contaminated by petroleum to enhance phytoremediation. Environ Sci Pollut Res Int 22(1):598–608

Yang Y, Liang Y, Ghosh A, Song Y, Chen H, Tang M (2015) Assessment of arbuscular mycorrhizal fungi status and heavy metal accumulation characteristics of tree species in a lead-zinc mine area: potential applications for phytoremediation. Environ Sci Pollut Res Int 22(17):13179–13193

Yang Y, Liang Y, Han X, Chiu T-Y, Ghosh A, Chen H, Tang M (2016) The roles of arbuscular mycorrhizal fungi (AMF) in phytoremediation and tree-herb interactions in Pb contaminated soil. Sci Rep 6:20469

Young D, Rice J, Martin R, Lindquist E, Lipzen A, Grigoriev I, Hibbett D (2015) Degradation of bunker C fuel oil by white-rot fungi in sawdust cultures suggests potential applications in bioremediation. PLoS One 10(6):e0130381. https://doi.org/10.1371/journal.pone.0130381

Zhuang W, Gao X (2013) Acid-volatile sulfide and simultaneously extracted metals in surface sediments of the southwestern coastal Laizhou Bay, Bohai Sea: concentrations, spatial distributions and the indication of heavy metal pollution status. Mar Pollut Bull 76:128–138

Zotti M, Piazza D, Simone, Roccotiello E, Lucchetti G, Mariotti MG, Marescotti P (2014) Microfungi in highly copper-contaminated soils from an abandoned Fe-cu sulphide mine: growth responses, tolerance and bioaccumulation. Chemosphere 117:471–476

Chapter 14
Microbial Biofilm Cell Systems for Remediation of Wastewaters

Muhammad Faisal Siddiqui, Lakhveer Singh, Farhana Maqbool, Ziaur Rahman, Abdul Rehman, Fazal Adnan, Sadia Qayyum, and Ajmal Khan

14.1 Introduction

The increasing pollution in water is of great alarm for the public about recalcitrant hazardous compounds from different sources. It is important to know that environment is contaminated via different pollutants like heavy metals, pesticides, phenolic compounds, dyes, nutrients and organic compounds which pose serious environmental issues (Mohamed et al. 2016; Rodgers-Vieira et al. 2015; Smułek et al. 2015; Bhat et al. 2017).

The majority of wastewater comes from different industries, and these wastewaters are deleterious and pose serious environmental and human health issues. Thus, it is of prime importance to control these pollutants via effective treatment technologies which must be efficient and cheap. Treated water must be recycled back to

M. F. Siddiqui (✉) · F. Maqbool · S. Qayyum
Department of Microbiology, Hazara University, Mansehra, Pakistan
e-mail: drfaisal@hu.edu.pk

L. Singh
Faculty of Engineering Technology, University of Malaysia Pahang, Pekan, Malaysia

Z. Rahman
Department of Microbiology, Abdul Wali Khan University Mardan, Mardan, Pakistan

A. Rehman
Department of Microbiology and Biotechnology, Abasyn University, Peshawar, Pakistan

F. Adnan
Atta-ur-Rahman School of Applied Biosciences, National University of Sciences & Technology, Islamabad, Pakistan

A. Khan
Department of Biotechnology, Bacha Khan University Charsadda, Charsadda, Pakistan

K. R. Hakeem et al. (eds.), *Bioremediation and Biotechnology*,
https://doi.org/10.1007/978-3-030-35691-0_14

the same process or reused. Various methods have been tested to treat industrial wastewaters such as chemical, biological and mechanical treatment methods. Biotechnological methods that use microbes for degradation of pollutants in wastewater have been largely used (Zhang et al. 2018). These biological treatment strategies are cheap and in these treatment processes no toxic chemical is added. Also, these methods have the ability to completely degrade pollutants (Sharma 2012). However, using microbes for the removal of pollutants from wastewater, there is also difficulty of separating microbes after treatment of wastewater (Tam et al. 2010).

Treatments of pollutant control methods have mainly two limitations: the microbial cells are difficult to separate after treatment process; their reuse and long-term stability. Thus, immobilized microbial cells via physical or chemical ways can overcome these limitations (Hartmeier 2012; Yiğitoğlu and Temoçin 2010). Immobilized microbial cells are gaining huge significance due to their numerous benefits compared to free cell methods for the treatment of wastewater. Immobilized microbial cells provide high mechanical strength, high mass, reuse, stability and resistance to toxic pollutants (Kadimpati et al. 2013).

Microbial cells can be attached or entrapped on/in different support materials. These support materials can be organic or inorganic or water-insoluble materials. Different immobilization cell systems and support materials were used for various wastewaters (Kadimpati et al. 2013). Whole cell immobilization for wastewater provides simple separation from treated water. Also, whole cell microbial immobilization makes available long-term stability of microbial stability of enzymes and activity of enzymes (Stolarzewicz et al. 2011).

In this chapter, we have comprehensively discussed the role of immobilized microbial cells and focused on immobilization methods and support materials for immobilization of microbial cells. Also, the role of microbial immobilized cell systems in the bioremediation of contaminated wastewater is discussed. Lastly, conclusions and future prospects in biofilm-based bioremediation are highlighted.

14.2 Microbial Immobilization

Microbial immobilization is the attachment or entrapment of microbial cells using support materials. Generally immobilization is the imprisonment or restriction of the movement of a cell (Zhang et al. 2004). Usually immobilization can be used for plant cells, animal cells, microbial cells or enzymes. In recent studies, whole microbial cells have been immobilized on support materials for the control of environmental pollutants. Immobilized microbial cells are of three types which consist of growing, dead and living. Therefore, it is important to select the suitable type of immobilized cells for a particular application (Rahman et al. 2006).

Microbial immobilized cells are more stable compared to immobilized enzymes. In this system, it is not necessary to extract enzymes from microbial cells. When using enzymes, they are prone to less stability in harsh conditions; moreover, enzyme systems' unnecessary reactions take place (Stolarzewicz et al. 2011). It is important to know that the area of whole cell immobilization is different from health sciences to food industries. Immobilized microbes on support materials can be reused in fresh bioprocess reactions for treatment of wastewater or production of a variety of products. Reuse of immobilized cells can reduce cost of the production or treatment process (Mrudula and Shyam 2012; Ohta et al. 1994).

14.3 Support Materials for Immobilization

For microbial cell immobilization, it is important to select a suitable support material. Support materials must meet the following norms (Zacheus et al. 2000):

(a) Immobilization support materials must have long shelf life.
(b) Support materials must be non-biodegradable and must be nonhazardous.
(c) Materials must be cheap and easily available.
(d) These support materials must be easily separated from cells.
(e) Support materials must have high chemical and mechanical stability.
(f) These materials can be sterilizable.
(g) Materials must be suitable for regeneration.

It is vital to know that the choice of support material for anoxic biomass immobilization can greatly affect the efficacy of a bioreactor or fermenter. Microbial cells attached on the surface depend on the support material which directly affects the number of microbes attached to it. Support materials are mainly classified in two main sets: organic and inorganic support materials (Lu and Toy 2009). Organic support materials are used such as dextran, celluloses, and agarose, while inorganic materials are porous glass, activated charcoal, clay, etc. (Lu and Toy 2009).

Organic support materials are available in large variety compared to inorganic support materials. Also, organic immobilization materials can be acquired with desired porosity. These materials are sensitive to pH, while inorganic support materials are resistant to chemicals, pH and microbial degradation. These support materials are also more feasible in scale-up process (Ispas et al. 2009; Magner 2013).

Also, organic support materials can be classified into synthetic and natural polymers. Different synthetic polymers such as polyvinyl, resins, acrylamide and polyurethane are also employed for microbial immobilization. Few examples of organic natural materials are agar, agarose and carrageenan (Hartmann 2005). In most of the studies, alginate polymers were used due to various benefits like they are environment-friendly and nontoxic to humans. Also, they are cheap and obtainable in huge amount. Moreover, immobilization in alginate also avoids changes in physiological condition (Buque et al. 2002).

14.4 Immobilization Methods

Recently, there is more focus on using immobilization methods for bioremediation of wastewater. Different types of immobilization methods have been employed for immobilization. Among these methods most important are encapsulation, adsorption, binding on surface and entrapment (Kourkoutas et al. 2004). This method is also elaborated briefly in Fig. 14.1.

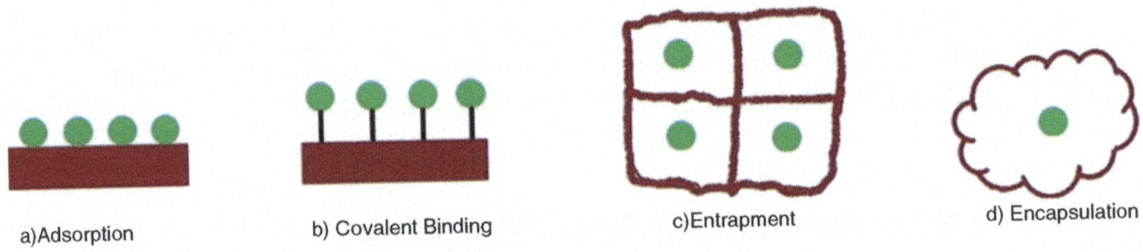

Fig. 14.1 Different immobilization methods (Bayat et al. 2015)

14.4.1 Adsorption

This method is simple and quick. This method is reversible. Adsorption is the most commonly used method for bioremediation of wastewater via immobilization technique. Adsorption can be defined as immobilization of enzymes or microbes via their physical interaction with surface of support materials. There is no need for addition of chemical additives. Adsorption method is cheap and environment-friendly. Adsorption completely is achieved via formation of weak bonds (hydrogen, ionic and van der waals forces, etc.). Also, these interactions are not strong and unstable; hence there is possibility of leakage in environment (Hou et al. 2014).

14.4.2 Covalent Binding

This method of immobilization is reversible due to covalent bond formation with the support material and microbial cell when a cross-linking material is present. This method is mostly used for immobilization of enzymes. In rare cases covalent binding method is applied owing to the toxicity which causes cell death. Covalent binding method is different from electrostatic binding (Groboillot et al. 1994).

14.4.3 Entrapment in Porous Matrix

Entrapment of microbial cells is widely used for pollutants treatment. Once microbes are entrapped, they can move inside the entrapped support material. Depending on the support materials, it can reduce the leakage of materials across the support material. Also, it can reduce the passage of nutrients. It is found that those microbes which stay near the surface have high activity and metabolic process, while other microbes have less activity (Bleve et al. 2011). There are numerous advantages of entrapment method such as it is environmentally friendly, cheap and nontoxic (Wojcieszyńska et al. 2012). This method helps in protection of microbes from the harsh environmental conditions. It is vital to understand that pore sizes of the support entrapment material must be smaller than the microbes. If these support materials have large pore size, there is possibility of leakage (Bleve et al. 2011).

14.4.4 Encapsulation

This method is quite similar to the method of entrapment. It is irreversible method of immobilization of microbial cells. In this method particles are isolated from external environment. The major benefit of this method is the protection of cells from toxic or extreme conditions. The protective barrier around the microorganism allows the passage of nutrients. This technique is cheap and fast. In addition to its benefits, there are also some limitations which hinder the large-scale application of this technique. Among the major limitations is injury can occur to the encapsulation material. Due to the limitations discussed here, this method is not widely used for bioremediation of wastewater. Pore size is also important, and if immobilization support material leaks, it can decrease the loading, and hence, it affects the efficiency of bioremediation process (Klein et al. 2012).

14.5 Role of Microbial Biofilm Cell Systems in Bioremediation of Wastewater

The increasing water pollution poses serious environmental concerns. It is important to find new ways to control water pollution. Recently, microbial biofilm-based strategies for the control of water pollution are increasing to cater this issue. In this section of the chapter, we will focus on the bioremediation of different wastewaters via immobilized microbial cells.

14.5.1 Bioremediation of Heavy Metals

Most of the industrial wastewaters contain hazardous metals such as copper, lead, cadmium, etc. (Jencarova and Luptakova 2017). These wastewaters produce free radicals which can cause serious environmental hazards and concerns (Gumpu et al. 2015). Hence, it is of prime important to treat heavy metals containing wastewater. Several methods have been tested for the treatment of heavy metal wastewater, but most of these methods are expensive and have various limitations. In few methods, several adsorbents are used but mostly are not efficient. Also their efficiency is usually augmented by increasing the surface area of the adsorbents (Ahmed et al. 2015). In one of the study reported, *Penicillium citrinum* was entrapped sodium alginate matrix. Sodium alginate beads containing *Penicillium citrinum* were produced, and beads were used for bioremediation of Cu(II) removal. It was observed that immobilized beads containing microbe remove Cu(II) up to 84.5%. However, for free cells it was 82.4%. From these results, it was exhibited that immobilized microbial cells are more efficient in removal of tested metal. Table 14.1 also demonstrates other methods for bioremediation of heavy metals using immobilized microbes. In another study, under batch conditions, 95% metal removal was achieved using

Table 14.1 Role of immobilized microbial cells in bioremediation of different pollutants

Pollutants	Microbes	Support materials	References
Phenol	*Bacillus thuringiensis J20*	Sodium alginate	Ereqat et al. (2018)
Phenol	*Bacillus* cells	Polyvinyl alcohol-sodium alginate	Ismail and Khudhair (2015)
Cu, Ni	*Chlorella vulgaris*	Alginate	Mehta and Gaur (2001)
Uranium	*Chlamydomonas reinhardtii*	Cellulose beads	Erkaya et al. (2014)
As(III), as(V)	*Corynebacterium glutamicum*	Neem leaves	Podder and Majumder (2015)
Cr(VI)	*Bacillus* species	Biomass of tea	Gupta and Balomajumder (2015)
Nitrite	Nitrite-oxidizing bacteria	Chitosan	Lertsutthiwong et al. (2013)
NH4-N	Nitrifier cells	Polyvinyl alcohol and sodium alginate	Wang et al. (2016)
Phosphorus and nitrogen	*Scenedesmus intermedius*	Alginate	Jimenez-Perez et al. (2004)
Nitrate	*Chlorella vulgaris*	Chitosan nanofibres	Eroglu et al. (2012)
Reactive dyes	*Pseudomonas putida* and *Bacillus Licheniformis*	Sodium alginate and polyacrylamide Gel beads	Suganya and Revathi (2016)
Methylene blue	*Bacillus subtilis*	Calcium alginate bead	Upendar et al. (2016)
Polyazo dye	*Bacillus firmus*	Tubular polymeric gel	Ogugbue et al. (2012)

sulphate-reducing bacteria in sodium alginate. A continuous removal strategy was also employed in which 99% of Cu(II) and 95.8% of Zn(II) were achieved (Kiran et al. 2018).

14.5.2 Bioremediation of Refractory Organic Wastewater

Bioremediation of phenolic- or aniline-based compounds is difficult to achieve (Luan et al. 2017). Most of the refractory compounds such as aniline or phenolic compounds are not degraded efficiently through present conventional treatment methods (Cesaro et al. 2013). It is owing to the long time needed for the microbial cells to grow and stay in the reaction process. However, immobilized microbial cells provide higher mass of cells and also microbial cells are more stable. In one of the study, *Bacillus* sp. SAS19 for phenol degradation was immobilized on porous carbonaceous gels. In this study it was exhibited that immobilized bacteria were more efficient in degradation of phenolics. It was found that immobilized bacteria can degrade phenol (1600 mg/L) up to 100% in 24 h (Ke et al. 2018). Lu and Toy (2009) also reported application of *Phanerochaete chrysosporium*. This fungus was immobilized on wood chips. Immobilized fungus was applied for biodegradation of phenolic compounds from coking wastewater. Immobilized fungus exhibited 87.05% degradation of phenol which was higher compared to non-attached fungus. The optimal removal of phenol was 84% in 3 days.

14.5.3 Bioremediation of Industrial Dyes

Different industries utilize dyes for different purposes including textile, leather, pharmaceutical, etc. After these processes, a huge amount of wastewater containing these dyes is discharged into the water bodies which pose different environmental concerns. Also, these dyes cause mutation and deleterious health concerns. Thus, it is important to treat these dyes containing wastewater. Many studies are performed to cater this issue and control dye wastewater before its discharge into the water streams. Different microorganisms are tested for dye treatment under different conditions in batch and continuous mode. Most common microbes are funguses which are more efficient in removal and degradation of dyes from wastewater (Couto 2009). In one of the study, *Brevibacillus parabrevis* was immobilized on coconut shell biochar. It was exhibited that under optimum conditions and with inoculum of 3 ml, removal of 95.7% of Congo red dye was achieved after 6 days (Talha et al. 2018). Hameed and Ismail (2018) found that using immobilized mix cells, reactive red dye (10 mg/L) was completely decolourized within 30 h under anaerobic conditions. For other studies, Table 14.1 shows different methods of bioremediation of dye wastewater.

14.5.4 Bioremediation of Nitrogen and Phosphorus

Nutrients are vital for the growth of microorganisms, but their increase in water bodies causes eutrophication (Tang et al. 2017).

Hence, it is necessary to treat water containing excessive nutrients before discharging it into rivers and lakes. Various techniques have been used for the control of nutrients (e.g. nitrogen and phosphorus). These conventional processes are adsorption, membrane processes, chemical precipitation, biological processes, etc. (Kumar et al. 2018). Due to the tremendous benefits of immobilized microbial cell technology, various studies have utilized immobilized cells for the bioremediation of nitrogen and phosphorus removal from wastewater. Shi et al. (2007) tested two green microalgae (*Chlorella vulgaris* and *Scenedesmus rubescens*) for phosphorus and nitrogen removal. Microalgae were immobilized via twin-layer system. It was exhibited that both algae tested removed nitrate from wastewater. When they have used secondary wastewater, both algae removed nitrate and phosphate to less than 10% in 9 days. Table 14.1 illustrates different immobilized microbial cells to control nutrients from wastewater.

14.6 Conclusions

It is exhibited from the studies that immobilized microbial cells have great potential for bioremediation of wastewater. In many studies it was found that immobilized cells are efficient compared to free cells for bioremediation of polluted water. Immobilized microbial cell systems provide numerous benefits compared to free cells for treatment of wastewater. In various studies, it was reported that immobilized cells have longer stability, lower cost, and higher degrading ability. Also, immobilized cells increase tolerance to harsh conditions which also makes them more suitable compared to free cells.

14.7 Future Prospects

Cloning of genes for biosurfactant synthesis and chemotactic ability of Genetically Engineered Microbes (GEMs) can further enhance the biodegradative capability of modified microbes. Nevertheless, the release and use of GEM in the nature and transmission is under much debate and controversial. However, the majority of organisms usually have other disabling mutations that will not permit the microbes to grow outside a given environment. Reengineering of secreted proteins in biofilm matrix is also an area for further development in the field of bioremediation. Bioremediation studies to test the effectiveness of biofilm under conditions similar to those encountered in natural environment still remain few. Many issues remain unclear such as the correlation between biofilm microstructure and biodegradation process, long-term behaviour of biofilms exposed to fluctuations in pollutant concentrations and the detail of the correlation between soil composition and biofilm behaviour. Solutions to these issues will provide a predictive and quantitative model for bioremediation using biofilm-based methods, so to improve this large scale of application of this environmentally sustainable technology.

References

Ahmed YM, Al-Mamun A, Al Khatib MFR, Jameel AT, AlSaadi MAHAR (2015) Efficient lead sorption from wastewater by carbon nanofibers. Environ Chem Lett 13(3):341–346

Bayat Z, Hassanshahian M, Cappello S (2015) Immobilization of microbes for bioremediation of crude oil polluted environments: a mini review. Open Microbiol J 9:48

Bhat RA, Shafiq-ur-Rehman, Mehmood MA, Dervash MA, Mushtaq N, Bhat JIA, Dar GH (2017) Current status of nutrient load in Dal Lake of Kashmir Himalaya. J Pharma Phytochem 6(6):165–169

Bleve G, Lezzi C, Chiriatti M, D'Ostuni I, Tristezza M, Di Venere D, Sergio L, Mita G, Grieco F (2011) Selection of non-conventional yeasts and their use in immobilized form for the bioremediation of olive oil mill wastewaters. Bioresour Technol 102(2):982–989

Buque EM, Chin-Joe I, Straathof AJ, Jongejan JA, Heijnen JJ (2002) Immobilization affects the rate and enantioselectivity of 3-oxo ester reduction by baker's yeast. Enzym Microb Technol 31(5):656–664

Cesaro A, Naddeo V, Belgiorno V (2013) Wastewater treatment by combination of advanced oxidation processes and conventional biological systems. J Bioremed Biodegrad 4(8):1–8

Couto SR (2009) Dye removal by immobilised fungi. Biotechnol Adv 27(3):227–235

Ereqat SI, Abdelkader AA, Nasereddin AF, Al-Jawabreh AO, Zaid TM, Letnik I, Abdeen ZA (2018) Isolation and characterization of phenol degrading bacterium strain Bacillus thuringiensis J20 from olive waste in Palestine. J Environ Sci Health A 53(1):39–45

Erkaya IA, Arica MY, Akbulut A, Bayramoglu G (2014) Biosorption of uranium (VI) by free and entrapped Chlamydomonas reinhardtii: kinetic, equilibrium and thermodynamic studies. J Radioanal Nucl Chem 299(3):1993–2003

Eroglu E, Agarwal V, Bradshaw M, Chen X, Smith SM, Raston CL, Iyer KS (2012) Nitrate removal from liquid effluents using microalgae immobilized on chitosan nanofiber mats. Green Chem 14(10):2682–2685

Groboillot A, Boadi D, Poncelet D, Neufeld R (1994) Immobilization of cells for application in the food industry. Crit Rev Biotechnol 14(2):75–107

Gumpu MB, Sethuraman S, Krishnan UM, Rayappan JBB (2015) A review on detection of heavy metal ions in water–an electrochemical approach. Sensors Actuators B Chem 213:515–533

Gupta A, Balomajumder C (2015) Simultaneous removal of Cr (VI) and phenol from binary solution using Bacillus sp. immobilized onto tea waste biomass. J Water Process Eng 6:1–10

Hameed BB, Ismail ZZ (2018) Decolorization, biodegradation and detoxification of reactive red azo dye using non-adapted immobilized mixed cells. Biochem Eng J 137:71–77

Hartmann M (2005) Ordered mesoporous materials for bioadsorption and biocatalysis. Chem Mater 17(18):4577–4593

Hartmeier W (2012) Immobilized biocatalysts: an introduction. Springer Science & Business Media, Berlin

Hou J, Dong G, Ye Y, Chen V (2014) Laccase immobilization on titania nanoparticles and titania-functionalized membranes. J Membr Sci 452:229–240

Ismail ZZ, Khudhair HA (2015) Recycling of immobilized cells for aerobic biodegradation of phenol in a fluidized bed bioreactor. Syst Cyber Informat 13(5):81–86

Ispas C, Sokolov I, Andreescu S (2009) Enzyme-functionalized mesoporous silica for bioanalytical applications. Anal Bioanal Chem 393(2):543–554

Jencarova J, Luptakova A (2017) The application of biogenically created sorbent for metal ions elimination. Inżynieria Mineralna 18

Jimenez-Perez M, Sanchez-Castillo P, Romera O, Fernandez-Moreno D, Pérez-Martınez C (2004) Growth and nutrient removal in free and immobilized planktonic green algae isolated from pig manure. Enzym Microb Technol 34(5):392–398

Kadimpati KK, Mondithoka KP, Bheemaraju S, Challa VRM (2013) Entrapment of marine microalga, Isochrysis galbana, for biosorption of Cr (III) from aqueous solution: isotherms and spectroscopic characterization. Appl Water Sci 3(1):85–92

Ke Q, Zhang Y, Wu X, Su X, Wang Y, Lin H, Mei R, Zhang Y, Hashmi MZ, Chen C (2018) Sustainable biodegradation of phenol by immobilized Bacillus sp. SAS19 with porous carbonaceous gels as carriers. J Environ Manag 222:185–189

Kiran MG, Pakshirajan K, Das G (2018) Heavy metal removal from aqueous solution using sodium alginate immobilized sulfate reducing bacteria: mechanism and process optimization. J Environ Manag 218:486–496

Klein S, Avrahami R, Zussman E, Beliavski M, Tarre S, Green M (2012) Encapsulation of Pseudomonas sp. ADP cells in electrospun microtubes for atrazine bioremediation. J Ind Microbiol Biotechnol 39(11):1605–1613

Kourkoutas Y, Bekatorou A, Banat IM, Marchant R, Koutinas A (2004) Immobilization technologies and support materials suitable in alcohol beverages production: a review. Food Microbiol 21(4):377–397

Kumar TP, Mandlimath TR, Sangeetha P, Revathi S, Kumar SA (2018) Nanoscale materials as sorbents for nitrate and phosphate removal from water. Environ Chem Lett 16(2):389–400
Lertsutthiwong P, Boonpuak D, Pungrasmi W, Powtongsook S (2013) Immobilization of nitrite oxidizing bacteria using biopolymeric chitosan media. J Environ Sci 25(2):262–267
Lu J, Toy PH (2009) Organic polymer supports for synthesis and for reagent and catalyst immobilization. Chem Rev 109(2):815–838
Luan M, Jing G, Piao Y, Liu D, Jin L (2017) Treatment of refractory organic pollutants in industrial wastewater by wet air oxidation. Arab J Chem 10:S769–S776
Magner E (2013) Immobilisation of enzymes on mesoporous silicate materials. Chem Soc Rev 42(15):6213–6222
Mehta SK, Gaur JP (2001) Removal of Ni and Cu from single and binary metal solutions by free and immobilized Chlorella vulgaris. Eur J Protistol 37(3):261–271
Mohamed A, El-Sayed R, Osman T, Toprak M, Muhammed M, Uheida A (2016) Composite nanofibers for highly efficient photocatalytic degradation of organic dyes from contaminated water. Environ Res 145:18–25
Mrudula S, Shyam N (2012) Immobilization of Bacillus megaterium MTCC 2444 by Ca-alginate entrapment method for enhanced alkaline protease production. Braz Arch Biol Technol 55(1):135–144
Ogugbue CJ, Morad N, Sawidis T, Oranusi NA (2012) Decolorization and partial mineralization of a polyazo dye by Bacillus firmus immobilized within tubular polymeric gel. 3. Biotech 2(1):67–78
Ohta T, Ogbonna J, Tanaka H, Yajima M (1994) Development of a fermentation method using immobilized cells under unsterile conditions. 2. Ethanol and L-lactic acid production without heat and filter sterilization. Appl Microbiol Biotechnol 42(2–3):246–250
Podder M, Majumder C (2015) Bacteria immobilization on neem leaves/$MnFe_2O_4$ composite surface for removal of As (III) and As (V) from wastewater. Arab J Chem. https://doi.org/10.1016/j.arabjc.2015.08.025
Rahman RNZA, Ghazali FM, Salleh AB, Basri M (2006) Biodegradation of hydrocarbon contamination by immobilized bacterial cells. J Microbiol 44(3):354–359
Rodgers-Vieira EA, Zhang Z, Adrion AC, Gold A, Aitken MD (2015) Identification of anthraquinone-degrading bacteria in soil contaminated with polycyclic aromatic hydrocarbons. Appl Environ Microbiol 81(11):3775–3781
Sharma S (2012) Bioremediation: features, strategies and applications. Asian J Pharm Life Sci 2231:4423
Shi J, Podola B, Melkonian M (2007) Removal of nitrogen and phosphorus from wastewater using microalgae immobilized on twin layers: an experimental study. J Appl Phycol 19(5):417–423
Smułek W, Zdarta A, Guzik U, Dudzińska-Bajorek B, Kaczorek E (2015) Rahnella sp. strain EK12: cell surface properties and diesel oil biodegradation after long-term contact with natural surfactants and diesel oil. Microbiol Res 176:38–47
Stolarzewicz I, Białecka-Florjańczyk E, Majewska E, Krzyczkowska J (2011) Immobilization of yeast on polymeric supports. Chem Biochem Eng Q 25(1):135–144
Suganya K, Revathi K (2016) Decolorization of reactive dyes by immobilized bacterial cells from textile effluents. Int J Curr Microbiol Appl Sci 5:528–532
Talha MA, Goswami M, Giri B, Sharma A, Rai B, Singh R (2018) Bioremediation of Congo red dye in immobilized batch and continuous packed bed bioreactor by Brevibacillus parabrevis using coconut shell bio-char. Bioresour Technol 252:37–43
Tam N, Chan M, Wong Y, Popov V, Itoh H, Mander U (2010) Removal and biodegradation of polycyclic aromatic hydrocarbons by immobilized microalgal beads. WIT Trans Ecol Environ 140:391–402
Tang C-J, Duan C-S, Yu C, Song Y-X, Chai L-Y, Xiao R, Wei Z, Min X-B (2017) Removal of nitrogen from wastewaters by anaerobic ammonium oxidation (ANAMMOX) using granules in upflow reactors. Environ Chem Lett 15(2):311–328
Upendar G, Dutta S, Chakraborty J, Bhattacharyya P (2016) Removal of methylene blue dye using immobilized bacillus subtilis in batch & column reactor. Mater Today Proc 3(10):3467–3472

Wang W, Ding Y, Wang Y, Song X, Ambrose RF, Ullman JL, Winfrey BK, Wang J, Gong J (2016) Treatment of rich ammonia nitrogen wastewater with polyvinyl alcohol immobilized nitrifier biofortified constructed wetlands. Ecol Eng 94:7–11

Wojcieszyńska D, Hupert-Kocurek K, Jankowska A, Guzik U (2012) Properties of catechol 2, 3-dioxygenase from crude extract of Stenotrophomonas maltophilia strain KB2 immobilized in calcium alginate hydrogels. Biochem Eng J 66:1–7

Yiğitoğlu M, Temoçin Z (2010) Immobilization of Candida rugosa lipase on glutaraldehyde-activated polyester fiber and its application for hydrolysis of some vegetable oils. J Mol Catal B Enzym 66(1–2):130–135

Zacheus OM, Iivanainen EK, Nissinen TK, Lehtola MJ, Martikainen PJ (2000) Bacterial biofilm formation on polyvinyl chloride, polyethylene and stainless steel exposed to ozonated water. Water Res 34(1):63–70

Zhang Y-Q, Tao M-L, Shen W-D, Zhou Y-Z, Ding Y, Ma Y, Zhou W-L (2004) Immobilization of L-asparaginase on the microparticles of the natural silk sericin protein and its characters. Biomaterials 25(17):3751–3759

Zhang W, Grimi N, Jaffrin MY, Ding L, Tang B, Zhang Z (2018) Optimization of RDM-UF for alfalfa wastewater treatment using RSM. Environ Sci Pollut Res 25(2):1439–1447

Chapter 15
Pollution Remediation by Way of Using Genetically Modified Plants (GMPs)

Fernanda Maria Policarpo Tonelli and Flávia Cristina Policarpo Tonelli

15.1 Introduction: Biotechnology and Phytoremediation

Environmental contamination with some harmful organic and inorganic contaminants is a consequence of some human economic activities that generate dangerous wastes (e.g., the ones from mine exploitation; petroliferous, fabric, and pharmaceutical industries; the agricultural use of herbicides) and pose a serious concern. These pollutants are hard to eliminate from nature and can cause serious damages to human and other forms of life. Among the organic ones, it is possible to highlight chlorinated solvents, halogenated hydrocarbons, and nitrogen compounds commonly present in explosives. Among the inorganic ones are the heavy metals and other elements such as the radioactive uranium (Jafari et al. 2013; Mendes et al. 2019; Pesantes et al. 2019; Pu et al. 2019; Rosculete et al. 2019; Vázquez-Luna and Cuevas-Díaz 2019; Zhang et al. 2019).

Heavy metal pollution is among the most serious environmental problems nowadays; once it is capable of bioaccumulating in living systems, is difficult to eliminate from contaminated water and soil, presents toxicity being able to cause poisoning and oxidative stress and also presents high carcinogenic potential (Alkorta et al. 2004; Ali et al. 2019). It is necessary to develop ways to extract them from contaminated environments, and the genetic manipulation of plants to perform this task is an elegant solution. In fact plants are more suitable to act in this sense once microorganisms can only convert metals into a less toxic form instead of removing them from a contaminated environment (Garbisu et al. 2002; Ojuederie and Babalola 2017).

F. M. P. Tonelli (✉)
Department of Cell Biology, Institute of Biological Science, Federal University of Minas Gerais, Belo Horizonte, Brazil

F. C. P. Tonelli
Department of Biochemistry, Federal University of São João del Rei, Divinópolis, Brazil

K. R. Hakeem et al. (eds.), *Bioremediation and Biotechnology*,
https://doi.org/10.1007/978-3-030-35691-0_15

Transgenic plants are the ones that underwent DNA manipulation with the intention to introduce a new trait to the organism, which does not occur naturally in the species. So, by applying methodologies to genetically modify plant DNA, it is possible to develop transgenic organisms to express or overexpress genes related to metal (or other contaminants) uptake/transportation/metabolization and apply these genetically modified organisms to perform phytoremediation with efficiency. By doing that it becomes possible to remove these contaminants from the environment, minimizing the risk they consist in our lives. Heavy metals were the first target of genetic manipulation of plants to perform remediation of contaminated soil (Misra and Gedamu 1989).

Plants offer some interesting characteristics that make its use advantageous when compared to the use of microorganisms for remediation. Working with plants is easier, especially when it comes to nutrient input due to the fact that as autotrophic systems they manage to provide their own nutrient sources. It is also an interesting feature the fact that plants can be controlled to avoid undesirable spreading, maintaining in situ remediation and also preventing the dispersion of contaminants. An eco-friendly system can be developed avoiding erosion, being suitable to application in a broad range of remediation sites, and presenting low costs associated with this renewable phytoremediation technology (Suresh and Ravishankar 2004; Abhilash et al. 2009; Lee 2013; Wan et al. 2016).

Transgenesis in plants for remediation commonly aims to insert or overexpress genes that codify proteins related to the uptake and/or sequestration of pollutants (Shukla et al. 2013; Mani and Kumar 2014; Das et al. 2016), for example, binding proteins and transporters (Table 15.1).

Among transporters it is interesting to highlight the members of ATP-binding cassette (ABC) family (proteins related to not only plant detoxification but also to important ion regulation process) (Martinoia et al. 2002), cation diffusion facilitator (CDF) family (e.g., the metal tolerance/transport protein (MTP) involved in metal storage) (Ricachenevsky et al. 2013), and metal ion transporters like the ones responsible for cytoplasmic transport of zinc and iron from the family of ZRT/IRT-related proteins (ZIP) (Ducic and Polle 2005).

Metals are important pollutants in the environment especially heavy ones, and it is common that plants possess genes related to their transportation to the organism's inner part as they consist in essential microelements for these organisms (Williams et al. 2000). Metal phytoremediation can be performed through different technologies such as phytoextraction (removing metals from soils and concentrating them in the shoots), phytostabilization (accumulating metals in roots or minimizing their mobility by causing their precipitation in rhizosphere), and phytovolatilization (Nascimento and Xing 2006).

Plants can be engineered to increase the accumulation of heavy metal in its shoots (phytoextraction); *Brassica juncea*, *Nicotiana tabacum*, *Arabidopsis thaliana,* and plants from *Populus* gender have already been modified with this purpose. Cd and Pb increased accumulation and tolerance in *B. juncea* shoots can be achieved by the overexpression of the ATP-binding cassette (ABC) family transporter AtATM3 (Bhuiyan et al. 2011); a transporter from the same family (in transgenic plants, the

Table 15.1 Examples of GMP developed to perform phytoremediation

Source of transgene	Transgene or desired gene product	Genetically modified plant (GMP) species generated	Desired characteristic presented by the GMP	Reference
Arabidopsis thaliana	*ATM3*	*Brassica juncea*	Cd and Pb increased accumulation and tolerance	Bhuiyan et al. (2011)
Saccharomyces cerevisiae	*YCF1*	*Populus tremula* and *Populus alba*	Cd, Zn, and Pb increased accumulation	Shim et al. (2013)
Psychotria gabriellae	*IREG1*	*Arabidopsis thaliana*	Ni tolerance and accumulation increased	Merlot et al. (2014)
Pseudomonas sp.	*copC*	*Arabidopsis thaliana*	Cu tolerance and accumulation increased	Rodríguez-Llorente et al. (2012)
Arabidopsis thaliana	*CAX2* and *CAX4*	*Nicotiana tabacum*	Increased biomass when grown in the presence of heavy metals and higher accumulation of Cd, Mn, and Zn	Korenkov et al. (2007)
Oryza sativa	*MTP1*	*Nicotiana tabacum*	Cd increased accumulation	Das et al. (2016)
Astragalus bisulcatus	*SMT*	*A. thaliana* and *B. juncea*	Se increased volatilization and tolerance	LeDuc et al. (2004)
Arabidopsis thaliana	*IRT1*	*Arabidopsis thaliana*	Cd and Zn increased accumulation	Connolly et al. (2002)
Noccaea caerulescens	ZNT1	*Arabidopsis thaliana*	Zn and Cd increased accumulation	Lin et al. (2016)
Escherichia coli	*gshI*	*Brassica juncea*	Cd, Cr, Cu, Pb, and Zn increased uptake	Zhu et al. (1999a, b)
Allium sativum and *Saccharomyces cerevisiae*	*PCS1 / GSH1*	*Arabidopsis thaliana*	Cd and As increased accumulation	Guo et al. (2008)
Thlaspi caerulescens	*PCS1*	*Nicotiana glauca*	Cd, Zn, and Pb increased accumulation	Martinez et al. (2006)
Arabidopsis thaliana	*PCS1*	*Brassica juncea*	Cd and As increased tolerance	Gasic and Korban (2007)
Elsholtzia haichowensis	*MT1*	*Nicotiana tabacum*	Cu increased tolerance and accumulation	Xia et al. (2012)
Sedum alfredii	*MT2*	*Nicotiana tabacum*	Cu increased tolerance and accumulation	Zhang et al. (2014)
Brassica campestris	*MT1* and *MT2*	*Arabidopsis thaliana*	Cd and Cu increased tolerance	Lu et al. (2015)
Oryza sativa	*MT2c*	*Arabidopsis thaliana*	Cu increased tolerance	Liu et al. (2015)

(continued)

Table 15.1 (continued)

Source of transgene	Transgene or desired gene product	Genetically modified plant (GMP) species generated	Desired characteristic presented by the GMP	Reference
Bacillus megaterium	*TnMERI1*	*Arabidopsis thaliana*	Cd and Pb increased accumulation and tolerance	Hsieh et al. (2009)
Enterobacter cloacae	*Onr*	*Nicotiana tabacum*	TNT and GTN increased tolerance	French et al. (1999)
Enterobacter cloacae	*NfsI*	*Nicotiana tabacum*	TNT increased tolerance	Hannink et al. (2007)
Arabidopsis thaliana	*743B4, 73C1*	*Arabidopsis thaliana*	TNT increased tolerance	Gandia-Herrero et al. (2008)
Rhodococcus rhodochrous	*XplA, XplB*	*Arabidopsis thaliana*	RDX phytoremediation	Jackson et al. (2007)
Escherichia coli	*NfsA*	*Arabidopsis thaliana*	TNT increased tolerance	Kurumata et al. (2005)
Homo sapiens	*CYP2E1*	*Arabidopsis thaliana*	Capacity to deal with residues of TCE	Doty et al. (2000)
Homo sapiens	*CYP1A1, CYP2B6*, and *CYP2C19*	*Oryza sativa*	Phytoremediation of the herbicides atrazine and metolachlor	Kawahigashi et al. (2006)
Homo sapiens	*CYP2C9, CYP1A1, CYP2B6*, and *CYP2C19*	*Solanum tuberosum*	Phytoremediation of herbicides including sulfonylureas	Inui and Ohkawa (2005)
Pseudomonas sp.	Modified bacterial *atzA* gene	*Medicago sativa* and *Nicotiana tabacum*	Atrazine-enhanced metabolism	Wang et al. (2005)
Zea mays	*gstI-6His*	*Nicotiana tabacum*	Phytoremediation of the herbicide alachlor	Karavangeli et al. (2005)

ABC transporters are commonly localized in the tonoplast, sequestrating metals in the vacuolar lumen (Song et al. 2014), the yeast cadmium factor 1 (YCF1), can be expressed in transgenic *Populus tremula* and *Populus alba* to increase the shoots' accumulation of Cd and Zn (Shim et al. 2013); *A. thaliana* can have its shoots' nickel and copper tolerance and accumulation increased by being genetically engineered to express, respectively, the metal transporter PgIREG1 (a gene originally expressed, e.g., in the hyperaccumulator shrub *Psychotria gabriellae*) (Merlot et al. 2014) and the copper-resistant protein (from *Pseudomonas* sp.) (Rodríguez-Llorente et al. 2012); *N. tabacum* can be genetically modified to overexpress the rice metal tolerance protein OsMTP1 increasing its capacity of Cd accumulation in shoots by high level of generation of thiol compounds that can chelate metals sequestrating them into vacuoles (Das et al. 2016). The *A. thaliana* CAX2 and CAX4 (low-affinity Ca^{2+}, heavy metal cation/H+ antiporters) when expressed in *N. tabacum* results in organisms with an increased biomass when grown in the presence of heavy metals and higher accumulation of Cd, Mn, and Zn (Korenkov et al. 2007).

N. tabacum and *A. thaliana* can be genetically modified to perform Hg phytovolatilization. Bacterial gene from the reductase merA and organomercurial lyase gene merB are interesting tools to achieve this goal. MerA transgenic plants can uptake Hg^{2+} through roots and convert it into Hg^0: a less toxic and volatile form. Mer B plants can convert the uptaken methylmercury into sulfhydryl-bound Hg^{2+}. And transgenic plants expressing both genes are able to convert not only Hg^{2+} but also methylmercury into the volatile form (Rugh et al. 1996, 1998, 2000; Heaton et al. 1998). The selenocysteine methyltransferase from *Astragalus bisulcatus* when overexpressed in *A. thaliana* and *B. juncea* leads to an increase in Se volatilization and tolerance (LeDuc et al. 2004).

Phytostabilization can be achieved by limiting the uptaken heavy metal transportation from plant's roots to the shoots. In transgenic *N. tabacum* expressing AtHMA4, the Cd transport is restricted by apoplastic barrier (Siemianowski et al. 2014); in *Manihot esculenta* the overexpression of the transporters AtZIP1 and AtMTP1 make possible to achieve Zn accumulation in the roots with restrict transportation to the shoots (Gaitán-Solís et al. 2015); in *A. thaliana* the overexpression of the metal transporter IRT1 can induce an increase in accumulation of Cd and Zn by the plant (Connolly et al. 2002), and the overexpression of Zn transporter ZNT1 from *Noccaea caerulescens* can increase in the transgenic plant the accumulation of Zn and Cd (Lin et al. 2016); and in the *Populus* gender species mentioned before, the same gene YCF1 when expressed also increases the accumulation of Pb in the plant roots (Shim et al. 2013).

When it comes to remediation of heavy metals by genetically modifying plants to express or overexpress binding proteins, it is necessary to highlight metal chelators like the peptides phytochelatins (Hirata et al. 2005), metallothioneins (Tripathi et al. 2015), and mercuric ion binding protein proteins (MerPs) (Huang et al. 2003).

When it comes to transgenic plants producing phytochelatins, the genetic modifications involve mainly two important enzymes that play a key role in their synthesis: phytochelatin synthase and *c*-glutamylcysteine synthetase (Hirata et al. 2005). *Brassica juncea* can extract more Cd, Cr, Cu, Pb, and Zn than wild plants when modified to overexpress γ-glutamylcysteine synthetase and glutathione synthetase (proteins involved in phytochelatin synthesis) (Zhu et al. 1999a, b). Bacterial and yeast glutathione synthetase expression in *A. thaliana* leads to increased accumulation of Cd and As in the transgenic organism (Guo et al. 2008). *Nicotiana glauca* genetically modified to overexpress TaPCS1 gene (from which product is a phytochelatin synthase) can accumulate high levels of Cd, Zn, and Pb (Martinez et al. 2006). The expression of phytochelatin synthase from *Arabidopsis* in *Brassica juncea* increases the transgenic tolerance not only to Cd but also to As (Gasic and Korban 2007).

Metallothionein genes can be introduced in target plant species to enhance heavy metal tolerance. These proteins rich in cysteine amino acid residues possess high affinity to cationic metals (Singh et al. 2003). *N. tabacum* can be modified using the EhMT1 (Xia et al. 2012) or SaMT2 (Zhang et al. 2014) gene to increase the Cu tolerance and accumulation; *A. thaliana* can be engineered using BcMT genes to increase the tolerance to Cd and Cu accumulating the latter in shoots (Lu et al. 2015); Cu tolerance can also be increased by OsMT2c gene (Liu et al. 2015).

A MerP from *Bacillus megaterium* strain MB1 transposon TnMERI1 when expressed in cell membrane and vesicles of transgenic *A. thaliana* can induce the increased accumulation and tolerance to Hg, Cd, and Pb (Hsieh et al. 2009).

In 1999 tobacco plants (*N. tabacum*) were engineered to also remediate other pollutants than heavy metals. By introducing the sequence responsible for codifying pentaerythritol tetranitrate reductase from *Enterobacter cloacae* into this plant species DNA makes it possible for it to increase the tolerance to TNT (2,4,6-trinitrotoluene) and glyceryl trinitrate (GTN); plants can also be engineered to remediate explosive residues that are persistent environmental cytotoxic pollutants (French et al. 1999). The attempts to reprogram plants to degrade toxic nitro-substituted compound continued allowing the development of other transgenic tobacco plant variants to remove TNT residues by using, for example, *E. cloacae* nitroreductase *NfsI* gene (Hannink et al. 2007); transgenic *A. thaliana* is able to deal well with TNT residues by overexpressing its own bifunctional O- and C-glucosyltransferases (Gandia-Herrero et al. 2008) or by being genetically modified to express *Escherichia coli* nitroreductase (Kurumata et al. 2005). *A. thaliana* can also be engineered to eliminate residues of the military explosive RDX (hexahydro-1,3,5-trinitro-1,3,5-triazine) by using genes of cytochrome P450 monooxygenases from *Rhodococcus rhodochrous* (Jackson et al. 2007).

In 2000 tobacco plants were also engineered to deal with residues of halogenated organic compound trichloroethylene (TCE), an industrial solvent. This substance can be metabolized up to 640-fold faster than in wild tobacco plants after receiving the DNA information to express mammalian cytochrome P450 2E1 enzyme. It also makes possible for the transgenic plant to increase not only the uptake but also the debromination of ethylene dibromide (Doty et al. 2000).

Herbicides are other important target for phytoremediation. These chemicals are applied to protect crop yields from weed; however to combat the resistant organisms it is necessary to increase the amount of chemicals used. By doing that the residues that remain on soil and water next to the plantations are worrying pollutants. The family of proteins most commonly used as tools to remediate this kind of residues is P450 family (Abhilash et al. 2009). By genetically modifying *Oryza sativa* to express human CYP1A1, CYP2B6, and CYP2C19, it is possible to program rice plants to phytoremediate the herbicides atrazine and metolachlor (Kawahigashi et al. 2006), and by using these genes and also CYP2C9, not only transgenic rice but also transgenic potato plants can be developed to deal with herbicide residues. CYP1A1, CYP2B6, and CYP2C19 make potato plants resistant to several herbicides, and transgenic rice plant expressing CYP2C9 presents resistance to sulfonylureas being both suitable for environment phytoremediation (Inui and Ohkawa 2005). Atrazine-enhanced metabolism can also be achieved by modifying plants to express bacterial atrazine chlorohydrolase; this commonly used herbicide can be efficiently degraded by transgenic *Medicago sativa* and *N. tabacum* (Wang et al. 2005). The chloroacetanilide herbicide alachlor can be efficiently remediated by using genetically modified tobacco plants overexpressing maize enzyme glutathione S-transferase I (Karavangeli et al. 2005).

15.2 Main Strategies of Plant Transgenesis

In order to manipulate the DNA from plants, there is a wide range of techniques that can be applied; these strategies can be divided in two main groups: biological and nonbiological methodologies.

The nonbiological genetic modification techniques most commonly applied include biolistics, gene delivery performed by different delivery vehicles (e.g., polymers, nanomaterials, and liposomes), electroporation, and microinjection. Biolistics consists in particle incorporated in a desirable DNA bombardment to deliver this DNA to plant target cells even in intact tissue fragment or to microspores. Proposed in the late 1980s (Sanford et al. 1987) and also known as gene gun and particle bombardment, it commonly uses tungsten particles of low cost or gold particles that offer higher efficiency in the process. Loaded particles accelerated by pressurized helium can penetrate cell efficiently to deliver the DNA making it possible to transform not only the nuclear genome but also the mitochondrial and plastidial ones; it is the most popular nonbiological technique to produce transgenic plants (Southgate et al. 1995; Baltes et al. 2017; Cunningham et al. 2018). Electroporation causes temporary opening of pores in cell membrane to allow DNA entrance into cells by submitting the sample to strong electric field pulses; it is commonly performed in protoplasm as target (Weaver 1995; Keshavareddy et al. 2018). Microinjection consists in injecting, with a glass microcapillary-injection pipette, the DNA sequence into protoplasts commonly immobilized by low melting point agarose (Mohanty et al. 2016). Polymers (Bart et al. 2006), nanoparticles (Cunningham et al. 2018), and liposomes (Wordragen et al. 1997) can also serve as gene delivery vehicles to plant transgenesis having as target most frequently the protoplasm.

When it comes to biological techniques, the most commonly applied is the use of *Agrobacterium tumefaciens* (most commonly or *Agrobacterium rhizogenes*), but it is also possible to use viral vectors (DNA or RNA virus) to deliver the transgene to target cells (Zaidi and Mansoor 2017). The use of *A. tumefaciens* to generate transgenic plants started in the 1970s and is up to date the most widely used method with this purpose. This soil bacterium naturally infects dicots, inserting in a stable way its DNA inside host's DNA causing crown gall disease. So, by engineering bacterial plasmid DNA replacing virulence genes by transgenes of interest, it is possible to generate transgenic plants by using this prokaryote (Cunningham et al. 2018) (Fig. 15.1).

15.3 Difficulties Associated with Phytoremediation and New Molecular Biology Strategies in Plant Transgenesis Field

There are plant species that can naturally hyperaccumulate metals such as some members of Brassicaceae family, and before plant genetic modification techniques fully developed, phytoremediation was performed using these species: able to uptake

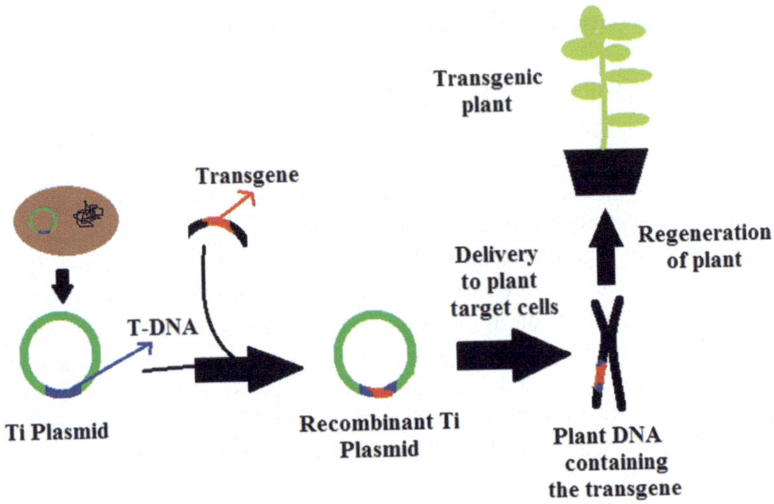

Fig. 15.1 Biological method of producing a transgenic plant using bacterium from *Agrobacterium* gender as transgene's delivery vehicle. The tumor-inducing plasmid (Ti-plasmid) from *A. tumefaciens*, for example, can receive the DNA sequence, in a site called T-DNA, which should be delivered to target plant cell to develop the transgenic organism. The recombinant plasmid containing the transgene can be processed and guided to the host genome. So, the desirable sequence can be integrated to the host's DNA making it possible to obtain the transgenic plant

a great amount of heavy metals and transport them from the root to the shoot. However some of these species, especially in the presence of a high level of contaminants, presented slow growth turning the decontamination process very time-consuming (Jafari et al. 2013).

The plant transgenesis presented solution to this kind of problem and other difficulties faced in this field of expertise. It made possible to implant desirable characteristics to some plant species and change undesirable ones. For example, plants that grow fast but were unable to survive in toxic environment or were unable to accumulate high levels of contaminants could be converted, for example, using genes from hyperaccumulators, into organisms suitable for phytoremediation (Ali et al. 2013). For example, as already mentioned, *Brassica juncea* (the Indian mustard) grows fast, and by adding to its DNA the codifying sequence for a selenocysteine methyltransferase from *Astragalus bisulcatus* (a selenium hyperaccumulator) made it possible to obtain a *B. juncea*, able to accumulate more Se, and tolerate it better than the wild type and also perform this element volatilization (e.g., suitable feature for soil decontamination) (LeDuc et al. 2004).

However inserting genes inside an organism DNA is not always an easy task. There are mainly three types of difficulties associated with plant genetic manipulation: undesirable effects associated with insertion (the transgene's insertion can occur in target cells' genome in a place different to the one previously planned resulting in undesirable results such as mutation and loss of function of important genes that can be interrupted by the insertion process), position (depending on the insertion place the regulatory sequences nearby can cause, e.g., unexpected transgene silencing), and somaclonal events (plant in vitro manipulation commonly induces somaclonal variation, impacting plant exhibited characteristics, changing some of them) (Ziemienowicz 2010). And these undesirable events become even more important when not only one gene insertion is aimed.

Sometimes not only one gene is necessary to be inserted in a plant genome in order to transform it into an efficient organism for phytoremediation (e.g., for Hg remediation). And the task of inserting multiple genes in nuclear genome can be laborious, elevate the study costs and may not lead to the desirable result. However, when multiple genes are involved, the target of genetic manipulation is commonly the chloroplast genome and the approach of homologous recombination to insert the transgene into DNA reduces problems related to off-target insertion (Hussein et al. 2007; Martret et al. 2011).

With advances in molecular biology field, more precise strategies for targeted genome editing have been developed (such as CRISPR/Cas9), and they are already being applied to obtain transgenic plants (Petolino 2015; Forsyth et al. 2016; Malzahn et al. 2017; Borrelli et al. 2018).

And it is also possible to make plants' present desirable characteristics without inserting an exogenous gene inside of their genome or without inserting more gene copies from the same species. RNA silencing can also be applied to obtain organism optimized to perform phytoremediation. Rice OsNRAMP5 Cd transporter, for example, can have its mRNA degraded by silencing methodology, resulting in enhanced cadmium translocation to the shoots, intensifying its pollutant phytoextraction from contaminated soil (Takahashi et al. 2014).

As genome projects from plant species continue to be performed and the identification of genes' function investigated, new possible sequences to be used to generate transgenes with desirable characteristics to phytoremediation continue to appear. But other important opportunity that genetic engineering of plants presents regarding dealing with environmental persistent pollution is the possibility to reduce the use of toxic chemical in crops, for example. It is possible to produce, for example, transgenic corn based on *Bacillus thuringiensis* as natural bioinsecticide (inducing, for example, the production of Cry protein endotoxins), offering a reduction of 56 million kilograms of insecticide use in USA from 1996 to 2011 (Benbrook 2012).

When it comes to transgenic plants developed to perform phytoremediation, there are few biosafety concerns once they are designed for one specific purpose (i.e., removing contaminants from the environment) and will not serve as food for human beings and animals. The major concerns would be related to gene flow from the transgenic plants used for phytoremediation in the environment to wild plants naturally present in that area (which chloroplast modification instead of nuclear DNA modification would help to avoid) and potential loss of diversity once transgenic plants would possess advantageous characteristics to survive in contaminated soil (Kotrba et al. 2009). However when it comes to the abovementioned strategies to reduce the use of chemicals in crops, as the use of insecticide in corn fields, concerns regarding biosafety stand out. Corn is used not only in human but also animals' food. So, food safety and allergenicity of new proteins produced in the transgenic plant, among other risks related to resistance genes introduced in the ecosystem, should receive attention.

15.4 Future Perspectives

The development of transgenic plants for phytoremediation is a promising and important tool in plant biotechnology field to deal with persistent and highly toxic environment pollution and also offers the opportunity to increase the knowledge regarding plant genomes and DNA manipulation, metabolism of heavy metals, and some organic substances that can be environmental contaminants.

Naturally, as the researches advance, advances also the comprehension over: metal uptake and elimination by plants, genetic manipulation of plant nuclear and chloroplastic genome, interaction plant-other forms of life and with the environment (also in contaminated areas) specially in rhizosphere, species that can be useful in phytoremediation providing genes to development of transgenic organisms or performing contaminants neutralization, and strategies to deal with mixed contamination in polluted sites. It is also expected that the numbers of field trials increase to enhance the understanding of transgenic plants' interaction and effects on the ecosystem and to evaluate the possible occurrence of negative economic and biological impacts.

Therefore it is expected that new techniques involving plant species can be developed by teams of researchers from diverse fields of expertise to offer interesting solutions to phytoremediation.

15.5 Conclusion

GMPs are an important tool when it comes to dealing with the increasingly evident global problem of pollution, especially the one related to heavy metals, explosives, industrial solvents, and herbicides. The transgenic plants make it possible to achieve, at low cost, removal of contaminant residues from the environment. Therefore, it is expected that new techniques involving plant species continue to be developed by teams of researchers from diverse fields of expertise to offer interesting and safe innovative solutions to phytoremediation.

References

Abhilash PC, Jamil S, Singh N (2009) Transgenic plants for enhanced biodegradation and phytoremediation of organic xenobiotics. Biotechnol Adv 27:474–488

Ali H, Khan E, Sajad MA (2013) Phytoremediation of heavy metals—concepts and applications. Chemosphere 91:869–881

Ali H, Khan E, Ilahi I (2019) Environmental chemistry and ecotoxicology of hazardous heavy metals: environmental persistence, toxicity, and bioaccumulation. J Chem 2019:1–14

Alkorta I, Hernandez-Allica J, Becerril JM, Amezaga I, Albizu I, Garbisu C (2004) Recent findings on the phytoremediation of soils contaminated with environmentally toxic heavy metals and metalloids such as zinc, cadmium, lead and arsenic. Rev Environ Sci Biotechnol 3:71–90

Baltes NJ, Gil-Humanes J, Voytas DF (2017) Genome engineering and agriculture: opportunities and challenges. Prog Mol Biol Transl Sci 149:1–26

Bart R, Chern M, Park CJ, Bartley L, Ronald PC (2006) A novel system for gene silencing using siRNAs in rice leaf and stem-derived protoplasts. Plant Methods 2:13–15

Benbrook CM (2012) Impacts of genetically engineered crops on pesticide use in the U.S.—the first sixteen years. Environ Sci Eur 24:24–29

Bhuiyan MSU, Min SR, Jeong WJ, Sultana S, Choi KS, Lee Y, Liu JR (2011) Overexpression of AtATM3 in *Brassica juncea* confers enhanced heavy metal tolerance and accumulation. Plant Cell Tissue Organ Cult 107:69–77

Borrelli VMG, Brambilla V, Rogowsky P, Marocco A, Lanubile A (2018) The enhancement of plant disease resistance using CRISPR/Cas9 technology. Front Plant Sci 9:1245–1249

Connolly EL, Fett JP, Guerinot ML (2002) Expression of the IRT1 metal transporter is controlled by metals at the levels of transcript and protein accumulation. Plant Cell 14:1347–1357

Cunningham FJ, Goh NS, Demirer GS, Matos JL, Landry MP (2018) Nanoparticle-mediated delivery towards advancing plant genetic engineering. Trends Biotechnol 36:882–897

Das N, Bhattacharya S, Maiti MK (2016) Enhanced cadmium accumulation and tolerance in transgenic tobacco overexpressing rice metal tolerance protein gene OsMTP1 is promising for phytoremediation. Plant Physiol Biochem 105:297–309

Doty SL, Shang QT, Wilson AM, Moore AL, Newman LA, Strand SE, Gordon MP (2000) Enhanced metabolism of halogenated hydrocarbons in transgenic plants contain mammalian P450 2E1. Proc Natal Acad Sci U S A 97:6287–6291

Ducic T, Polle A (2005) Transport and detoxification of manganese and copper in plants. Braz J Plant Physiol 17:103–112

Forsyth A, Weeks T, Richael C, Duan H (2016) Transcription activator-like effector nucleases (TALEN)-mediated targeted DNA insertion in potato plants. Front Plant Sci 7:1572–1576

French CJ, Rosser SJ, Davies GJ, Nicklin S, Bruce NC (1999) Biodegradation of explosives by transgenic plants expressing pentaerythritol tetranitrate reductase. Nat Biotechnol 17:491–494

Gaitán-Solís E, Taylor NJ, Siritunga D, Stevens W, Schachtman DP (2015) Overexpression of the transporters AtZIP1 and AtMTP1 in cassava changes zinc accumulation and partitioning. Front Plant Sci 6:492–496

Gandia-Herrero F, Lorenz A, Larson T, Graham IA, Bowles J, Rylott EL (2008) Detoxification of the explosive 2,4,6-trinitrotoluene in *Arabidopsis*: discovery of bi-functional O and C-glucosyltransferases. Plant J 56:963–974

Garbisu C, Hernandez-Allica J, Barrutia O, Alkorta I, Becerril JM (2002) Phytoremediation: a technology using green plants to remove contaminants from polluted areas. Rev Environ Health 17:173–188

Gasic K, Korban SS (2007) Transgenic Indian mustard (*Brassica juncea*) plants expressing an *Arabidopsis* phytochelatin synthase (AtPCS1) exhibit enhanced As and Cd tolerance. Plant Mol Biol 64:361–369

Guo J, Dai X, Xu W, Ma M (2008) Overexpressing gsh1 and AsPCS1 simultaneously increases the tolerance and accumulation of cadmium and arsenic in *Arabidopsis thaliana*. Chemosphere 72:1020–1026

Hannink NK, Subramanian M, Rosser SJ, Basran A, Murray JAH, Shanks JV, Bruce NC (2007) Enhanced transformation of TNT by tobacco plants expressing a bacterial nitroreductase. Int J Phytoremediation 9:385–401

Heaton ACP, Rugh CL, Wang N, Meagher RB (1998) Phytoremediation of mercury- and Methylmercury-polluted soils using genetically engineered plants. J Soil Contam 7:497–509

Hirata K, Tsuji N, Miyamoto K (2005) Biosynthetic regulation of phytochelatins, heavy metal-binding peptides. J Biosci Bioeng 100:593–599

Hsieh JL, Chen CY, Chiu MH, Chein MF, Chang JS, Endo G, Huang CC (2009) Expressing a bacterial mercuric ion binding protein in plant for phytoremediation of heavy metals. J Hazard Mater 161:920–925

Huang CC, Su CC, Hsieh JL, Tseng CP, Lin PJ, Chang JS (2003) Polypeptides for heavy-metal biosorption: capacity and specificity of two heterogeneous MerP proteins. Enzym Microb Technol 33:379–385

Hussein HS, Ruiz ON, Terry N, Daniell H (2007) Phytoremediation of mercury and organomercurials in chloroplast transgenic plants: enhanced root uptake, translocation to shoots, and volatilization. Environ Sci Technol 41:8439–8446

Inui H, Ohkawa H (2005) Herbicide resistance in transgenic plants with mammalian P450 monooxygenase genes. Pest Manag Sci 61:286–291

Jackson EG, Rylott EL, Fournier D, Hawari J, Bruce NC (2007) Exploring the biochemical properties and remediation applications of the unusual explosive-degrading P450 system XplA/B. Proc Natl Acad Sci U S A 104:16822–16827

Jafari M, Danesh YR, Goltapeh EM, Varma A (2013) Bioremediation and genetically modified organisms. In: Goltapeh E, Danesh Y, Varma AM (eds) Fungi as bioremediators. Springer, Berlin, pp 433–450

Karavangeli M, Labrou NE, Clonis YD, Tsaftaris A (2005) Development of transgenic tobacco plants overexpressing maize glutathione S-transferase I for chloroacetanilide herbicides phytoremediation. Biomol Eng 22:121–128

Kawahigashi H, Hirose S, Ohkawa H, Ohkawa Y (2006) Phytoremediation of herbicide atrazine and metolachlor by transgenic rice plants expressing human CYP1A1, CYP2B6 and CYP2C19. J Agric Food Chem 54:2985–2991

Keshavareddy G, Kumar ARV, Ramu VS (2018) Methods of plant transformation—a review. Int J Curr Microbiol App Sci 7:2656–2668

Korenkov V, Hirschi K, Crutchfield JD, Wagner GJ (2007) Enhancing tonoplast Cd/H antiport activity increases Cd, Zn, and Mn tolerance, and impacts root/shoot Cd partitioning in *Nicotiana tabacum* L. Planta 226:1379–1387

Kotrba P, Najmanova J, Macek T, Ruml T, Mackova M (2009) Genetically modified plants in phytoremediation of heavy metal and metalloid soil and sediment pollution. Biotechnol Adv 27:799–810

Kurumata M, Takahashi M, Sakamoto A, Ramos JL, Nepovim A, Vanek T, Hirata T, Morikawa H (2005) Tolerance to, and uptake and degradation of 2,4,6-trinitrotoluene (TNT) are enhanced by the expression of a bacterial nitroreductase gene in *Arabidopsis thaliana*. Z Naturforsch C 60:272–278

LeDuc DL, Tarun AS, Montes-Bayon M, Meija J, Malit MF, Wu CP, Abdel Samie M, Chiang CY, Tagmount A, deSouza M, Neuhierl B, Bock A, Caruso J, Terry N (2004) Overexpression of selenocysteine methyltransferase in *Arabidopsis* and Indian mustard increases selenium tolerance and accumulation. Plant Physiol 135:377–383

Lee JH (2013) An overview of phytoremediation as a potentially promising technology for environmental pollution control. Biotechnol Bioprocess Eng 18:431–439

Lin YF, Hassan Z, Talukdar S, Schat H, Aarts MG (2016) Expression of the ZNT1 zinc transporter from the metal Hyperaccumulator *Noccaea caerulescens* confers enhanced zinc and cadmium tolerance and accumulation to *Arabidopsis thaliana*. PLoS One 11:e0149750

Liu J, Shi X, Qian M, Zheng L, Lian C, Xia Y, Shen Z (2015) Copper-induced hydrogen peroxide upregulation of a metallothionein gene, OsMT2c, from *Oryza sativa* L. confers copper tolerance in *Arabidopsis thaliana*. J Hazard Mater 294:99–108

Lu Y, Deng X, Quan L, Xia Y, Shen Z (2015) Metallothioneins BcMT1 and BcMT2 from *Brassica campestris* enhance tolerance to cadmium and copper and decrease production of reactive oxygen species in *Arabidopsis thaliana*. Plant Soil 367:507–519

Malzahn A, Lowder L, Qi Y (2017) Plant genome editing with TALEN and CRISPR. Cell Biosci 7:21–25

Mani D, Kumar C (2014) Biotechnological advances in bioremediation of heavy metals contaminated ecosystems: an overview with special reference to phytoremediation. Int J Environ Technol 11:843–872

Martinez M, Bernal P, Almela C, Velez D, Garcia-Agustin P, Serrano R (2006) An engineered plant that accumulates higher levels of heavy metals than *Thlaspi caerulescens*, with yields of 100 times more biomass in mine soils. Chemosphere 64:478–485

Martinoia E, Klein M, Geisler M, Bovet L, Forestier C, Kolukisaoglu U, Müller-Röber B, Schulz B (2002) Multifunctionality of plant ABC transporters—more than just detoxifiers. Planta 214:345–355

Martret B, Poage M, Shiel K, Nugent GD, Dix PJ (2011) Tobacco chloroplast transformants expressing genes encoding dehydroascorbate reductase, glutathione reductase, and glutathione-S-transferase, exhibit altered anti-oxidant metabolism and improved abiotic stress tolerance. Plant Biotechnol J 9:661–673

Mendes KF, Régo APJ, Takeshita V, Tornisielo VL (2019) Water resource pollution by herbicide residues. IntechOpen. https://doi.org/10.5772/intechopen.85159

Merlot S, Hannibal L, Martins S, Martinelli L, Amir H, Lebrun M, Thomine S (2014) The metal transporter PgIREG1 from the hyperaccumulator *Psychotria gabriellae* is a candidate gene for nickel tolerance and accumulation. J Exp Bot 65:1551–1564

Misra S, Gedamu L (1989) Heavy metal tolerant transgenic *Brassica napus* L. and *Nicotiana tabacum* L. plants. Theor Appl Genet 78:161–168

Mohanty D, Chandra A, Tandon R (2016) Germline transformation for crop improvement. In: Rajpal VR, Rao SR, Raina SN (eds) Molecular breeding for sustainable crop improvement. Springer, Cham, pp 178–183

Nascimento CWA, Xing B (2006) Phytoextraction: a review on enhanced metal availability and plant accumulation. Sci Agr 63:299–311

Ojuederie OB, Babalola OO (2017) Microbial and plant-assisted bioremediation of heavy metal polluted environments: a review. Int J Environ Res Public Health 14:1504–1511

Pesantes AA, Carpio EP, Vitvar T, López MMM, Menéndez-Aguado JM (2019) A multi-index analysis approach to heavy metal pollution assessment in river sediments in the ponce enríquez area, Ecuador. Water 11:590–596

Petolino JF (2015) Genome editing in plants via designed zinc finger nucleases. In Vitro Cell Dev Biol Plant 51:1–8

Pu Q, Sun JQ, Zhang FF, Wen XY, Liu WH, Huang CM (2019) Effects of copper mining on heavy metal contamination in a rice agrosystem. Acta Geochim 2019:1–21

Ricachenevsky FK, Menguer PK, Sperotto RA, Williams LE, Fett JP (2013) Roles of plant metal tolerance proteins (MTP) in metal storage and potential use in biofortification strategies. Front Plant Sci 4:144–149

Rodríguez-Llorente ID, Lafuente A, Doukkali B, Caviedes MA, Pajuelo E (2012) Engineering copper hyperaccumulation in plants by expressing a prokaryotic cop C gene. Environ Sci Technol 46:12088–12097

Rosculete CA, Bonciu E, Rosculete E, Olaru LA (2019) Determination of the environmental pollution potential of some herbicides by the assessment of cytotoxic and genotoxic effects on *Allium cepa*. Int J Environ Res Public Health 16:75–85

Rugh CL, Wilde HD, Stack NM, Thompson DM, Summers AO, Meagher RB (1996) Mercuric ion reduction and resistance in transgenic *Arabidopsis thaliana* plants expressing a modified bacterial merA gene. Proc Natl Acad Sci U S A 93:3182–3187

Rugh CL, Senecoff JF, Meagher RB, Merkle SA (1998) Development of transgenic yellow poplar for mercury phytoremediation. Nat Biotechnol 16:925–928

Rugh CL, Bizily SP, Meagher RB (2000) Phytoreduction of environmental mercury pollution. In: Raskin I, Ensley BD (eds) Phytoremediation of toxic metals—using plants to clean up the environment. Wiley, New York, pp 151–171

Sanford JC, Klein TM, Wolf ED, Allen N (1987) Delivery of substances into cells and tissues using a particle bombardment process. Particulate Sci Technol 5:27–37

Shim D, Kim S, Choi YI, Song WY, Park J, Youk ES, Jeong S-C, Martinoia E, Noh E-W, Lee Y (2013) Transgenic poplar trees expressing yeast cadmium factor 1 exhibit the characteristics necessary for the phytoremediation of mine tailing soil. Chemosphere 90:1478–1486

Shukla D, Kesari R, Tiwari M, Dwivedi S, Tripathi RD, Nath P, Trivedi PK (2013) Expression of *Ceratophyllum demersum* phytochelatin synthase, CdPCS1, in *Escherichia coli* and *Arabidopsis* enhances heavy metal(loid)s accumulation. Protoplasma 250:1263–1272

Siemianowski O, Barabasz A, Kendziorek M, Ruszczyńska A, Bulska E, Williams LE, Antosiewicz DM (2014) AtHMA4 expression in tobacco reduces Cd accumulation due to the induction of the apoplastic barrier. J Exp Bot 65:1125–1139

Singh OV, Labana S, Pandey G, Budhiraja R, Jain RK (2003) Phytoremediation: an overview of metallic ion decontamination from soil. Appl Microbiol Biotechnol 61:405–412

Song WY, Park J, Eisenach C, Maeshima M, Lee Y, Martinoia E (2014) ABC transporters and heavy metals. In: Geisler M (ed) Plant ABC transporters, Signaling and communication in plants, vol 22. Springer, Cham, pp 1–17

Southgate EM, Davey MR, Power JB, Marchant R (1995) Factors affecting the genetic engineering of plants by microprojectile bombardment. Biotechnol Adv 13:631–651

Suresh B, Ravishankar GA (2004) Phytoremediation—a novel and promising approach for environmental clean-up. Crit Rev Biotechnol 24:97–124

Takahashi R, Ishimaru Y, Shimo H, Bashir K, Senoura T, Sugimoto K, Ono K, Suzui N, Kawachi N, Ishii S, Yin YG, Fujimaki S, Yano M, Nishizawa NK, Nakanishi H (2014) From Laboratory to Field: OsNRAMP5-Knockdown Rice Is a Promising Candidate for Cd Phytoremediation in Paddy Fields. PLoS ONE 9(6):1–7. https://doi.org/10.1371/journal.pone.0098816

Tripathi P, Singh PK, Mishra S, Gautam N, Dwivedi S, Chakrabarty D, Tripathi RD (2015) Recent advances in the expression and regulation of plant metallothioneins for metal homeostasis and tolerance. In: Chandra R (ed) Environmental waste management. CRC Press, New York, pp 551–564

Vázquez-Luna D, Cuevas-Díaz MC (2019) Soil contamination and alternatives for sustainable development. In: Vázquez-Luna D, Cuevas-Díaz MC (eds) Soil contamination and alternatives for sustainable development. IntechOpen, London. https://doi.org/10.5772/intechopen.83720

Wan X, Lei M, Chen T (2016) Cost-benefit calculation of phytoremediation technology for heavy-metal-contaminated soil. Sci Total Environ 564:796–802

Wang L, Samac DA, Shapir N, Wackett LP, Vance CP, Olszewski NE, Sadowsky MJ (2005) Biodegradation of atrazine in transgenic plants expressing a modified bacterial atrazine chlorohydrolase (atzA) gene. Plant Biotechnol J 3:475–486

Weaver JC (1995) Electroporation theory. In: Nickoloff JA (ed) Methods in molecular biology. Plant cell electroporation and electrofusion protocols. Humana, Totowa, pp 3–28

Williams LE, Pittman JK, Hall JL (2000) Emerging mechanisms for heavy metal transport in plants. Biochim Biophys Acta Biomembr 1465:104–126

Wordragen MV, Shakya R, Verkerk R, Peytavis R, Kammen AV, Zabel P (1997) Liposome-mediated transfer of YAC DNA to tobacco cells. Plant Mol Biol Rep 15:170–178

Xia Y, Qi Y, Yuan Y, Wang G, Cui J, Chen Y, Zhang H, Shen Z (2012) Overexpression of *Elsholtzia haichowensis* metallothionein 1 (EhMT1) in tobacco plants enhances copper tolerance and accumulation in root cytoplasm and decreases hydrogen peroxide production. J Hazard Mater 234:65–71

Zaidi SS, Mansoor S (2017) Viral vectors for plant genome engineering. Front Plant Sci 8:539–545

Zhang J, Zhang M, Tian S, Lu L, Shohag MJI, Yang X (2014) Metallothionein 2 (SaMT2) from *Sedum alfredii* Hance confers increased Cd tolerance and accumulation in yeast and tobacco. PLoS One 9:e102750

Zhang Q, Yu R, Fu S, Wu Z, Chen HYH, Liu H (2019) Spatial heterogeneity of heavy metal contamination in soils and plants in Hefei, China. Sci Rep 1049:2045–2051

Zhu Y, Pilon-Smits EA, Tarun AS, Weber SU, Jouanin L, Terry N (1999a) Cadmium tolerance and accumulation in Indian mustard is enhanced by overexpressing γ-glutamylcysteine synthetase. Plant Physiol 121:1169–1177

Zhu Y, Pilon-Smits EAH, Jouanin L, Terry N (1999b) Overexpression of glutathione synthetase in *Brassica juncea* enhances cadmium tolerance and accumulation. Plant Physiol 119:73–79

Ziemienowicz A (2010) Plant transgenesis. Methods Mol Biol 631:253–268

Index

K. R. Hakeem et al. (eds.), *Bioremediation and Biotechnology*,
https://doi.org/10.1007/978-3-030-35691-0

The manufacturer's authorised representative in the EU is Springer Nature Customer Service Centre GmbH, Europaplatz 3, 69115 Heidelberg, Germany. If you have any concerns regarding our products, please contact ProductSafety@springernature.com

Printed and bound by CPI Group (UK) Ltd, Croydon, CR0 4YY
22/09/2026
02221057-0015

The manufacturer's authorised representative in the EU is Springer Nature Customer Service Centre GmbH, Europaplatz 3, 69115 Heidelberg, Germany. If you have any concerns regarding our products, please contact ProductSafety@springernature.com

Printed and bound by CPI Group (UK) Ltd, Croydon, CR0 4YY
25/09/2026
02222855-0012